The Culture of Time and Space

1880–1918

THE CULTURE OF TIME AND SPACE 1880–1918

STEPHEN KERN

Harvard University Press Cambridge, Massachusetts

Printed in the United States of America
10 9 8 7 6 5 4 3 2

Library of Congress Cataloging in Publication Data

Kern, Stephen.
The culture of time and space 1880–1918.

Includes index.
1. Technology and civilization. 2. Space and time. 3. Civilization, Modern—19th century. 4. Civilization, Modern—20th century. I. Title.
CB478.K46 1983 303.4'83 83-303
ISBN 0-674-17972-2 (cloth)
ISBN 0-674-17973-0 (paper)
Designed by Gwen Frankfeldt

To Rudolph Binion

ACKNOWLEDGMENTS

Permission to quote a fragment of "Spring Strains," by William Carlos Williams, in *Collected Earlier Poems of William Carlos Williams,* copyright 1938 by New Directions Publishing Corporation, has been granted by the publisher. Editions Gallimard has given its permission to quote a fragment from Valéry Larbaud's "Poèmes par un rich amateur," in *Oeuvres de Valéry Larbaud,* Bibliothèque de la Pléiade, © Editions Gallimard 1957.

I wish to thank the Rockefeller Foundation for a Humanities Fellowship during 1977–78, the Center for European Studies of Harvard University for an Honorary Research Fellowship during the same year, and Northern Illinois University for a sabbatical leave in the spring of 1980. Thanks also to Keith Cohen, Michael Gelven, and Paul Starr, who commented on parts of the manuscript, and to Mary Kay Damer, Lewis Erenberg, Donald Lowe, George Roeder, and Sean Shesgreen, who read it in its entirety. In a series of conversations George Mosse suggested many sources, and Martin Sklar was a sounding board for ideas from the inception of the book to its completion. Aida Donald and Anita Safran deftly edited the final version.

This book is dedicated to Rudolph Binion for the extraordinary attention he gave to my work when I was his student, for the interpretive boldness that his own work has inspired, for countless suggestions and a critical reading of this book, and for his abiding friendship.

S.K.
DeKalb, Illinois

CONTENTS

ILLUSTRATIONS

INTRODUCTION

From around 1880 to the outbreak of World War I a series of sweeping changes in technology and culture created distinctive new modes of thinking about and experiencing time and space. Technological innovations including the telephone, wireless telegraph, x-ray, cinema, bicycle, automobile, and airplane established the material foundation for this reorientation; independent cultural developments such as the stream-of-consciousness novel, psychoanalysis, Cubism, and the theory of relativity shaped consciousness directly. The result was a transformation of the dimensions of life and thought. This book is about the way Europeans and Americans came to conceive of and experience time and space in those years.

The idea for this kind of interpretation came to me from reading the works of phenomenologically oriented psychiatrists, who viewed their patients' mental lives in these terms. They used a categorical frame of reference to reconstruct their patients' experience of time, space, causality, materiality, and other essential categories. The work of the French psychiatrist Eugène Minkowski, elaborated in a collection of case studies published as *Le Temps vécu* in 1933, was especially useful. While Minkowski explored other categories, the focus of his attention was on time, especially on how his patients experienced past, present, and future. He applied the phenomenological method to understand patients who had acute psychotic disorders and could not reconstruct their lives genetically or historically as the psychoanalytic method required. His method is particularly suited for psychotics, because it is often not feasible to link their prior personality with their current pathological personality, which is generally too fragmented and disorganized. I adapt that aspect of the phenomenological method informally, for it is possible to identify many origins or "causes" of changing ideas about time and space, such as the scheduling requirements of railroads that directly necessitated the institution of World Standard Time, or the telephone that immediately and directly changed the sense of space. For all its diversity, the culture of an age hangs together more coherently than does the mind of a psychotic. My primary object, however, is to survey significant changes in the experience of time and space, including some for which I am able to identify no specific "cause." Hence I do not explain why the telephone was invented or why the stream-of-consciousness novel began to appear.

As basic philosophical categories, time and space are particularly suitable as a framework for a general cultural history, because they are comprehensive, universal and essential.

Since all experience takes place in time and space, the two categories provide a comprehensive framework that can include such wide-ranging cultural developments as Cubism, simultaneous poetry, and ragtime music along with the steamship, skyscraper, and machine gun. To avoid the crazy-quilt effect that such an assemblage of sources might create, I select only material that conforms to the essential nature of each of the subtopics that make up the first nine chapters—The Nature of Time, The Past, The Present, The Future,

Speed, The Nature of Space, Form, Distance, and Direction—and emphasize those developments that differ significantly from earlier periods.

I followed two lines of thinking to arrive at these subtopics. The three modes of time—past, present, and future—came from philosophy and were part of Minkowski's conceptual framework. Even Henri Bergson (who insisted that the division of the flux of time into three discrete parts distorted its essentially fluid nature) used the terms repeatedly in his analyses. These modes of time seemed to be natural, compelling, and comprehensive subdivisions for all possible human experiences of time. Subtopics for space were more difficult to determine. In a discussion with Alan Henrikson, I learned that map makers identify four aspects of space that plane maps can show—shape, area, distance, and direction. These categories suggested a framework that was as comprehensive as those for time that I had already decided on, so I combined "shape" and "area" into "form" and added an introductory chapter on the nature of space as I did on the nature of time. My categories thus encompass a wide range of human activities and are mutually exclusive—except for the material on speed. I treat speed in a separate chapter because it was widely discussed around the turn of the century as a topic in its own right, because the material on it would have been impossible to classify as either exclusively temporal or spatial, and because as a juncture of time and space it formed a natural transition between them.

To avoid repetition I have dispersed single corpuses among these chapters. The work of Marcel Proust, for example, appears in chapters on the nature of time, the past, the nature of space, and distance. Prominent figures such as Proust have been interpreted with such uniformity that their contributions to the cultural landscape have tended to become as solid and fixed as a rock. By cracking into such routine interpretations, identifying different contributions from various parts of their corpus, and distributing my discussion of them among the subtopics of this study, I attempt to expose fresh surfaces and attribute those contributions to the precise modes of time or space that are appropriate.

To illustrate further the comprehensive range of these topics, in the two concluding chapters I survey how these changes shaped the diplomatic crisis and the actual fighting of World War I. Individuals behave in distinctive ways when they feel cut off from the flow of time, excessively attached to the past, isolated in the present, without a future, or rushing toward one. Nations also demonstrate distinctive attitudes toward time. For example, the contrast between Austria-Hungary, convinced its time was running out, and Russia, which felt it had time to spare, is striking and is revealed repeatedly in diplomatic documents. The experience of space also varies considerably along national lines: some countries, like Germany, believed they needed more; Austria-Hungary thought that its space was excessively heterogeneous and distintegrating; Russia was universally viewed (and feared) as the country with boundless space. These final chapters illustrate how the changes in thinking about and experiencing these abstract philosophical categories were manifested in a concrete historical situation. The categories of time and space thus provide a comprehensive theoretical framework that allows not only the integration of many areas across the cultural spectrum but also integration along a theoretical vertical axis from "high culture" to popular culture and the material aspects of everyday life.

Not every society has kings, parliaments, labor unions, big cities, bourgeoisies, Christian churches, diplomats, or navies. I do not mean to question the significance of histories of such entities but only to point out that they are not universal. Time and space are. All people, everywhere, in all ages, have a distinctive experience of time and space and, however unconscious, some conception of it. It is possible to interpret how class structures, modes of production, patterns of diplomacy, or means of waging war were manifested historically in terms of changing experiences of time and space. Thus class conflict is viewed as a function of social distance, assembly lines are interpreted in conjunction with Taylorism and time management studies, the diplomatic crisis of July 1914 is seen to have a historically unique temporality, and World War I can be interpreted under a Cubist metaphor. The phonograph and cinema are evaluated in terms of the way they modified the sense of the past, the telephone and World Standard Time are seen restructuring the experience of the present, the steamship and the Schlieffen Plan reflect a desire to control the future, urbanism is viewed as a process of diminishing living space, the politics of imperialism is seen as a universal impulse to claim more space, wealth is conceived as the power to control time and space.

Such interpretations are reductionistic. But if one is to make generalizations about the culture of an age, one must be able to show how a wide variety of phenomena have certain common features in their essential nature or function, and one must also be able to interpret these features in a common language. The interpretation of phenomena such as class structure, diplomacy, and war tactics in terms of modes of time and space makes possible the demonstration of their essential similarity to explicit considerations of time and space in literature, philosophy, science, and art. Put together, they create the basis for generalizations about the essential cultural developments of the period. And by interpreting the culture as a function of time and space, it becomes possible to compare different ages and different cultures topic by topic with less confusion than would be involved in trying to compare historically and culturally specific interpretative categories such as parliaments, unions, families, or bourgeoisies. It should be possible, therefore, to compare the experience of time and space in the Renaissance or the Enlightenment with that of the *fin de siècle* to discover what essential changes occurred in the intervening years. This study provides a contribution to such a larger historical project.

By arguing that my topics are essential, I run the risk of implying that cultural histories with other foci are unessential. Researching the culture of an age, over a number of years, with two topics constantly in mind, one inevitably begins to see everything in that context, even the work of others who have classified and interpreted sources differently. General cultural histories of the period, including some that focus on single nations or cities, have been inspirational and suggestive in drawing my attention to sources and offering a variety of interpretations of them. While I am mindful of the natural bias that comes to any researcher, I must nevertheless venture the claim that my focal topics are more essential from a strictly philosophical point of view. The topics of Roger Shattuck, H. Stuart Hughes, and Carl E. Schorske are framed according to conventional academic disciplines and artistic genres. While I used those frames for subdivisions within my chapters, my basic categories derive from two essential philosophical categories—essential in that they are, as Kant argues, the necessary foundation of all experience. Shattuck focused on four themes in French culture of the period: childhood, humor, dream, and ambiguity as they were expressed in four genres of art, music, drama, and poetry. Schorske interpreted the politics of literature, architecture, city planning, psychiatry, art, and music in Vienna; and Hughes examined a specific discovery in social thought, which can be interpreted as one aspect of the new sense of space—perspectivism.[1] The more limited focus of these studies enabled their authors to go into greater detail, but they did not attempt to analyze the essential foundations of experience, as I have tried to do.

I originally planned to organize the new thinking according to traditional artistic genres and academic disciplines, however much of it cut across those dividers. I finally decided to base the theoretical framework on philosophical concepts, because that allowed me to treat concepts such as simultaneity as a whole and not scatter them throughout various genre and discipline chapters; it forced me to break up large corpuses, which sharpened my assessment of their various contributions; and it necessitated thinking through the historical significance of the culture of the period in fresh terms. This approach obliged me to decide upon a suitable subtopic to cover simultaneity—a concept that cut across traditional divisions—and that posed problems. Instantaneous electronic communication, which made simultaneity a reality, affected the sense of the present, speed, form, and distance. I concluded that its most distinctive effect was on the sense of the present.

Technological developments are temporally specific events that often affect great numbers of people, and as such they are a compelling source for historical explanation. To avoid a monocausal technological determinism in cultural history, it is essential to clarify precisely how technology and culture interact.

Some cultural developments were directly inspired by new technology. James Joyce was fascinated by the cinema, and in *Ulysses* he attempted to recreate in words the montage techniques used by early film makers. The Futurists worshipped modern technology and celebrated it in manifestos and art. Several poets wrote "simultaneous" poetry as a response to the simultaneity of experience made possible by electronic communication. Many conceptions of time and space, however, were altered independently of technology, in response to pressures within various genres and disciplines. Paul Cézanne revolutionized the treatment of space in art as he concentrated on the eternal form of *Mont Sainte-Victoire* and the arrangement of bottles and apples in his still lifes. Einstein's challenge to Newton was suggested by the results of an experiment made possible by a new machine—the inferometer—but relativity was largely a revision of theoretical problems that physics had been struggling with for years. The thematic similarity between developments inspired by technology and those independent of it suggests that a cultural revolution of the broadest scope was taking place, one that involved essential structures of human experience and basic forms of human expression.

Other technics provided metaphors and analogies for changing structures of life and thought. The opening up of the interior anatomical terrain of the human body by x-ray was part of a general reappraisal of what is properly inside and what is outside in the body, the mind, physical objects, and nations. Thomas Mann's hero in *The Magic Mountain* remarked that he felt as though he were peering into the grave when he observed his cousin's insides by means of x-ray. Edmund Husserl challenged the Cartesian idea that perception takes place in the mind and argued instead that it is a relation between a perceiver and a thing perceived. The Cubists rendered both the interior and exterior of objects from a variety of perspectives on a single canvas, thereby transcending traditional spatial and temporal limits in art. The airplane altered the significance of national boundaries and traditional geographical barriers between peoples.

In the process of integrating such an array of sources, I use a working principle of *conceptual distance.* Thus, there is greater conceptual distance between the thinking of an architect and that of a philosopher on a given subject than there is between the thinking of two philosophers, and I assume that any generalization about the thinking of an age is the more persuasive the greater the conceptual distance between the sources on which it is based. However the distance must not be too great or the juxtaposition becomes forced. Mindful of that problem I have at times used metaphor and analogy to link material from especially "distant" sources to extend interpretations beyond the confines of strict academic disciplines and their exacting requirements for evidence and argumentation. And so, for example, a discussion of the discovery of the constituent function of negative space juxtaposes evidence all across Western culture including field theory in physics, architectural spaces, sculpted voids, Cubist positive negative space, the pauses and blanks in Mallarmé's poetry, and silence in literature and music. Such broad cross-discipline and cross-genre constructions involve a radical gerrymandering of traditional cultural areas.

This method of grouping thematically related developments without an apparent causal link occasionally led to the discovery of a link. The connection between Cubism and camouflage, for example, was suggested by Picasso's remark to Gertrude Stein, upon seeing the first camouflaged trucks parading in Paris in 1915, that the Cubists had invented camouflage. For a number of reasons the historical significance of these two phenomena was strikingly similar, but, as neither Picasso nor Gertrude Stein documented the connection, I at first assumed that he was just pointing out that significant similarity, much as I have done with other cultural developments throughout this book. But further inquiry revealed that the man who invented camouflage was inspired by the Cubists and explicitly acknowledged that debt. This discovery tightened my interpretation of the major changes in the actual fighting of World War I within a Cubist metaphor. Some analogies, however, remain mere analogies, and although I did not discover any actual connection between their elements (as, for example, between field theory in physics and Futurist "force lines"), the similarity between the two was sufficiently strong to link them in the only way justified by my research—analogically—having a similar structure or function within their respective disciplines or genres and possibly related in fact by processes of communication that I was unable to discover. These analogies constitute the open end of my thinking, but they do not make up the bulk of my argument, which is based on developments of similar cultural function that were causally or, at least, consciously related at that time.

It is impossible to identify a single thesis that properly encompasses all changes in the experience of time and space that occurred in this period. Indeed, one major change was the affirmation of a plurality of times and spaces. Nevertheless, it is possible to indicate the most important development for each of the two major topics—the affirmation of the reality of private time and the leveling of traditional spatial hierarchies. Bergson's philosophy forms the theoretical core of the argument for private time, and Cubism graphically negates the traditional notion that the subject of a painting, for example, is more important than the background. This leveling of hierarchy in various areas of Western culture, it will be seen, parallels the leveling of aristocratic society, the rise of democracy, and the dissolution of the distinction between the sacred and profane space of religion. Although there is some evidence for direct, conscious connection between these parallel developments, such as Louis Sullivan's affirmation of a new "democratic" architecture, the connection remains largely one of analogy, based on compelling similarity.

While I do not mean to present the "relevance" of this study to current problems in a simplistic way, its very conception is associated with the energy crisis of recent years. Contemplating the disastrous consequences of a long-range depletion of energy sources, especially those that affect transportation, it struck me that in the period I wanted to analyze, new energy sources had revolutionized the experiences of time and space. The age thus had an energy crisis of its own—a crisis of abundance. The tremendous development of railroads and steamships and the invention of the automobile and airplane greatly accelerated transportation and proliferated the places where people could travel at new high speeds. The petroleum industry began to supply combustible fuels on a large scale for the automobiles, and power stations distributed electricity to light up the night and drive electric motors. It was the reverse of the current energy crisis, since its alarmists were generally concerned about a surfeit of new energy sources and its possible nefarious consequences. There was little talk of running out. And unlike the current crisis that has caused panic, the crisis of the prewar period generally inspired hope.

Each chapter begins with the technological or institutional developments that shaped the mode of time and space that is its subject, and then surveys the cultural record, following traditional academic disciplines and artistic genres. Within each subsection I have reconstructed events in chronological order. The concluding date of this study is a natural historical marker; the beginning date is approximate. Some events, such as the publication of Jules Verne's *Around the World in Eighty Days* in 1873 or the invention of the telephone in 1876 precede it, but the bulk of the changes cluster in the turn-of-the-century period and constitute a generally coherent cultural unit.

1

THE NATURE OF TIME

In the preface to a collection of essays on the history of ideas, the cultural historian Arthur O. Lovejoy complained that many studies overunified the views of an author in order to present his thinking "all-of-a-piece." In his own essays he sought to correct that weakness and present the "inner tensions—the fluctuations or hesitancies between opposing ideas or moods, or the simple and more or less unconscious embracing of both sides of an antithesis."[1] Lovejoy was referring to an individual's thinking, but the warning applies even more to the thought of an age. The great variety of views in any particular age do not all line up on one side of the issues. I present the critical concepts dramaturgically in accord with the theory that knowledge is essentially dialectical, that ideas are generated in opposition to other ideas and have a basic polemical nature. The development of a body of thought involves a selection from, and an occasional resolution of, contrasting views. The ideas of this period on the nature of time will be organized around three pairs of opposing views: whether time was homogeneous or heterogeneous, atomistic or a flux, reversible or irreversible.

∞

As every child quickly learns, there is only one time. It flows uniformly and may be divided into equal parts anywhere along the line. This is the time Isaac Newton defined in 1687: "Absolute, true, and mathematical time, of itself, and from its own nature, flows equally without relation to anything external." In *The Critique of Pure Reason* (1781) Immanuel Kant rejected the Newtonian theory of absolute, objective time (because it could not possibly be experienced) and maintained that time was a subjective form or foundation of all experience. But even though it was subjective, it was also universal—the same for everybody. No doubt Newton and Kant experienced different paces of private time, but before the late nineteenth century no one (with the possible exception of Laurence Sterne, who explored private time in *Tristram Shandy*) systematically questioned the homogeneity of time. The evidence for it was written on the faces of the millions of clocks and watches manufactured every year.

The most momentous development in the history of uniform, public time since the invention of the mechanical clock in the fourteenth century was the introduction of standard time at the end of the nineteenth century. A pioneer in promoting uniform time was the Canadian engineer Sanford Fleming, who in 1886 outlined some reasons for its adoption. The use of the telegraph "subjects the whole surface of the globe to the observation of civilized communities and leaves no interval of time between widely separated places proportionate to their distances apart." This system mixes up day and night as "noon, midnight, sunrise, sunset, are all observed at the same moment," and "Sunday actually commences in the middle of

Saturday and lasts until the middle of Monday."[2] A single event may take place in two different months or even in two different years. It was important to be able to determine local times and to know precisely when laws go into effect and insurance policies begin. The present system, he concluded, would lead to countless political, economic, scientific, and legal problems that only the adoption of a coordinated world network could prevent.

The most famous supporter of standard time, Count Helmuth von Moltke, in 1891 appealed to the German Parliament for its adoption. He pointed out that Germany had five different time zones, which would impede the coordination of military planning; in addition there were other time zones, he protested, that "we dread to meet at the French and Russian boundaries."[3] When Fleming sent Moltke's speech to the editor of *The Empire* for publication, he did not dream that in 1914 the world would go to war according to mobilization timetables facilitated by standard time, which he thought would rather engender cooperation and peace.

Despite all the good scientific and military arguments for world time, it was the railroad companies and not the governments that were the first to institute it. Around 1870, if a traveler from Washington to San Francisco set his watch in every town he passed through, he would set it over two hundred times. The railroads attempted to deal with this problem by using a separate time for each region. Thus cities along the Pennsylvania Railroad were put on Philadelphia time, which ran five minutes behind New York time. However, in 1870 there were still about 80 different railroad times in the United States alone.[4] The day the railroads imposed a uniform time, November 18, 1883, was called "the day of two noons," because at mid-day clocks had to be set back in the eastern part of each zone—one last necessary disruption to enable the railroads to end the confusion that had so complicated their functioning and cut into their profits. In 1884 representatives of twenty-five countries that convened at the Prime Meridian Conference in Washington proposed to establish Greenwich as the zero meridian, determined the exact length of the day, divided the earth into twenty-four time zones one hour apart, and fixed a precise beginning of the universal day. But the world was slow to adopt the system, for all its obvious practicality.

Japan coordinated railroads and telegraphic services nine hours ahead of Greenwich in 1888. Belgium and Holland followed in 1892; Germany, Austria-Hungary, and Italy in 1893; but in 1899, when

John Milne surveyed how countries throughout the world determined their time and its relation to Greenwich, there was still a great deal of confusion. Telegraph companies in China used a time that was approximately the same as in Shanghai; foreigners in coastal ports used their own local time taken from solar readings; and all other Chinese used sundials. In Russia there were odd local times such as that of St. Petersburg—two hours, one minute, and 18.7 seconds ahead of Greenwich. In India hundreds of local times were announced in towns by gongs, guns, and bells.[5]

Among the countries in Western Europe, France had the most chaotic situation, with some regions having four different times, none of which had a simple conversion to Greenwich time. Each city had a local time taken from solar readings. About four minutes behind each local time was astronomical time taken from fixed stars. The railroads used Paris time, which was nine minutes and twenty-one seconds ahead of Greenwich. A law of 1891 made it the legal time of France, but the railroads actually ran five minutes behind it in order to give passengers extra time to board: thus the clocks inside railway stations were five minutes ahead of those on the tracks.[6] In 1913 a French journalist, L. Houllevigue, explained this "retrograde practice" as a function of a national pride, expressed in the wording of a law of 1911 promoting the system that other countries of Europe had adopted twenty years earlier. The French law declared that "the legal time in France and Algeria is the mean Paris time slowed nine minutes and twenty-one seconds." Houllevigue pointed out the Anglophobic intent of the wording: "By a pardonable reticence, the law abstained from saying that the time so defined is that of Greenwich, and our self-respect can pretend that we have adopted the time of Argentan, which happens to lie almost exactly on the same meridian as the English observatory."[7] In spite of their previous isolation the French finally took the lead in the movement for unified world time based on the guidelines of 1884. If the zero meridian was to be on English soil, at least the institution of world time would take place in France. So President Raymond Poincaré had Paris host the International Conference on Time in 1912, which provided for a uniform method of determining and maintaining accurate time signals and transmitting them around the world.

The wireless telegraph made it all possible. As early as 1905 the United States Navy had sent time signals by wireless from Washington. The Eiffel Tower transmitted Paris time in 1910 even before it was legally declared the time of France. By 1912 the system was expanded with installations in Nancy, Charleville, and Langres so that the entire country could receive the same signals simultaneously. Houllevigue boasted that Paris, "supplanted by Greenwich as the origin of the meridians, was proclaimed the initial time center, the watch of the universe."[8] The observatory at Paris would take astronomical readings and send them to the Eiffel Tower, which would relay them to eight stations spaced over the globe. At 10 o'clock on the morning of July 1, 1913, the Eiffel Tower sent the first time signal transmitted around the world. The independence of local times began to collapse once the framework of a global electronic network was established. Whatever charm local time might have once had, the world was fated to wake up with buzzers and bells triggered by impulses that traveled around the world with the speed of light.

Around the time of the International Conference on Time various proposals for calendar reform were made. Nothing concrete came of them, but they reveal a parallel effort to rationalize public time. In 1912 an American reformer noted that while the year, month, and day have a basis in nature, the week and the hour are entirely artificial. The "stupid" arrangement of the calendar, he argued, should be simplified by dividing the year into four equal seasons of 91 days each and leaving out New Year's day and one day every four years.[9] In an introduction to a proposal of 1913 by Paul Delaporte for calendar reform, the French scientific writer Camille Flammarion applauded the achievements of the International Conference on Time, observed that the unequal divisions of the year should be modified, and endorsed Delaporte's proposal to shorten every month to twenty-eight days, with an intercalary period added in the middle of the year so that workers could be paid every four weeks and rent would be due and interest computed for the same length of time every month. The year would always begin on the same day, thus obviating the reprinting of calendars.[10] In 1914 an Englishman emphasized difficulties in scheduling for business and government and recommended a calendar in which each quarter would be composed of two thirty-day months and one thirty-one day month, with leap year not counted at all.[11] A German reformer proposed a "hundred-hour day" composed of units approximately equivalent to a quarter-hour. Just as the introduction of the decimal system in spatial measurements had enabled the German people to make rapid economic development, he contended, so might the introduction of a temporal decimal system liberate resources for other pursuits.[12]

A science-fiction novel of 1893 about life on Mars incorporated some of the developments in standard time made in the previous decade. In Henry Olerich's *A Cityless and Countryless World,* every dwelling and working place was furnished with clocks that were astronomically regulated and electronically synchronized. The standard for money was time: "In business, when you say I want so many dollars, cents and mills for an article, we say I want so many days, hours, minutes and seconds for it."[13] Martian currency consisted of paper bills stamped with units of time. This time money was perhaps inspired by the introduction of time-recording machines for workers. The same year that Olerich's book was published, an article in *Scientific American* described a machine, in service since 1890, that stamped an employee's card with the time he entered and left.[14] Though he was paid in dollars, the time-stamped tape determined the amount. Olerich had only to make the slightest alteration to create a utopian world where time *is* money.

Punctuality and the recording of work time did not originate in this period, but never before had the temporal precision been as exact or as pervasive as in the age of electricity.[15] From the outset there were critics. Some pathological effects were noted in that catalog of medical alarmism, George Beard's *American Nervousness.* He blamed the perfection of clocks and the invention of watches for causing nervousness wherein "a delay of a few moments might destroy the hopes of a lifetime."[16] Every glance at the watch for these nervous types affects the pulse and puts a strain on the nerves. There were many other alarmists who reacted adversely to the introduction of standard time, but the modern age embraced universal time and punctuality because these served its larger needs. That prerevolutionary, pastoral image in Arthur Koestler's *Darkness at Noon,* of Russian peasants coming to the railroad station at dawn to wait for a train that might not arrive until the late afternoon, suggested a life style more frustrating and wasteful than it was idyllic.

The proponents of world time were few, and none of them (aside from Moltke) were well known beyond the narrow circle of fellow reformers. Nevertheless the concept of public time was widely accepted as a proper marker of duration and succession. There were no elaborate arguments on its behalf because there seemed to be no need. The passion in the debate about homogeneous versus heterogeneous time was generated rather by those novelists, psychologists, physicists, and sociologists who examined the way individuals create as many different times as there are life styles, reference systems, and social forms.

∞

Of all the assaults on the authority of uniform public time that appeared in the imaginative literature of this period, the most direct was the one assigned to the Russian anarchist in Joseph Conrad's *The Secret Agent* (1907). His task as an *agent provocateur* in England was to blow up the Greenwich Observatory. Conrad could not have picked a more appropriate anarchist objective, a more graphic symbol of centralized political authority.

The heterogeneity of private time and its conflict with public time was explored in a number of literary works. In 1890 Oscar Wilde imagined a sinister discordance between body time and public time for his Dorian Gray, whose portrait aged in his place while he stayed young. When Dorian kills the portraitist, the magic ends and the two times race back to their proper positions: the portrait changes back to innocent youth, and Dorian's face registers the corruption that the portrait had concealed.

Marcel Proust's *Remembrance of Things Past* takes place in a clearly identifiable public time from the Dreyfus affair to World War I. But the private time of its narrator, Marcel, moves at an irregular pace that is repeatedly out of phase with that of the other characters and defies reckoning by any standard system. Marcel reflected that his body kept its own time while he slept, "not on a dial superficially marked but by the steadily growing weight of all my replenished forces which, like a powerful clockwork, it had allowed, notch by notch, to descend from my brain into the rest of my body."[17] In the search for lost time, mechanical timepieces will be utterly useless as Proust learns to listen for the faint stirrings of memories implanted in his body long ago and destined to recur to him in unpredictable and enchanting ways.

The dials that superficially mark time for Proust are virtual enemies in the troubled lives of Franz Kafka's heroes. When Gregor Samsa awakens in *The Metamorphosis* and discovers himself to be a great insect, his distress is intensified by the discovery that he is going to miss his train. This first break with the routine of public time is symbolic of the complete breakdown of his relationship with the world. In *The Trial* (1914–15) Josef K. tells his employer about the summons to his first hearing: "I have been rung up and asked to go somewhere, but they forgot to tell me when." He assumes he should arrive at nine but oversleeps and arrives over an hour late. A few minutes later the Examining Magistrate reproaches him: "You should have been here an hour and five minutes ago."[18] The next week when he returns he is on time but no one shows up. This confusion mirrors his larger problems with the world. He eventually loses the ability to differentiate inner and outer sources of guilt just as he was unable to determine who was responsible for his missed appointments. In a diary entry of 1922 Kafka commented on the maddening discordance between public and private time. "It's impossible to sleep, impossible to wake, impossible to bear life or, more precisely, the successiveness of life. The clocks don't agree. The inner one rushes along in a devilish or demonic—in any case, inhuman—way while the outer one goes, falteringly, its accustomed pace."[19] His heroes feel absurd when they arrive too early and guilty when late.

The public time that Proust found superficial and Kafka terrifying, Joyce found to be arbitrary and ill-suited to order the diverse temporal experiences of life. In *Ulysses* he modified traditional treatment of time by compressing Odysseus's twenty years of travel into sixteen hours in the life of Leopold Bloom as he meandered about the shops and pubs of downtown Dublin. During that day we are given a microscopic account of everything Bloom does, thinks, and feels, but within the limited duration of the story Joyce widens the temporal range with interior monologues and authorial comments about Bloom's unique experience of time and its relation to the infinite expanses of cosmic time.

The heterogeneity of time is presented formally by means of the specific rhythm of the prose of each chapter.[20] In the "Aeolus" episode the rhythm varies like the unpredictable winds which blew Odysseus off course and which in *Ulysses* blow like the windbag newspapermen whose views are chopped up into newspaper-length articles. In "Lestrygonians" Bloom goes for lunch and the rhythm is the peristaltic motion of digestion. Bloom looks into the river and reflects on the way everything flows: food through the alimentary canal, the foetus through the birth canal, the traffic of Dublin, his bowels, thought, language, history, and time itself. The prose in "Oxen of the Sun" approximates the long cadences of a woman in labor. In "Ithaca" Joyce describes the journey home of Stephen and Bloom as a catechism in which their thoughts, like their footsteps, alternate in a series of questions and answers. And in the final episode the rhythm is that of the flow of Molly's stream of consciousness.

In the midst of telling how Bloom flopped over the back fence to get into his home, Joyce suddenly breaks into the narrative with a list of possible ways of describing when Bloom last weighed himself. It was "the twelfth day of May of the bissextile year one thousand nine hundred and four of the christian era (jewish era five thousand six hundred and sixty-four, mohammedan era one thousand three hundred and twenty-two), golden number 5, epact 13, solar cycle 9, dominical letters C B, Roman indication 2, Julian period 6617, MXMIV."[21] We are told that Bloom walked around Dublin precisely on June 16, 1904, only Joyce leaves us wondering exactly when that is.

Joyce's reminder that time is relative to the system by which it is measured also points to Einstein's theory that all temporal coordinates are relative to a specific reference system. In a textbook of 1883 Ernst Mach raised some questions about classical physics that anticipated one of the greatest scientific revolutions ever. Mach rejected Newton's views of absolute space and absolute motion and dismissed his absolute time as an "idle metaphysical conception."[22] This passing shot at classical mechanics triggered a series of modifications that eventually culminated in the bold dismantling of it by Einstein. The next blow to absolute time came from an experiment intended to show the existence of a luminiferous ether through which light was propagated. According to classical mechanics the speed of light perpendicular to the ether flow generated by the passage of the earth through it ought to have been faster than the speed of light in line with it, but the famous experiment of Michelson and Morley of 1888 showed no detectable difference. This troublesome result led to several hypotheses about a slowing down of time from its movement through ether.

In 1895 Hendrick Lorentz speculated that perhaps time was dilated by motion through the ether just enough to account for the observed equality of the two speeds of light.[23] This position was midway between classical physics and relativity theory. It looked forward to relativity by suggesting that time measurements are modified by motion, that there is a plurality of "local times," each dependent on the relative motion of the clock and observer. But it adhered to the traditional concept of absolute time by insisting that the change actually took place *in the object* as a result of motion through the ether, similar to the way other elastic bodies contract in the direction of their motion through a gas or fluid. Lorentz believed that the dilation of time was real, and he thus retained the concept of absolute time. Einstein would argue that the dilation of time was only a perspectival effect created by relative motion between an observer and the thing observed. It was not some concrete change inherent in an object but merely a consequence of the act of measuring. Such an interpretation rejected absolute time, because time only existed when a measurement was being made, and those measurements varied according to the relative motion of the two objects involved.

With the special theory of relativity of 1905 Einstein calculated how time in one reference system moving away at a constant velocity appears to slow down when viewed from another system at rest relative to it, and in his general theory of relativity of 1916 he extended the theory to that of the time change of accelerated bodies. Since every bit of matter in the universe generates a gravitational force and since gravity is equivalent to acceleration, he concluded that "every reference body has its own particular time."[24] In a subsequent popularization of his theory he contrasted the older mechanics, which used only one clock, with his theory which requires that we imagine "as many clocks as we like."[25] The general theory of relativity had the effect, figuratively, of placing a clock in every gravitational field in the universe, each moving at a rate determined by both the intensity of the gravitational field at that point and the relative motion of the object observed. Einstein, who could not afford to have a clock on the wall of his room when he was working in the patent office in Berne, had filled the universe with clocks each telling a different correct time.

Although several investigations of the social origin of time were made in the late nineteenth century,[26] the prodigious work of Emile Durkheim constitutes the first one of major significance. The sociology and anthropology of that age was full of information about primitive societies with their celebration of the periodic processes of life and the movement of heavenly bodies, their vital dependence on seasonal change and the rhythmic activity of plants and animals, their exotic commemorations of ancestral experience, and their cyclic and apocalyptic visions of history. It is no wonder that Durkheim came to believe in the social relativity of time.

In *Primitive Classification* (1903) Durkheim mentioned in passing that time is closely connected with social organization, and in *The Elementary Forms of the Religious Life* (1912) he explored the subject in detail. There he distinguished between private time and "time in general," which has a social origin: "the foundation of the category of time is the rhythm of social life." More concretely, "the divisions into days, weeks, months, and years, etc., correspond to the periodical recurrence of rites, feasts, and public ceremonies." Societies organize their lives in time and establish rhythms that then come to be uniformly imposed as a framework for all temporal activities. Thus "a calendar expresses the rhythm of the collective activities, while at the same time its function is to assure their regularity."[27]

Arguments for a relativity of time were also made by psychiatrists and philosophers. Karl Jaspers's work in phenomenological psychiatry outlined different modes of perceiving time and space that can occur in mental illness.[28] In a history of the idea of memory and time Pierre Janet recounted the contributions of "a whole generation" of experimental psychologists and clinicians in the late nineteenth century who investigated subjective time. Citing his own account of a distorted sense of time among the mentally ill in *Névroses et idées fixes* (1898), he then characterized Jean Guyau's essay of 1890 as opening "a new era in the psychology of time."[29] Janet also discussed Charles Blondel's *La Conscience morbide* of 1914, which examined the diverse temporal worlds of the mentally ill. One patient lived "from day to day, like an animal, in a kind of retreat from the past and the future," with time appearing interminable. A few days in the past seemed like years, and all events in time were mixed in nightmarish confusion. For another patient, "Gabrielle," time contracted and dreaded future events were transposed into the past and generated anxiety as if they had already occurred and would remain forever present. It was as if her mind constantly surveyed the entire temporal range to collect and condense all morbid thoughts into a present and inescapable experience of anxiety.[30]

∞

The argument on behalf of the atomistic nature of time had a variety of sources. Perhaps most influential was Newton's calculus, which conceived of time as a sum of infinitesimally small but discrete units. Clocks produced audible reminders of the atomistic nature of time with each tick and visible representations of it with their calibrations. The modern electric clock with the sweeping fluid movement of its second hand was invented in 1916. Until then clocks could offer no model for time as a flux.[31] Experimental psychologists attempted to determine the precise intervals of human responses and the shortest duration one can detect. In the laboratories of Gustav Fechner and Wilhelm Wundt metronomes and watches were used to study human life as a construction of measurable bits of time.

In the late 1870s two pioneers of the cinema studied atomized movement by means of a series of still photographs. Eadweard Muybridge recorded the motion of a galloping horse by setting up some cameras in line along the course with a thin wire strung across the track that triggered the shutter as the horse ran by. He went on to make sequential photographic studies of human and animal movements. In 1882 the French physician E. J. Marey began to study movement with a technique he called chronophotography—literally, the photography of time: "a method which analyzes motions by means of a series of instantaneous photographs taken at very short and equal intervals of time."[32] Marey was particularly interested in the aerodynamics of flight and developed an apparatus for photographing birds simultaneously from three different points of view. He believed that the best way to understand motion was to break it up into parts and then reassemble them into a composite picture or plastic model.

When the cinema was improved to permit the first public showing in 1896, it also broke up motion into discrete parts. The Futurist photographer Anton Bragaglia proposed a technique he called photodynamism, which involved leaving the shutter open long enough to record the blurred image of an object in motion.[33] This, he believed, offered the only true art of motion in contrast to both chronophotography and cinematography, which broke up the action and missed its "*intermovemental* fractions." Bragaglia's photographs look more like the errors of a beginner than an artistic solution of the problem of motion, and they offer a vivid, if somewhat ludicrous, illustration of the difficulty that all the visual arts had in capturing the fluid nature of movement or time.

The difficulty painters have rendering the movement of an object in time has always been a frustrating limitation of the genre. That limitation, formalized in the eighteenth century with Gotthold Lessing's division of the arts as temporal and spatial, came to haunt painters of the late nineteenth century.[34] Artists had often attempted to imply a past and future by painting a moment that pointed beyond the present. The Impressionists attempted to render time more directly with a sequence of paintings of the same motif at different times of the day, seasons, and climatic conditions, as in Claude Monet's haystacks and his series depicting the Rouen Cathedral. Monet himself explained, "One does not paint a landscape, a seascape, a figure. One paints an impression of an hour of the day."[35] The Impressionists also tried to portray their impression of motion, but no matter how well they suggested the luminous shifting caused by a passing cloud or the ripple of the wind on water, everything was fixed in a single moment.

The Cubists attempted to go beyond the instant with multiple perspectives, at least so a group of early commentators argued. In 1910 Leon Werth wrote that Picasso's Cubist forms show "the sensations and reflections which we experience with the passage of time." In the same year the Cubist painter Jean Metzinger suggested that in Braque's paintings "the total image radiates in time." In 1911 Metzinger explained how he thought the multiple perspective of the Cubists added the temporal dimension. "They have allowed themselves to move round the object, in order to give a concrete representation of it, made up of several successive aspects. Formerly a picture took possession of space, now it reigns also in time." In 1910 Roger Allard noted that a Metzinger painting was a "synthesis situated in the passage of time."[36] These arguments are all overstated. The multiple and successive perspective that the Cubists did integrate into a single painting does not justify the conclusion that they radiate *in* time. No matter how many successive views of an object are combined, the canvas is experienced in a single instant (aside from the time necessary for the eye to scan the surface). The Cubists toyed with the limitations of their genre, perhaps even with some intended mockery. Their inventions presented time in art in a new way, but that did not constitute the experience of time as it passes.

In 1899 the Dutch critic Ernst Te Peerdt observed in *The Problem of the Representation of Instants of Time in Painting and Drawing* that our visual field is not composed of a series of timeless unities. Each instant of perception synthesizes a sequence of numerous perceptions. "It is precisely those moments that are put together as a simultaneity, a *Nebeneinander,* which constitute a sequence, a *Nacheinander,* in the seeing of an object."[37] Unlike a still photograph the eye is able to integrate a succession of observations. The task of the painter is to integrate temporal sequence with forms in space. Despite Te Peerdt's argument that good visual art can suggest a sequence, it nevertheless cannot portray the movement of an object or the passage of time.

No motif gives as graphic a reminder of the atomized nature of time as a clock, and there are few clocks in the art of this period. Around 1870 Paul Cézanne painted a still life dominated by a massive black clock without hands—symbol of the timelessness he sought to create in his painting. I have not been able to find clocks again in any major work of Western art until 1912 with *The Watch* by

Juan Gris. Here time is out of joint in several respects. The watch is rotated ninety degrees, making a first quick reading of it difficult. It is broken into four quadrants, only two of which are visible. The other two are obscured and the minute hand is missing, making an exact reading impossible even after some contemplation. On this Cubist watch, time is fragmented, discontinuous, and ambiguous, but fixed forever by the hand that points to XI on the visible quarter-face. In 1913 Albert Gleizes put a clock in a Cubist portrait and effaced half the numbers.[38] The time is precisely 2:35, but the clock is useless for readings on the effaced portion. Gleizes broke up time as easily as he fragmented objects and space.

In *Enigma of the Hour* of 1912 Giorgio de Chirico painted a clock with plainly visible time towering over a small figure that looks up at its imposing grandeur. De Chirico included prominent clocks looming like Van Gogh's suns in a number of paintings: *The Delights of the Poet* (1913), *The Soothsayer's Recompense* (1913), *The Philosopher's Conquest* (1914), and *Gare Montparnasse* (*The Melancholy of Departure*) (1914). In all but the first of these a railroad train chugs by, which suggests that he may have deliberately connected the railroads and the standard time that began to be imposed on a global scale precisely in 1912. Although the titles of the paintings suggest transcendence in space and time, the clocks fix the action in a single and immutable moment. There is a rigid, static quality to them that no train journey or soothsayer's vision could undo. Unlike Cézanne, Gris, and Gleizes, de Chirico chose to concede that the plastic arts are condemned to a single moment, and he celebrated the dominating power of clock time by making its universal symbol so prominent.

As if that concession to round and wholly visible clocks were too much to endure, some years later Salvador Dali painted three melting watches in *The Persistence of Memory* (1931). One is hanging from a tree in a reminder that the duration of an event may be stretched in memory. Another with a fly on it suggests that the object of memory is some kind of carrion that decays as well as melts. The third deformed watch curls over a hybrid embryonic form—symbol of the way life distorts the geometrical shape and mathematical exactness of mechanical time. The one unmelted watch is covered with ants that seem to be devouring it as it devours the time of our lives.

Aside from de Chirico, who placed readable clocks clearly in view, all the other painters deformed, obscured, or defaced these reminders that their genre is incapable of representing time. Lessing's iron law was challenged but never surmounted. The argument on behalf of the flux of time would be carried through more effectively by the philosophers and novelists who could give it an extended formulation.

The theory that time is a flux and not a sum of discrete units is linked with the theory that human consciousness is a stream and not a conglomeration of separate faculties or ideas. The first reference to the mind as a "stream of thought" appears in an essay by William James in 1884, which criticized David Hume's view of the mind as an "agglutination in various shapes of separate entities called ideas" and Johann Herbart's representation of it as the result of "mutual repugnancies of separate entities called *Vorstellungen.*"[39] His descriptions of this "vicious mode of mangling thought's stream," this "illegitimate" and "pernicious" treatment of atoms of feeling, anticipate Bergson's characterizations of the spatial representation of time as a "vice." James distinguishes between the separate "substantive parts" and the fluid "transitive parts" which have been neglected by sensationist psychologists. Utilizing his favorite metaphor for the activity of the mind, James ridicules associationist psychology as saying that the river is composed of "pailfuls" of water. Rather "every image in the mind is steeped and dyed in the free water that surrounds it." Each mental event is linked with those before and after, near and remote, which act like a surrounding "halo" or "fringe." There is no single pace for our mental life, which, "like a bird's life, seems to be made of an alternation of flights and perchings." The whole of it surges and slows, and different parts move along at different rates, touching upon one another like the eddies of a turbulent current.

In 1890 James repeated these arguments in a popular textbook of psychology and added a formulation that subsequently became famous. "Consciousness does not appear to itself chopped up in bits. Such words as 'chain' or 'train' do not describe it fitly . . . It is nothing jointed; it flows. A 'river' or a 'stream' are the metaphors by which it is most naturally described. In talking of it hereafter, let us call it the stream of thought, of consciousness."[40] Although James and Bergson tended to use somewhat different metaphors to characterize thought, they agreed that it was not composed of discrete parts, that any moment of consciousness was a synthesis of an ever changing past and future, and that it flowed.

In *An Introduction to Metaphysics* (1903) Henri Bergson approached the subject of the fluid nature of time by distinguishing two ways of knowing: relative and absolute. The former, impoverished kind is achieved by moving around an object or by coming to know it through symbols or words that fail to render its true nature. Absolute knowledge is achieved by experiencing something as it is from within. This absolute knowledge can only be given by intuition, which he defined as "the kind of intellectual sympathy by which one places oneself within an object in order to coincide with what is unique in it and consequently inexpressible." Here we encounter a major difficulty. If absolute knowledge, the goal of his philosophy, is inexpressible, how can we write about it usefully? Bergson strives to communicate this kind of knowing, and the existence that comes from it, by a series of analogies and metaphors, all of which, he is quick to admit, can never fully express it, but the metaphors succeed in part because we all share one experience of intuition: "our own personality in its flowing through time—our self which endures." When he contemplated his inner self he found "a continuous flux, a succession of states, each of which announces that which follows and contains that which precedes it." This inner life is like the unfolding of a coil or a continual rolling of a thread on a ball. But as soon as he suggested these similes he conceded that they were misleading, because they referred to something spatial, whereas mental life is precisely that which is not extended in space but in time. In a final effort to provide an approximate analogy Bergson directed the reader to imagine "an infinitely elastic body [which cannot be imagined], contracted, if it were posssible [which it is not] to a mathematical point." Imagine a line drawn out of that point, and then focus not on the line but on the action by which it is traced. Then "let us free ourselves from the space which underlies the movement in order to consider only the movement itself, the act of tension or extension, in short pure mobility. We shall have this time a more faithful image of the development of our self in duration."[41] Bergson thus asks us to imagine something which is unimaginable, conceive of an action of that unimaginable image which is inconceivable, and then effect a limitation of our attention to an aspect of that action which is impossible. The effect of this trying analogy is to underline the difficulty of expressing in words the true nature of our existence in time, which he called "duration" (*durée*).

Bergson became incensed at the way contemporary thought, especially science, tended to distort the real experience of *durée* and represent it spatially, as on a clock. A quarter of an hour *becomes* the 90-degree arc of the circle that is traversed by the minute hand. In another argument against the translation of time into space he refuted Zeno's "proofs" that motion or change is impossible. Zeno concluded that if an arrow in flight passes through the various points on its trajectory, it must be at rest when at them and therefore can never move at all. Bergson countered that the mistake was in assuming that the arrow can be *at* a point. "The arrow never is in any point of its course. The most we can say is that it might be there, that it passes there and might stop there."[42] Movement, like time, is an indivisible flux. Zeno founders on the assumption that such a division is possible and that "what is true of the line which traces the path followed is true of the movement." The line may be divided but the movement may not. And so with time: we cannot consider movement as a sum of stoppages nor time as a sum of temporal atoms without distorting their essentially fluid nature.

Bergson's theory of duration generated a broad and varied cultural response ranging from passionate support to frantic condemnation. In the 1890s Georges Sorel developed a blueprint for socialist revolution that was intended to create an "intuition" of socialism for the workers by having them participate in a general strike. In Bergsonian language Sorel argued that the scientific analysis of revolutionary socialism is static and misses the essential nature of historical change, which must be intuited in its durational flux. He found the European working class stopped in its revolutionary course like Zeno's arrow was stopped in flight—artificially frozen by analyses that obscured the essential indivisibility of change and movement.[43] Charles Péguy used Bergson's philosophy to attack the Cartesian tradition that he believed locked French thought in unproductive rigidity.[44] Péguy explained the spiritual death of modern Christianity by its mindless repetition of fixed ideas: layers of habit stifle the dynamic energies of true faith.

In the concluding paragraph of *Creative Evolution* Bergson outlined the proper aim of the philosopher who dispenses with all fixed symbols. "He will see the material world melt back into a single flux, a continuity of flowing, a becoming." This vision horrified some of Bergson's critics. Perhaps the most colorful of his detractors, and certainly the most hysterical, was the English artist Wyndham Lewis, who in 1927 concluded that Bergson's romance with flux was the start of a most unfortunate development in the modern world which cooked up all the articulate distinctions of clear analysis into a murky durational stew. Lewis accused Bergson of putting the hyphen between space and time, and he registered his passionate disapproval:

> As much as he enjoys the sight of things 'penetrating' and 'merging' do we enjoy the opposite picture of them standing apart—the wind blowing between them, and the air circulating freely in and out of them: much as he enjoys the 'indistinct,' the 'qualitative,' the misty, sensational and ecstatic, very much more do we value the distinct, the geometric, the universal, non-qualitied—the clear light, the unsensational. To the trance of music, with its obsession of *Time,* with its emotional urgency and visceral agitation, we prefer what Bergson calls 'obsession of Space.'

Lewis viewed Bergson's philosophy and Einstein's physics as well as a good deal of literature of the period as "one vast orthodoxy" that conspired to remove clean lines from art and separate faculties from human perception. He found another example of Bergsonian fluidity in the "softness, flabbiness, and vagueness" of James Joyce's *Ulysses.*[45]

Joyce's treatment of the stream of consciousness is the culmination of a literary development first explored by Sterne and revived in 1888 by Edouard Dujardin in a novel in which the protagonist's thoughts about past and future are presented along with his current perceptions. This technique has been identified as the direct interior monologue, because the inner workings of the mind are given directly without authorial clarifications such as "he thought" or explanations of what is happening.[46] Many writers before Dujardin had attempted to *analyze* the thoughts of a character, and occasionally they had narrated as if through a character's consciousness, but none had made prespeech levels of consciousness the subject of an entire novel. Although the technique is intended to recreate the entirety of consciousness, it is especially well suited to deal with its temporal fluidity, as the following passage reveals.

> The hour is striking, six, the hour I waited for. Here is the house I have to enter, where I shall meet someone; the house; the hall; let's go in. Evening has come; good the air is new; something cheerful in the air. The stairs; the first steps. Supposing he has left early; he sometimes does; but I have got to tell him the story of my day. The first landing; wide, bright staircase; windows. He's a fine fellow, friend of mine; I have told him all about my love affair. Another pleasant evening coming on. Anyway he can't make fun of me after this. I'm going to have a splendid time.[47]

Although the narrative time of the segment lasts only a few seconds, the private time extends over a large duration and shifts erratically in it. Dujardin's direct interior monologue expressed the inner workings of the mind with its brief span of attention, its mixture of thought and perception, and its unpredictable jumps in space and time.

The term "stream of consciousness" came into literary use after 1890, following William James's famous definition. Although *Ulysses* was no mere application of either Dujardin's direct interior monologue or James's stream of consciousness, it provides a superb embodiment of a generation of developments in literature and philosophy on the nature of human consciousness and its life in time. Sections of direct interior monologue are scattered throughout the novel, with a final uninterrupted flow of it as Molly Bloom fades into sleep at the end. The different verb tenses in one passage reveal her widely ranging leaps about the temporal spectrum.

> . . . my belly is a bit too big Ill have to knock off the stout at dinner or am I getting too fond of it the last they sent from ORourkes was as flat as a pancake he makes his money easy Larry they call him the old mangy parcel he sent at Xmas a cottage cake and a bottle of hogwash he tried to palm off as claret that he couldnt get anyone to drink God spare his spit for fear hed die of the drouth or I must do a few breathing exercises I wonder is that antifat any good might overdo it thin ones are not so much the fashion now garters that much I have the violet pair I wore today thats all he bought me out of the cheque he got on the first . . .[48]

The metaphor of "stream" is not entirely appropriate to describe this mental activity, because it suggests a steady flow in a fixed course, while Molly's mind revolves about her universe in defiance of conventional calculations of its pace or direction. In this final episode Joyce achieves the fullest expansion of the time of Molly's world as it is experienced in her consciousness. It is the only episode to which Joyce assigned no particular hour of the day and its symbol is that of eternity and infinity—"∞."[49] The rigid dimensions of conventional time with its sharp dividers are useless to plot the action of her mind. It is as irrelevant to ask when Molly is having these thoughts as it is silly to ask where. They are an endless rewriting of the story of her life that change with every passing reflection and every flickering of sexuality. Her memory is not a faculty for bringing fixed ideas out of the past; it is one that enables her to transform them repeatedly in the endless creativity of her present consciousness, where all is fluid without separate thoughts or isolated moments of time.

∞

The structure of history, the uninterrupted forward movement of clocks, the procession of days, seasons, and years, and simple common sense tell us that time is irreversible and moves forward at a steady rate. Yet these features of traditional time were also challenged as artists and intellectuals envisioned times that reversed themselves, moved at irregular rhythms, and even came to a dead stop. In the *fin de siècle,* time's arrow did not always fly straight and true.

This challenge had a basis in two technological developments: the electric light and the cinema. The first commerically practical incandescent lamp was invented by Thomas Alva Edison in 1879, and three years later he opened the first public electric supply system at the Pearl Street district of New York that made possible the widespread use of the electric light. The eminent historian of architecture Rayner Banham has called electrification "the greatest environmental revolution in human history since the domestication of fire." One of the many consequences of this versatile, cheap, and reliable form of illumination was a blurring of the division of day and night. Of course candles and gas lamps could light the darkness, but they had not been able to achieve the enormous power of the incandescent light bulb and suggest that the routine alternation of day and night was subject to modification. One of many such observations occurs in a novel of 1898, where a Broadway street scene at dusk is illuminated by a flood of "radiant electricity" which gave the effect of an "immortal transformation" of night into day.[50]

From another perspective, the cinema portrayed a variety of temporal phenomena that played with the uniformity and the irreversibility of time. A pioneer of the cinema in France, Georges Méliès, recalled an incident that inspired a series of tricks of motion picture photography. One day in 1896 he was filming a street scene at the Place de l'Opéra and his camera jammed. After a few moments he got it going and continued filming, and when he projected the entire sequence it created the illusion that an omnibus had suddenly changed into a hearse.[51] This suggested to Méliès several other effects he could achieve by stopping the camera and changing the scene. He used this technique in *The Vanishing Lady* (1896) where a skeleton suddenly becomes a living woman, implying both a jump in time and its reversal.

Méliès stopped the camera to effect these tricks. The American film maker Edwin S. Porter discovered that time could be compressed, expanded, or reversed in a more versatile way by editing the film. Intervals of time could be literally cut out of a sequence and temporal order could be modified at will. He applied those techniques in *The Life of an American Fireman* (1902), where we see first someone setting off a fire alarm and then the sleeping firemen just before the alarm sounds. David Griffith developed the technique of parallel editing to expand time by showing simultaneous action in response to a single event. In *A Corner on Wheat* (1909) Griffith first used the freeze-frame technique by having his actors hold still to create the illusion of stopping time.[52] In 1916 Hugo Münsterberg noted that several contemporary playwrights attempted to imitate the cinema and use time reversals on stage as in Charlotte Chorpenning's *Between the Lines,* where "the second, third, and fourth acts lead up to the three different homes from which the letters came and the action in the three places not only precedes the writing of the letters, but goes on *at the same time.*"[53]

An even more striking representation of time reversal was produced by running film backwards through the projector, first tried by Louis Lumière in *Charcuterie mécanique* (1895). One cinema critic described these amazing effects: boys fly out of water feet first and land on the diving board, firemen carry their victims back into a burning building, and eggs unscramble themselves. His account of a mass of broken glass ascending through space and reforming on a table into the perfect original suggests a Cubist decomposition in reverse.[54]

Several prominent novelists commented on the problems they faced in presenting the passage of time; some found solutions unmistakably parallel to, if not directly inspired by, the innovative temporal manipulations of the cinema. Conrad's method was to isolate a particular moment and hold it up for extended scrutiny as if suspended in time.[55] Ford Madox Ford summarized a view that he and Conrad shared.

> It became very early evident to us that what was the matter with the Novel, and the British Novel in particular, was that it

> went straight forward, whereas in your gradual making acquaintanceship with your fellows you never do go straight forward . . . To get . . . a man in function you could not begin at his beginning and work his life chronologically to the end. You must first get him with a strong impression, and then work backwards and forwards over his past.
>
> Life does not say to you: in 1914 my next door neighbor, Mr. Slack, erected a greenhouse and painted it with Cox's green aluminum paint . . . If you think about the matter you will remember in various unordered pictures, how one day Mr. Slack appeared in his garden and contemplated the wall to his house.[56]

In *Ulysses* Joyce created a dramatic interruption in the forward movement of narrative time. As Bloom approaches a brothel he steps back to avoid a street cleaner and resumes his course forty pages and a few seconds later. In those few seconds of his time the reader is led through a long digression that involves dozens of characters and covers a period of time far exceeding the few seconds that elapsed public time would have allowed. Virginia Woolf believed that it was the writer's obligation to go beyond "the formal railway line of a sentence." "This appalling narrative business of the realist: getting on from lunch to dinner, it is false, unreal, merely conventional." She also recorded Thomas Hardy's observation about the new way of rendering time in literature: "They've changed everything now. We used to think there was a beginning and a middle and an end. We believed in the Aristotelian theory."[57]

Psychologists and sociologists observed modifications of the continuity and irreversibility of time in dreams and psychoses and in religion and magic. In a letter of 1897 Freud commented on the temporal distortions he observed in dreams and fantasies. There occurs a distortion of memory that comes from "a process of fragmentation in which chronological relations in particular are neglected."[58] In *The Interpretation of Dreams* Freud surveyed how the sequence of experiences in the course of our conscious life is rearranged to suit the needs of the dreaming mind. The psychic forum of our instinctual life, primary process, entirely disregards the demands of logic and space as well as time. In 1920 he summarized his theory that unconscious mental processes are "timeless," for the passage of time does not change them in any way and "the idea of time cannot be applied to them."[59]

"Summary Study of the Representation of Time in Religion and

Magic" (1909), by Henri Hubert and Marcel Mauss, argued that time in religion and magic serves a social function and provides a framework for the qualitative rather than the quantitative experience of succession. They viewed time as heterogeneous, discontinuous, expandable, and partially reversible. With Durkheim they contended that the social origin of time insures its heterogeneity. In contrast to most views of the calendar as quantitative, they proposed that it is qualitative, composed of special days and seasons. Sacred time is also discontinuous, for such events as the appearance of a deity interrupt ordinary continuity. Periods separated in chronology can be linked in their sacred function to give time a "spasmodic character." Some special moments may "contaminate" the entire interval that follows, and instants may be united if they have the same religious significance. Time can also be expanded, as "heroes can live years of magical life in an hour of ordinary human existence." Their observation that rites of entry and exit may be united over time implies a partial reversibility as end is joined with beginning.

Following Bergson, Hubert and Mauss believed that time is dynamic, and they endorsed his substitution for time-images of position and succession the concept of time as an "*active tension* by which consciousness realizes the harmony of independent durations and different rhythms." But they differed from Bergson in the extent to which they were willing to allow public time to be part of the inner consciousness of time. They maintained that public time is one pole of the "scale of tensions of consciousness." "The play of notions, which distinguishes the psychological reality of successive images, consists in the adjustment of two series of representations. The one is constant and periodic: it is the calendar . . . The other constructs itself perpetually by the action of generating new representations. The mind works constantly to associate in a single tension certain elements of these two series." The time of magic and religion is a compromise among interpsychic tensions set up by our private experience of a uniform and homogeneous time. The celebration of an anniversary, especially one associated with magical and sacred happenings, is an integration of our own unique rhythm of living with the uniform rhythm observed by a social community.[60]

Their argument that the divisions of time "brutally interrupt the matter that they frame" parallels a revolutionary theory of Einstein's about the interaction of time and matter that further challenged the classical theory of the irreversibility of time. Newton believed that no occurrence in the material world could affect the flow of time, but

Einstein argued that the relative motion between an observer and an object makes the passage of time of the object appear to go more slowly than if it were observed from a point at rest with respect to it. Therefore it is possible for event A to be observed from one point and seen as occurring before event B, and after it when observed from another point, if relative motion is involved. However, the succession of events that occur at the same place and the succession of causally related events are not reversible from any conceivable conditions of observation, and thus remain absolute in relativity theory.[61]

We all learn to tell time with ease, but to tell what it is remains as baffling as it was to Saint Augustine over fifteen hundred years ago. "What, then, is time?" he asked. "If no one asks me, I know what it is. If I wish to explain it to him who asks me, I do not know." In the period we are looking at, the question was taken up repeatedly and with a determination to break through the impasse that had stopped Augustine. There was a sharp rise in the quantity of literature about time, and contemporary observers thought that this was of historical significance. Already in 1890 the British philosopher Samuel Alexander hailed Bergson as the "first philosopher to take time seriously." A French critic saw Proust as the first to discover "that our body knows how to measure time," and Wyndham Lewis bemoaned the preoccupation with time by so many of that generation.[62]

Contrasting views about the number, texture, and direction of time were complicated by the fact that generally two kinds of time were being considered: public and private. The traditional view of a uniform public time as the one and only was not challenged, but many thinkers argued for a plurality of private times, and some, like Bergson, came to question whether the fixed and spatially represented public time was really time at all or some metaphysical interloper from the realm of space. The introduction of World Standard Time created greater uniformity of shared public time and in so doing triggered theorizing about a multiplicity of private times that may vary from moment to moment in the individual, from one individual to another according to personality, and among different groups as a function of social organization. Similarly, thinkers about the texture of time were divided between those who focused on its public or its private manifestations. The popular idea that time is made up of discrete parts as sharply separated as the boxed days on a calendar continued to dominate popular thinking about public time, whereas the most innovative speculation was that private time was the real time and that its texture was fluid. The argument about time going in one direction also separated along the lines of public and private time. Only Einstein challenged the irreversibility of public time, and even then for a special kind of event series that occurs in different sequences when viewed from different moving reference systems. All others left public time to flow irreversibly forward but insisted that the direction of private time was as capricious as a dreamer's fancy. The temporal reversals of novelists, psychiatrists, and sociologists further undermined the traditional idea that private time runs obediently alongside the forward path of public time.

The thrust of the age was to affirm the reality of private time against that of a single public time and to define its nature as heterogeneous, fluid, and reversible. That affirmation also reflected some major economic, social, and political changes of this period. As the economy in every country centralized, people clustered in cities, and political bureaucracies and governmental power grew, the wireless, telephone, and railroad timetables necessitated a universal time system to coordinate life in the modern world. And as the railroads destroyed some of the quaintness and isolation of rural areas, so did the imposition of universal public time intrude upon the uniqueness of private experience in private time. It was a subtle intrusion, one that appears sharper in historical perspective than it did around the turn of the century. Conrad dramatized the tension between authoritarian world time and the freedom of the individual by having the anarchist leader ask Mr. Verloc to prove himself by "blowing up the meridian." Most spokesmen for private time, however, did not identify the connection between the new world time and urban clustering, monopoly, bureaucracy, or big government, though it seems likely that their statements were energized in part by a reaction to the intrusion of a variety of collectivizing forces in this period, including World Standard Time.

The technology of communication and transportation and the expansion of literacy made it possible for more people to read about new distant places in the newspaper, see them in movies, and travel more widely. As human consciousness expanded across space people could not help noticing that in different places there were vastly different customs, even different ways of keeping time. Durkheim's insistence on the social relativity of time challenged the temporal ethnocentrism of Western Europe, in the same way as the literary explorations of private time challenged the authoritarian and overbearing tendencies of world time.

All across the cultural record we have identified polarities: three divisions about the nature of time—its number, texture, and direction—as well as a basic polarity about the reality of private as opposed to public time. The sense of time throughout this period emerged from tensions and debates in physical science, social science, art, philosophy, novels, plays, and concrete technological change. In tracing its various modes of past, present, and future we will see other polarities over different issues, structuring the culture through conflict.

2
THE PAST

The experience of the past varies considerably among individuals. For some it stretches far back and memories in it are coherently ordered. Some lose track of events almost as soon as they have occurred and confuse the sequences of the little that remains. Others cannot forget and dwell on the past at the expense of the present and future.[1] Every age also has a distinctive sense of the past. This generation looked to it for stability in the face of rapid technological, cultural, and social change. Its thinkers developed a keen sense of the historical past as a source of identity in an increasingly secular world and investigated the personal past with a variety of purposes. For Bergson it was a source of freedom, for Freud a promise of mental health, for Proust a key to paradise. Others viewed the past as a source of remorse, an excuse for resignation and inaction, a burden of guilt. Thinking about the past centered on four major issues: the age of the earth, the impact of the past on the present, the value of that impact, and the most effective way to recapture a past that has been forgotten.

∞

The first line of the Old Testament had always been able to anchor the inquiring mind and keep it from spinning out of control when contemplating the infinite expanse of the past. Even if the exact moment was not known, one could, with a little bit of faith, find comfort in the thought that there had been a beginning. But the comfort of revealed truth comes in part from its imprecision; in 1654, when Bishop Ussher calculated the year of creation as 4004 B.C., he invited scientific challenge. In the 1770s the Comte de Buffon determined that the earth was at least 168,000 years old, and by 1830 Charles Lyell estimated it as "limitless," time enough for the geological formations to be created by gradual processes still in action. Lyell formulated his uniformitarian theory by substituting time for catastrophic upheavals, and in 1859 Darwin, working on the assumption that he had time to burn, stretched out his theory of ever so slight variations and estimated the age of one area that he studied at over 300 million years. This steady appropriation of time by geological theory came to a sudden halt in 1862, when the eminent English physicist Lord Kelvin argued from calculations based on the rate at which the earth had cooled that it probably was not more than a hundred million years old and was possibly not more than twenty million.[2] With such a vastly reduced time scale geologists and biologists were forced to make theoretical modifications by postulating catastrophic surges of great force that accelerated the evolution of geological and living forms. Darwin made a concession to catastrophism and speculated that evolution must have occurred more rapidly in earlier periods, and for over forty years the debate about the age of the earth continued to center on the validity of Kelvin's calculations. In 1895 the British geologist Archibald Geikie complained about the reduction of the time scale: "the physicists have been insatiable and inexorable. As remorseless as Lear's daughters, they have cut down their grant of years by successive slices, until some of them have brought the number to something less than ten millions."[3] On the other side, in 1897 the American geologist Thomas Chamberlain praised Kelvin for "restraining the reckless drafts on the bank of time" by earlier geologists.[4] The argument was settled in favor of an expanded time scale when another scientific discovery provided generous funding for the time bank. In March of 1903 Pierre Curie and Albert Laborde announced that radium salts constantly release heat, and geologists were quick to apply this finding to extend the age of the earth. They reasoned that the release of heat must have slowed the rate at which the earth cooled. Already in October of 1903 John Joly wrote that "the hundred million years which the doctrine of uniformity requires may, in fact, yet be gladly accepted by the physicists."[5] In just over a century the age of the earth had oscillated from the cramped temporal estimates of biblical chronology to the almost unlimited time scale of Lyell, down to Kelvin's meager twenty million years, and then back up to hundreds of millions of years. While geologists and biologists tried to work out patterns of development through those vast stretches of time, the history of man came to appear increasingly as a parenthesis of infinitesimal brevity.

∞

If the past of the geologists seemed to rush away from the present, the past of human experience seemed to rush toward it, and a second focus of discussion emerged about the force of its impact.

Two inventions brought the past into the present more than ever before, changing the way people experienced their personal past and the collective past of history. The phonograph, invented by Edison in 1877, could register a voice as faithfully as the camera could a form. The two provided direct access to the past and made it possible to exercise greater control over what would become the historical past. An article of 1900 explained that the phonograph had already been used to record the voices of singers and orators and facilitate the study of psychology, language, and folklore.[6] In the same year the Anthropology Society of Paris founded a Musée glossophonographique and the Vienna Academy of Science created a phonographic archive. In 1906 the American critic G. S. Lee wrote a rapturous eulogy of technology, *The Voice of the Machines.* He took his title from the phonograph which enabled man to speak "forward" in time to the unborn and listen "backwards" to the dead.[7] In *Ulysses* James Joyce reflected on a similar use of the phonograph. While observing a funeral Bloom fantasizes about a way to continue to hear someone long after he has died: "Have a gramophone in every grave or keep it in the house. After dinner on a Sunday. Put on poor old greatgrandfather Kraahraark! Hellohellohello amawfullyglad kraark awfully gladaseeragain hellohello amarawf kopthsth. Remind you of the voice like the photograph reminds you of the face."[8]

The cinema was also used to record events and even to shape the course of history. Hugo Münsterberg commented on the unique ability of the cinema to create a direct vision of the past. While in the theater we must recall past events to give present action its full force, in the motion picture we can be shown the past. With the cinema we experience an "objectivation of our memory function."[9] Both still and motion picture photography preserved the past with all the clutter of detail that painting and the theater leave out. The early movie houses with their one-reel slices of recent history were an important institution that increased the detail and accuracy of memories for the millions of customers who watched.

Although the still camera was not unique to this period, the formation of photographic record societies was. In *The Camera as Historian* (1916) H. D. Gower surveyed the history of these societies, which began in 1890 with the Scottish Photographic Survey. The English followed in 1897 with the National Photographic Record Association, which was to collect photographic records of scenes of interest and deposit them in the British Museum. The United States, Belgium, and Germany soon developed similar societies. Analogous to these were institutions formed to preserve or restore architectural structures of historical interest that were threatened with destruction from reckless urban growth. In 1895 the English established the National Trust to look after places of historic interest or natural beauty, and an Act of Parliament of 1907 gave it legal power to protect sites and hold them in trust for the nation. The French passed a law in 1905 that protected national monuments for posterity. In 1903 Ferdinand Avenarius founded the Dürerbund in Germany, and a year later the Heimatschutz was created to protect historical monuments as well as natural areas.[10]

Marcel Proust's vignette of an old church and Georg Simmel's sociological essay on "The Ruin" reveal the sensitivity of high culture to the function of architecture in preserving the past in solid form. Proust described the village church as an embodiment of the passion and faith of his ancestors. The angles of its ancient porch were smoothed and hollowed "as though the gentle brushing of generations of countrywomen entering and dipping their fingers had, through the centuries, acquired a power of destruction and had carved in the long suffering flint furrows like those made by wagons on a milestone against which they daily bump."[11] This church, the church of Combray in *Swann's Way*, evoked precious memories of the narrator's childhood and the drama of history: the thickness of its walls hid the "rugged barbarities of the eleventh century," its gravestones were "softened and sweetened" by time, the stained glass windows sparkled "with the dust of centuries." These features made it "a building which occupied, so to speak, four dimensions of space—the name of the fourth being Time—which had sailed the centuries with that old nave, where bay after bay, chapel after chapel, seemed to stretch across and hold down and conquer not merely a few yards of soil, but each successive epoch from which the whole building emerged triumphant."[12] What Proust saw in a church Simmel found in a ruin, which intensifies and fulfills the past in the present. The peace we experience in the presence of a ruin comes from the resolution of the tension between two moments in time: "the past with its destinies and transformations has been gathered into this instant of an aesthetically perceptible present."[13] All the uncertainties of change in time and the tragedy of loss associated with the past find in the ruin a coherent and unified expression.

The phonographic cylinders, the motion pictures, and the preservation societies constituted silent arguments for the persistence of the past and its impact on the present. Contemporary psychologists and philosophers followed a path parallel to Proust and Simmel in making explicit what these cultural artifacts implied. Some of them thought memories were locked in living tissue as a cumulative residue of voluntary movements and bodily processes. Henry Maudsley introduced the concept of organic memory in 1867 with his observation that memory exists in every part of the body, even in "the nervous cells which lie scattered in the heart and in the intestinal walls."[14] This was reformulated in 1870 in a classic essay by the German psychologist Ewald Hering, who concluded that every living cell contains the memory of the experience of the entire series of its parent cells and even those of former generations.[15] Samuel Butler argued essentially the same in *Life and Habit* (1878) as did Bergson in

Matter and Memory (1896), although Bergson left out inherited memories. For Bergson every movement leaves traces that continue to affect all subsequent physical or mental processes. The past collects in the fibers of the body as it does in the mind and determines the way we walk and dance as well as the way we think. A ghoulish elaboration of the idea of an organic persistence of the past was Bram Stoker's *Dracula* (1897). The blood of several centuries of victims flowed in the veins of the four-hundred-year-old hero along with the blood of his ancestors—more ancient, the Count boasted, than the Hapsburgs or the Romanovs.

In 1881 the psychological concept of memory was revised in three countries. In Vienna Joseph Breuer's hysterical patient, Anna O., had a variety of symptoms including paralytic contractures, anaesthesias, tremors, and disturbances of speech, hearing, and vision. In the course of treatment he discovered that the symptoms disappeared after she gave verbal descriptions of the particular episode in her past that had precipitated them. Thus her fear of drinking water disappeared after she vented her suppressed anger and described having seen a dog drinking out of a glass. From such successes Breuer generalized a technique that rested on a recognition of the chronic pathogenic effect of past traumatic experiences.[16] At approximately the same time, the French psychiatrist Théodule Ribot was working out the laws according to which events come to be forgotten. He concluded that memories disappear according to a law of regression—"from the more recent to the older, from the complex to the simple, from the voluntary to the automatic, from the less to the more organized."[17] This regression works in reverse when memory is restored: memories of childhood are the most securely fixed, the last to disappear in amnesia. And in Germany the psychologist Wilhelm Preyer published the first systematic child psychology, which led to further work on how childhood experiences determine adult professional choices, love relations, artistic work, mental illness, and dreams.[18]

Many of these findings on memory, forgetting, and the role of childhood were incorporated in the great retrospective system of the period—psychoanalysis. Like an archaeologist excavating for lost structures in the earth's crust, Freud dug through layers of defensive structure to uncover the sources of his patients' neuroses. In his first theory of neurosis of 1892 he speculated that the foundation for later pathology was established before the "age of understanding." That limit soon receded to the second dentition (eight to ten years), to the age of the Oedipus complex (five to six years), and finally to the age of three. On several occasions he dramatically retold the story of the gradual edging back to childhood: "In my search for the pathogenic situations in which the repressions of sexuality had set in and in which the symptoms had their origin, I was carried further and further back into the patient's life and ended by reaching the first years of his childhood. What poets and students of human nature had always asserted turned out to be true: the impressions of that early period of life, though they were for the most part buried in amnesia, left ineradicable traces."[19]

Freud made five distinctive contributions to thinking about the impact of the past. He argued that the most distant past, that of our early childhood, is the most important; that the crucial experiences are sexual in nature; that the most important memories are repressed and not just forgotten; that all dreams and neuroses have their origin in childhood; and that all experiences leave some lasting memory trace. He repeatedly emphasized the universality of the childhood factor; in a letter of 1898 he wrote that "dream life seems to me to proceed directly from the residue of the prehistoric stage of life (one to three years), which is the source of the unconscious and alone contains the aetiology of all the psychoneuroses."[20] Perhaps most provocative was his idea that all memories are somehow retained: "all impressions are preserved, not only in the same form in which they were first received, but also in all the forms which they have adopted in their further developments . . . Theoretically every earlier state of mnemic content could thus be restored to memory again, even if its elements have long ago exchanged all their original connections for more recent ones."[21] As Darwin assumed that remnants of the past are indelibly inscribed in organic matter and triggered miraculously in the proper order to allow embryos to recapitulate all that has gone before, so Freud maintained that every experience, however insignificant, leaves some trace that continues to shape psychic repetitions and revisions throughout life. In 1920, in a final concession to the relentless action of the past, Freud concluded that there is an instinct to repeat in every organism, and he introduced the "repetition compulsion" as a principle governing all behavior.

In *The Critique of Pure Reason* Kant had argued that even the simplest act of perception has a temporal structure, which is a synthesis of the immediate presentation and memory.[22] So this notion was not entirely new, but the philosophers of time in the turn-of-the-century period elaborated it as though it was of historical signficance. Bergson was the most explicit when he announced in 1896: "The moment has come to reinstate memory in perception." Although William James credited others before him with an understanding of the way the past persists, he presented the argument as if it were pathbreaking. Husserl maintained that his phenomenological method provided a new scientific basis for all philosophical inquiry, including the subject of memory. They all assumed that any moment must involve consciousness of what has gone before, otherwise it would be impossible to hear a melody, maintain personal identity, or think. The melody would appear as a series of discrete sounds unrelated to what had gone before, understanding of ourselves would be chopped into unconnected fragments, and it would be impossible to learn a language or follow an argument. The philosophical problem was to explain how it was possible in a single moment to be aware of events that have occurred at different times.

Bergson wrote: "Either you must suppose that this universe dies and is born again miraculously at each moment of duration, or you must make of its past a reality which endures and is prolonged into its present."[23] He opted for continuity and concluded that the past survives as motor mechanism and as recollection. The body is "an ever advancing boundary between the future and the past"; it integrates past action and points ahead. The past can also be manifested in mental images, which he described with a variety of metaphors to suggest its impact on the present. In 1889 he wrote of a self whose former states "permeate," "melt," or "dissolve" into one another as do notes in a tune. Another metaphor has the conscious states of our inner self effecting a "mutual penetration, an interconnection."[24] In 1896 he described that action as a mingling, an interlacing, a "process of osmosis," and another aggressive metaphor described the present as the "invisible progress of the past gnawing into the future."[25] Years later he repeated that predatory image of duration as "the continuous progress of the past which gnaws into the future and which swells as it advances." He pushed the metaphor even further by elaborating on the masticatory action of the past on the present: "Real duration gnaws on things and leaves on them the mark of its tooth."[26] Human consciousness is not the tranquil passage of discrete ideas imagined by the associationist psychologists; rather, it is a thunderous action of memories that interlace, permeate, melt into, drag down, and gnaw on present experience.

William James saw the persistence of the past as a function of the fluid nature of human consciousness and, like Bergson, believed that the past remained in a dynamic relation with the present—with one essential difference. James saw a sharp distinction between recent memories that are part of the present and distant memories that are recollected as something separate; Bergson emphasized the constant interconnection of all past experiences with the present regardless of how far back they may have occurred. He would not allow any differentiation between two kinds of memories, especially if they were characterized, as they were by James, in spatial terms as near and distant. Nothing is "far away" in Bergson's *durée.*

Edmund Husserl, like Bergson and James, began by considering how we can know in one moment something that occurred before. With them he reasoned that in listening to a melody, if the past sound were entirely to disappear, one would hear only unconnected notes and not be able to make out the melody. But in order for the past to integrate with the present, it must diminish in intensity from its original form; otherwise the crescendo of sounds in a melody would soon become a hopeless jumble. The past must remain in consciousness but in a changed form.

Husserl shared with James the idea that there are two kinds of past experience—a recent one called "retention" and a more distant one called "recollection." As a perception fades away from the present it becomes one and then the other. We first experience a "now-point," which then becomes a fresh retention that remains attached to the next now-point. In time the retention fades away entirely and ceases to be part of the present as immediately given. To be experienced again, it must be reconstituted as a recollection. Retentions and recollections have a different vividness, a different relation to the present, and a different nature as parts of the past. A recollection may change the order or the rate of the original events, whereas the order and rate of a rentention is always fixed in experience. When I originally experience A and then B, I have no knowledge of B when I first experience A. In recollection I experience the whole interval "A and then B" at once, or "B follows A"; in either case my experience of A is mixed with that of B. In recollection the pace of events is also more malleable, because in it we can "accommodate larger and smaller parts of the presentified event with its modes of running-off and consequently run through it more quickly or more slowly."[27] Husserl believed that the simplest perception, even one of a single note of a melody, has a temporal structure and is "constituted" in consciousness in a manner that depends upon the retention or recollection of a receding trail of memories.

The three philosophers shared a conviction that the past had an enormous impact on the present. Bergson's past gnawed into the present, James's streamed into it, and Husserl's clung to it. They differed, however, in the value they attached to that impact. Husserl's phenomenological method formally eschewed any such evaluations, while James suggested that the full life was elaborated with the fringes and halos of memory. Bergson's metaphysics of time addressed itself most explicitly to the question of value.

∞

The debate about the value of the impact of the past on the present ranged between those who argued that the past had a positive effect as a source of meaning, freedom, identity, or beauty and those who viewed it critically as an excuse for inaction, a deadening force of habit and tradition. The work of Wilhelm Dilthey and Bergson provided the philosophical foundation for a number of the positive views that followed.

For Dilthey the past is a source of knowledge and meaning. All understanding is historical because man is a historical being. The individual life, like that of a society, develops in time and can only be understood by us because we are able to experience directly our own temporal nature. "Autobiography," he wrote, "is the highest and most instructive form in which understanding of life confronts us." "The only complete, self-contained and clearly defined happening encountered everywhere in history and in every concept that occurs in the human studies, is the course of a life." The language in which we think and the concepts we employ all originate in time. "Thus to impenetrable depths within myself," he wrote, "I am a historical being."[28] Dilthey then considered how we experience the temporal structure of our lives. Memory enables us to integrate experience in a series of ongoing syntheses which become understandable as we interpret the past and future in a changing present, in the same way as we understand a sentence whose meaning comes from words grasped sequentially in time. In an age of spectacular achievements in the natural sciences, his insistence on the historical nature of all knowledge provided some badly needed philosophical support for all the sciences of man.

Bergson based his theory of knowledge on the way we know ourselves in time. In his dissertation of 1889, he characterized the representation of time in terms of space as a "bastard concept" and dismissed such metaphysical extravagance as a "vice." This is strong language for such an abstract subject, but Bergson was passionately convinced that time is the heart of life: his metaphysics implies an epistemology and an ethic. It is not enough that we properly understand time—we must learn to live it; on it everything else turns. The absolute knowledge acquired by intuition is not merely a better way of knowing reality; it is essential to living the good life in it, and our ability to integrate the past in the present is one source of our freedom.[29]

To clarify the virtues of a life fully open to duration Bergson commented on the meager lives of those cut off from it. To live only in the present and respond only to immediate stimuli is suitable for the lower animals; for men it constitutes a life of impulse. On the other extreme, the man who lives only in the past is a visionary. Between these poles lies the life of good sense, at all times poised effectively in the present with easy access to past and future. Bergson cast certain human characteristics in time-related terms. Sorrow is a "facing towards the past"; grace is a "mastering of the flow of time," possible only when the future flows immediately from present attitudes. Jerky movements result when the future is cut off from the present, so that movements are "self-sufficient" and "do not announce those to follow."[30]

The heart of Bergson's evaluation of the past is his notion that duration is a source of freedom, which we must seek in the dynamics of experience. "It is into pure duration that we plunge back, a duration in which the past, always moving on, is swelling unceasingly with a present that is absolutely new . . . We must, by a strong recoil of our personality on itself, gather up our past which is slipping away, in order to thrust it, compact and undivided, into a present which it will create by entering."[31] The freest individual has an integrated past and is capable of utilizing the greatest number of memories to respond to the challenges of the present. As the dancer is free to move because he can integrate a complex network of past motor experiences, so is the effective individual free to coordinate a vast stream of past experiences in a present action or thought that reaches out towards an ever fuller future.

One passage of Bergson's dissertation reads like an invitation for Proust's novel:

> If some bold novelist, tearing aside the cleverly woven curtain of our conventional ego, shows us under this juxtaposition of

> simple states an infinite permeation of a thousand different impressions which have already ceased to exist the instant they are named, we commend him for having shown us better than we knew ourselves. The very fact that he spreads out our feeling in a homogeneous time and expresses its elements by words shows that he is only offering us its shadow: but he has arranged this shadow in such a way as to make us suspect the extraordinary and illogical nature of the object which projects it.[32]

The bold novelist took up the challenge twenty years later. In the hour when Proust discovered his life's vocation as a novelist, he responded with a metaphysics and an aesthetics that approximated the views of Bergson and shared his intrigue with the power of the past to produce beauty and joy.

On a snowy January evening in 1909 Marcel Proust had a cup of tea with dry toast that overwhelmed him with a sensation of "extraordinary radiance and happiness." As he concentrated on the feeling, the shaken screens of memory suddenly gave way and he recalled the happy years of his childhood in a country house at Auteuil and remembered the nibbles of rusk soaked in tea that his grandfather used to give him in those long lost summer days. In his first account of this episode Proust observed: "the taste of a rusk soaked in tea was one of the shelters where the dead hours—dead as far as the intellect knew—hid themselves away."[33] What no effort of the intellect could recapture, the organic memory of the tea and biscuit could. This episode was transposed into his great novel, *Remembrance of Things Past,* which he began to write in July of that year.

In the opening pages Marcel relates an episode from his childhood that marked the beginning of his loss of the past. One night he contrived to have his mother kiss him goodnight and unexpectedly got her to spend the entire night with him. That night he realized that his recent loneliness and suffering were part of life, not an accidental misfortune; it was the beginning of the relentless erosion of childhood happiness, which is the content of time lost. The process of recovery may be delayed for many years, until it is set in motion by an experience like the taste of tea and rusk that had so shaken the author when it first occurred in real life. In the fictional version Marcel's mother served him tea with some madeleines. The taste sent a shudder of pleasure through him and he ceased to feel mediocre or mortal. Another sip enabled him to recall that the taste was that of a madeleine soaked in lime-flowered tea, which his aunt Léonie used to give him when he was a child. The elegant simplicity of this episode contrasts with the massive complexity of the story that unfolds out of it like Japanese paper flowers in water. Marcel informs us that at the time he did not know why the memory made him so happy and that he must "long postpone" his discovery of why it did. The reader must share the search for the time that he began to lose the night of his mother's kiss and to recapture many years later in her cup of tea.

Marcel interrupts his story to relate one that took place before he was born, when his neighbor, Swann, courted Odette. Marcel had his first *moment bienheureux* while sipping tea; Swann's opening into the past was triggered by a musical passage. During his courtship Swann heard it played, and years later when jealousy had obliterated the earlier tender feelings, the passage set off a flood of painful memories. A striking contrast to Marcel's pleasure at recapturing his past is provided by Swann's shattering discovery that Odette had a homosexual encounter on the evening that he had believed to be the supreme moment of their love. The news cracked the time of his life in two, and the memory of that evening became the nucleus of a pain "which radiated vaguely round about it, overflowing into all the preceding and following days."[34] For both men the truth about the past became known years later when the confusion of the present had disappeared. But Swann's discovery of the past led to misery, to thoughts of death, and to an immersion in time that was endlessly revealed to him as the obsessive repetition of this trauma, while Marcel's past eventually became a source of happiness and a deliverance from the relentless course of time.

In a later volume we learn that the moment when Swann searched for Odette in fear that she was with another man continued to dominate his love for all subsequent women. "Between Swann and her whom he loved this anguish piled up an unyielding mass of already existing suspicions . . . allowing this now aging lover to know his mistress of the moment only in the traditional and collective phantasm of the 'woman who made him jealous,' in which he had arbitrarily incarnated his new love."[35] His past experience with Odette created this iron law of love that compelled him forever to repeat his first dismal experience. Swann strives to divest himself of the past while Marcel searches for its meaning, as contrapuntally they make their respective ways into and out of oblivion.

In the last volume, when Marcel moves toward the final spectacular revelations, we know that much time has passed. His love for Albertine had disappeared from his conscious mind, but in his limbs there lingered an "involuntary memory" that still reached out for her in the old way. In a recapitulation of the opening pages of the novel when the child awakens in bed and reaches out for the world, the older Marcel, lying in bed lost in a dreamy nebula of memories, responds physically to the ghostlike presence of his dead mistress. His legs and arms were full of "torpid memories" of her body. A memory in his arm made him fumble for the bell as he had formerly done with her at his side. They are portents of the powerful recollective energies of bodily sensation that will reveal to Marcel by means of several involuntary memories the secret of time lost.

When Marcel returns to Paris after the war, he attends a party at the Princess de Guermantes; there, upon his arrival, he experiences five involuntary memories which finally make clear to him their mysterious sources of power and joy. The first, triggered by the feeling of an uneven paving stone underfoot and recalling a similar sensation he had had years earlier in the baptistry at St. Mark's in Venice, removes all anxiety about the future and reminds him of the happiness he had felt when he tasted the madeleine soaked in tea. The involuntary memories that follow reveal to him their tremendous potency: they give him back *le temps perdu.* The simplest act is associated with colors, scents, and temperatures that words cannot recreate, but when an involuntary memory recalls the past in the sensuous fullness of those associations we experience the intense pleasure of the *moments bienheureux.* The present is too confusing to allow us to discern the essentials of reality, and the intellect is useless to grasp it. Only through the perspective of time passed and time regained can we come to understand the past and enjoy its retrieval: "The only true paradises are the paradises that we have lost."[36]

Like metaphors, these moments of recollection unite what is separate to illuminate and give pleasure. In them Marcel experienced the present "in the context of a distant moment, so that the past was made to encroach upon the present."[37] They revealed the essence of things "outside time," and being freed from the rigid order of time he became momentarily indifferent to the idea of death. Previously there had been an urgency about capturing these moments, which slipped away so quickly and dissolved under close scrutiny, but this time he discovered a way to sustain them by making them the subject of a novel.

Proust resolved to write the story of his life, not in the traditional "two-dimensional psychology" but in a different sort of "three-dimensional psychology" that would reconstruct the movement of life in time. In a letter of 1912 he explained: "There is a plane geometry and a geometry of space. And so for me the novel is not only plane psychology but psychology in space and time. That invisible substance, time, I try to isolate."[38] Most novels take place in time, but they fail to capture the experience of time passing, the emptiness of time lost, and—the only real pleasure in life—time regained. The final sentence and statement of purpose of the novel underlines the priority of the temporal dimension of life: to "describe men first and foremost as occupying a place, a very considerable place compared with the restricted one which is allotted to them in space . . . in the dimension of Time."

If there is a single illusion that Proust most wanted to dispel it is that life takes place primarily in space. The spaces in which we live close about us and disappear like the waters of the sea after a ship passes through. To look for the essence of life in space is like trying to look for the path of the ship in the water: it only exists as a memory of the flow of its uninterrupted movement in time. The places where we happen to be are ephemeral and fortuitious settings for our life in time, and to try to recapture them is impossible. But in our bodies time is preserved in memories of tea and cake that give us back the days of our childhood and the vast stream of events that have since overlaid them and make up our life.

The fact that Bergson, Proust, and Freud, all of Jewish heritage, insisted that the past was an essential source of the full life—of freedom, beauty, and mental health—suggests a possible connection between their lives and their theories about time. There are some striking similarities between the temporal experience of the Jews and these works. Both Judaism and Christianity share a reverence for the past and argue their validity partly from tradition. The implicit ethic is that old is good. Judaism is the older of the two, and in the search for identity its longer history would set it above Christianity on the time scale. It is possible that the insistence of these men that the past alone is real, that only the recapture of the past can inspire art or cure neurosis, is linked to this feature of the Jewish experience.

A second similarity that might be attributed to the Jewish experience derives from their respective arguments that life in time is more important than life in space. Bergson became angry when he contemplated the spatialization of time, Freud strove to reconstruct his patients' lives out of the past, and Proust created his characters' lives in the dimension of time that occupied "a very considerable place compared with the restricted one which is allotted to them in space." This shared feature of their work parallels the experience of the Jews, who did not have a space of their own except the cramped enclaves of the ghettoes. Their spatial existence was always a tenuous and painful reminder of their isolation from the surrounding world and was far less important to them than their existence in time. Thus the Wandering Jew is at home only in time. The Jewish religion also eschewed all spatial representations of the deity whose reality and goodness became known through his action in history. In modern Europe the history of the Jews had no surviving physical landmarks. They had to internalize their landmarks and preserve them in memory in written and oral form, whereas in the Christian world the past was tangibly preserved in monuments and could easily be seen and reconstructed in the imagination.[39] This experience of the Jews may have shaped the three thinkers' evaluations of the primacy of our existence in time.

There was also a special interest in the way the past can shape modes of thinking, social forms, and organic structures. The beginnings of this historicization of thought go back to Locke's idea that all knowledge comes from sensation. The Enlightenment philosophers worked this principle hard to reject innate ideas and an *a priori* human nature and to prove that man is entirely shaped by history and society. In the nineteenth century Comte, Hegel, Darwin, Spencer, and Marx shared the idea that philosophies, nations, social systems, or living forms become what they are as a result of progressive transformations in time, that any present form contains vestiges of all that has gone before. With the decline of the religious conception of man in the late nineteenth century, many drew from these systems to give meaning to life in a world without God. If man could no longer believe he had a place in eternity, he could perhaps find one in the movement of history.

While the great historicist systems of the nineteenth century as well as the work of Dilthey, Bergson, Proust, and Freud celebrated the historical or genetic approach, many of their contemporaries rejected it with passion and condemned the way the past can overwhelm the present. This view of the past as a burden was forcefully presented by a German philosopher, a Norwegian dramatist, an Irish novelist, an Austrian architect, and a group of Italian writers. It was not unambiguous, as they all had a profound sense of the past and some appreciation of its positive value, but their most distinctive work was a strong negative evaluation of the paralyzing and destructive action of memories, habits, and traditions.

In an essay of 1874, *The Use and Abuse of History,* Friedrich Nietzsche reacted sharply to the domination of historicism. As the title implies, his treatment is at least partly balanced, and he concedes that "every man and nation needs a certain knowledge of the past," but the central message warns about excessive pondering over what has gone before. For the acutely miserable, dwelling on history is a deliverance, a "cloak under which their hatred of the present power and greatness masquerades as an extreme admiration of the past." The conservatives find comfort in the past—the old house, the portrait gallery provides meaning and stability in a changing world. This "antiquarian history" hinders the impulses for action and "greedily devours all the scraps that fall from the bibliographical table." It inclines an individual to capitulate to circumstance and renounce his inner resources. It teaches that the present is the old age of mankind, that people are "late survivals," born with grey hair. It creates cynicism about the possibility of changing anything and paralyzes the energies of art. The entire age was suffering from a "malignant historical fever," and Nietzsche was particularly incensed by those who are chained to precedent and bowed under the weight of an ever heavier accumulation of memory and tradition. "One who wished to feel everything historically would be like a beast who had to live by chewing a continual cud." This chronic rumination, this hypertrophy of the historical sense "finally destroys the living thing, be it man or a people or a system of culture."[40]

Ten years later Nietzsche wrote a critical analysis of the effect of an overbearing sense of the personal past on the individual will. In *Thus Spoke Zarathustra* (1883–1885) Nietzsche introduced his notion of the will to power, which is thwarted in most people by a load of memories, regrets, and guilt. In the chapter "On Redemption" he insisted that the only true redemption was the victory of the will to power over time and the obstacles that the past puts in its way. "Powerless against what has been done, [the will] is an angry spectator to all that is past. The will cannot will backwards; and that he cannot break time and time's covetousness, that is the will's loneliest melancholy." The will strives to liberate an individual from the residue of the past, but " 'that which was' is the name of the stone he cannot move . . . This, indeed this alone, is what *revenge* is: the will's ill will against time and its 'it was'." The frustration created by the indelible nature and dead weight of the past causes resentment and guilt and the destructive responses of punishment and revenge. To free itself the will must "recreate all 'it was' into 'thus I willed it'."[41] Then the past becomes an appropriation of the will, which can function properly as a creative force, a liberator, a joy-bringer, and a bridge to the future.

The destructive effect of the past that Nietzsche analyzed, Henrik Ibsen dramatized in a series of plays in which inheritance, a sudden disclosure about their past, or a persistent memory works upon his characters and leaves them crippled or dead.[42] The characters of the early plays might overcome their past through special effort, but with *A Doll's House* in 1879 he grew pessimistic. In the concluding scene Nora tells her husband that it will take "a miracle of miracles" for their marriage to work, but her final gesture as she walks out on him seems to slam the door on that possibility as well. With *Ghosts* (1881) the past is unequivocally triumphant, and thereafter his characters struggle in vain against its power as evoked by hereditary diseases, haunting memories, and spirits of the dead.[43] This is powerfully summed up in Mrs. Alving's exclamation: "I almost think we are all ghosts—all of us . . . It isn't just what we have inherited from our father and mother that walks in us. It is all kinds of dead ideas and all sorts of old and obsolete beliefs . . . and we can never rid ourselves of them."[44]

In two other plays a sudden disclosure provides the dramatic focus. The contented family life of Hjalmar Ekdal in *The Wild Duck* (1884) is destroyed when he learns that years before his wife had an affair with a man who fathered her daughter, Hedvig, arranged for Hjalmar to meet his wife to be, and then set him up in business. In the anguish of these revelations Hjalmar condemns his wife for spinning a web of deceit and disowns Hedvig, who then kills herself in a desperate effort to regain his love. As they all contemplate the senselessness of Hedvig's sacrifice, a drunken character explains that most people need a "life lie" to sustain them and that nothing is more dangerous than to strip it away and reveal the naked truth about the past. The parents of *Little Eyolf* (1894) try to suppress the guilty memory reminding them that the boy was crippled when he fell off a table while they were making love. This comes out after Eyolf drowns. Their bereavement, the father realizes, is the "gnawing of conscience," and they resolve to devote themselves to helping other children, a lifelong project of forgetting what they never can and remembering what they think they must.

In other plays by Ibsen the past torments as a persistent memory. The heroine of *The Lady from the Sea* (1888) is so fixated on the memory of a dashing sailor who charmed her long ago that she will never be able to escape from the "horrible, unfathomable power" he has over her, as she tells her landlubber suitor Arnhold in rejecting his proposal. In *When We Dead Awaken* (1899) a sculptor meets his former model, who tells him that she left him years ago because when she undressed to pose and offered herself to him naked he remained unmoved and merely thanked her for a "very happy episode." The meeting years later allows them to awaken only to discover that they are dead, and in the final scene as they climb a mountain to a supposed undoing of that event, they are swept off by an avalanche—symbol of the cumulative weight and overpowering force of the past. And for the hero of *Rosmersholm* (1886) the dead live on at his estate, symbolized by the white horses that stand on its grounds, memorialized by the portrait gallery that lines the walls, and sustained by memories of treachery.[45]

In *Hedda Gabler* (1890) Ibsen indicted the historical profession, represented by the heroine's husband, Tesman, who was so preoccupied with completing a study of domestic industry in Brabant during the Middle Ages that he took his research along on their honeymoon. At one climactic moment Hedda bemoans the fact that she is condemned "To hear nothing but the history of civilization, morning, noon, and night!" Her frustrations—social, sexual, and existential—are accented by her husband's routine fulfillment from doting on the past. His neglect of her, compounded with other disappointments, finally drives her to suicide.

The thoughts that haunt Ibsen's characters provide the theme for James Joyce's story "The Dead."[46] Gabriel gives an after-dinner speech in which he evokes the memory of the dead and "thoughts of the past, of youth, of changes, of absent faces that we miss here tonight." As the party ends he sees his wife listening intently to a song and feels a special attraction for her, but for his wife the song triggered memories of a youth who sang it years before. While Gabriel had been reminiscing about their moments together, she had been thinking about her life apart, and as she slept that night he agonized over the years she had secretly savoured the image of her lover. The revival of the past transforms the meaning of the party, makes a mockery out of the sentimental gush of his speech, reveals the emptiness of their life together, and leaves Gabriel alone and ashamed at the side of his wife.

In *Ulysses* Joyce's views on memory range from an appreciation of the potency of recollection to a condemnation of the deathly paralysis of a life wholly immersed in the past. In an unmistakably Proustian perception he noted that some memories "are hidden away by man in the darkest places of the heart but they abide there and wait until a chance word will call them forth suddenly and they will rise up to confront one in the most varied circumstances." The kinds of chance circumstances he mentions could also have come from Proust: "a shaven space of a lawn one soft May evening" or "the well remembered grove of lilacs." However, Joyce's general view of memory focused on its limitations and dangers: its potential to paralyze the artist, its power to sustain guilt, and its abuse by pedants.

Stephen Dedalus recognizes that memory preserves identity: "I, entelechy, form of forms, am I by memory because under everchanging forms." Memory holds him together but he is desperate to break the hold it has on his free expression as an artist. His uniqueness is swamped; his self is too much part of everything as if attached by an enormous umbilical cord that stretches back to Eve: "The cords of all link back, strandentwining cable of all flesh." His identity flows into the world about him and with the past that it conjures up. The remains of a shipwreck he sees at the shore take him back to the days of the Spanish galleys that ran to beach there, to the Danevikings and ancient whalers. "Their blood is in me," he thinks, "their lusts my waves." He feels caught up in the teeming life about him, entangled with everybody in the relentless passing of time. Most painful is the persistent recollection of his mother's death and his refusal to attend her funeral. Joyce refers to this nagging guilt with the phrase "Agenbite of inwit" that echoes in Stephen's mind with each recollection and fires his contempt for history.

Stephen's famous outburst that history is "a nightmare from which I am trying to awake" is the culmination of his frustration with the cheap anti-Semitism of a pedantic colleague whose history is a "dead treasure," but it refers more directly to his own struggle to break the hold that history has on himself. When a student of his fails to answer a question about the battle of Pyrrhus, Stephen reflects on the panorama of disaster that all history has recorded: "I hear the ruin of all space, shattered glass and toppling masonry, and time one livid final flame. What's left then?" Man has been fighting senselessly from the battle of Pyrrhus to World War I. The imagery undoubtedly refers to the effect of an artillery shell exploding on a building, but beyond the ruinous action of the war and the prospect of Europe going up in flames, Stephen is troubled by history itself—the illusion of reality that it offers and its distortion of the living present. What is left for man after the disaster of World War I, and what is left of truth if history is always "fabled by the daughters of memory?"

Stephen's long struggle to realize himself as an artist reaches a climax towards the end of the novel as he strikes out against history, time, and guilt by smashing a chandelier in a brothel just as his dead mother appeared "green with grave mold" and, choking in the agony of her death rattle, begged the Lord to have mercy on him. Finally he had had enough. "Time's livid final flame leaps and in the following darkness, ruin of all space, shattered glass and toppling masonry." He smashes the light to obscure the image of his mother and put out the fire of his guilt. Stephen has stopped time in order to become an artist. The present is the only reality, especially for the artist, and Joyce has Stephen articulate his credo: "Hold to the now, the here, through which all future plunges to the past."[47] An effort must be made to hold the present because it is always slipping away, always threatening to have its uniqueness swamped by the old patterns of the past.

No group of artists was more acutely aware of the dead weight of the past than the architects, who quite literally could see it lining the streets of European cities. One of the most palpable monuments to traditional architecture was the Ringstrasse in Vienna—a ring of public buildings around the Austrian capital that was constructed from the 1860s to the 1890s, each designed in a historical style that was considered to be appropriate to its function. Thus the Parliament building had a classical Greek architecture, the City Hall was Gothic, the University was Renaissance, and the Burgtheater, Baroque. In 1893 Otto Wagner won a competition for a plan to expand urban development beyond the Ringstrasse. Although he had built a number of apartments in the Ringstrasse area and had contributed to some of its public buildings, his writings and architecture from the 1890s sharply repudiated the historicism of the Ringstrasse and of much nineteenth-century architecture generally. His urban development proposal focused on transportation and the needs of a modern industrial city rather than on the blocks of buildings that had been constructed to beautify the city and memorialize the past. In a textbook of 1895, *Modern Architecture,* Wagner speculated about what had produced such deadly eclecticism and slavish devotion to the past and concluded that while most ages had been able to adapt artistic forms to changing techniques and needs, in the latter half of the nineteenth century social and technological change had proceeded too rapidly for artists to keep pace, and the architecture fell back on earlier styles. Wagner assailed this capitulation before the forms of the past that had brought artistic innovation to a halt, and in a specific recommendation for the training of architects, he condemned the Italian journey that had traditionally been the culmination of a classic *beaux arts* education.[48]

The nagging effect of the past that so angered Nietzsche, overwhelmed Ibsen, and threatened Joyce drove the Italian Futurists into a frenzy. They voiced the most passionate repudiations of the past in manifestos that recommended burning the Louvre and filling the canals of Venice. Their most energetic spokesman was Filippo Marinetti, whose manifesto of February 1909 contained the essentials of their *antipasséiste* project to destroy the museums and the academies and to free the land from "its smelly gangrene of old professors, archaeologists, ciceroni, and antiquarians." In April 1909 he vowed "to mock everything consecrated by time." In *Against Past-Loving Venice* he envisioned the city turned into a modern commercial port. He ridiculed the English as victims of traditionalism who carefully preserve every remnant of the past and attacked the Symbolists, who "swam the river of time with their heads always turned back toward the far blue spring of the past" and who had a contemptible passion for the eternal. In contrast, his fellow Futurists celebrated the here and now and created an art that was perishable and ephemeral. In 1914 he announced the funeral of all *passéiste* beauty including its nefarious ingredients of memory, legends, and ruins.[49]

∞

Although there was considerable division about the value of the past, there was general agreement that it could not be entirely forgotten, that the complete artist, the wise statesman, the healthy individual must come to terms with it somehow. The thinkers who wanted to retrieve the past differed over the best means to do so. Proust's search emphasized the passive approach and contrasted with the active methods of Bergson, Freud, and Henry James.

In *Swann's Way* Proust argued that the past cannot be recaptured by any conscious effort. "The past is hidden somewhere . . . beyond the reach of the intellect, in some material object (in the sensation which that material object will give us) which we do not suspect," and we must wait until we chance upon that object to be able to recapture what it holds.[50] Marcel stumbles onto the first clue with the tea and madeleines but is unable to sustain or understand the pleasure it gave him. In a weak moment some time later he attempts to recapture the lost happiness of a childhood love by returning to the spot in the Bois de Boulogne where it had occurred, but, like all efforts of conscious intention, it fails and only intensifies his longing. Years later Marcel finally discovers the signficance of his involuntary memories and how the lost past can be recaptured. The involuntary memory is entirely passive; however, once it has occurred, one can work to make it last by embodying it in art. Even though that is active work, Proust's distinctive contribution is his emphasis on the passive, involuntary memory that springs up out of experience by chance.

In contrast to Bergson, who believed that we experience a continuous gnawing of the past into the present, Proust valued the shock and pleasure of being suddenly immersed in time which has been experienced *dis*continuously. The ecstasy and the feeling of immortality that Proust's character Marcel has "outside time" comes from a sudden reentry into time and a deliverance from its relentless movement. For Proust duration is a series of isolated moments that produce such pleasure upon retrieval precisely because they are so remote from each other, while Bergsonian duration is at every moment a composite of each successive moment and therefore continuous. If Bergson's duration is like a stream, Proust's is like a series of steep cataracts where the mind recaptures intermittent surges of memory out of oblivion. Their views on the process of recollection also differ. Although Bergson believed that the recapture of the past was a difficult undertaking that required a "vigorous effort," he thought that one could make that effort at any time. In contrast Proust argued that the first step must occur fortuitiously. He commented on this difference in a letter of 1912, denying that his novels were Bergsonian because they were "dominated by a distinction which not only doesn't figure in the philosophy of Bergson but which is even contradicted by it."[51] The crucial distinction was between voluntary and involuntary memory: Proust insisted that Bergson's involved the intellect, while his could not be summoned up by any volitional activity. Although Bergson did speak of the spontaneous activity of memory, Proust's distinction between the two is valid if for no other reason than his greater emphasis on chance.

There is an even sharper contrast in the methods of recollection of Proust and Freud. In the first place Proust insists on the solitary nature of recollection. "As for the inner book of unknown symbols . . . if I tried to read them no one could help with any rules, for to read them was an act of creation in which no one can do our work for us or even collaborate with us."[52] In the search for *le temps perdu* psychiatrists are as distracting as lovers or friends. Second, for Proust the crucial phase is passive: waiting for an involuntary memory. Once it has occurred the active search for its meaning may begin, as in the prodigious work he put into embodying it in a novel. In Freud's psychotherapy a systematic procedure predominates. It is hard to imagine a more active search for the past than arranging punctual sessions for its exploration in which the therapist's experience with former patients serves as a model. Psychoanalysis also has an arsenal of theories about unconscious processes, stages of development, and potential points of fixation to help decipher the disguises that might retard the discovery of a patient's history. This search for the past is as active as a fox hunt, and the repressed memories are the object of a continuous hounding that may go on for years.[53]

Another kind of active search takes place in Henry James's novel of 1917, *The Sense of the Past.* An American gentleman, Ralph Pendrel, is failing in the here and now with his courtship of Aurora Coyne, when he learns of the imminent death of an English relative and begins to fantasize that he will be left something "ancient and alien." The bitterness of Aurora's rejection is partially softened by her dubious compliment about his "natural passion for everything old." He had written *An Essay in Aid of Reading History,* which had prompted her remark and inspired the relative to will to Ralph a house built in 1710. Ralph travels to England to take possession in the hope that there he will be able to "remount the stream of time" and "bathe in its upper and more natural waters." Once he gets settled, the house becomes an enclave in time that separates him from the present as much as its isolation cuts him off from the surrounding world. The objects in it were, like the walls of Proust's church of Combray, "smoothed with service and charged with accumulated messages." The house embodied the entire line of Ralph's ancestors, and "as the house was his house, so the time, as it sank into him, was his time."[54] This identification with the past turns into a hallucination when he sees his own face on a portrait in the house and comes to believe that he and his ancestor had exchanged personalities. His sense of the past, which first appeared as a scholarly inclination, finally took over his personality and became his identity. Though cast in fiction, the story is nevertheless one of the complete appropriation of a man's ancestral past by direct, though psychologically unbalanced, means.

Proust's emphasis on the passive recovery of the past is alone among the numerous active projects. Some stretching must be done to weave Bergson's philosophy, Freud's therapy, and Henry James's imagination into a single cultural statement, but these works cluster unmistakably on the opposite side of this issue. The comparison with James is a bit forced because Proust was not concerned with ancestral past, but he nevertheless would have found Ralph's journey, his possession of the old house, and his conscious effort to go back in time to be a futile effort of the intellect and doomed to failure.

Of the four debates about the length, force, value, and method of retrieval of the past the one on the age of the earth had the most limited cultural impact. The difference between Kelvin's stingy estimate of twenty million years and the several hundred million allowed following the discovery of radioactivity interested primarily a small circle of geologists. Nevertheless the eons opened up to popular consciousness since Darwin's time continued to influence writers such as Joyce, who portrayed the petty punctuality of Leopold Bloom against the vast ages of the universe. This vertiginous extension of the time scale dealt yet another blow to the egocentrism of man, whose tenure on earth seemed to shrink to minuscule proportions.

The cultural record lined up consistently behind the idea that the past has a powerful influence on the present. The phonograph and motion picture camera provided a historically unique technology for the preservation of the past, and photographic archives and preservation societies provided new institutional support for it. After almost two millennia of Christianity that belittled the significance of human history by holding that the ultimate meaning or purpose of life was realized in the timeless and unchanging gaze of God, nineteenth-century thinkers sought to find meaning and justification for life in human history. Historians found new sources, dug up buried civilizations, raised standards for accuracy and documentation, and generally professionalized the discipline. Evolutionary theory overwhelmed the biological sciences; historicist systems such as those of Hegel and Marx revolutionized philosophy and social sciences; liberal and socialist faith in history dominated political thinking. Toward the end of the century Dilthey claimed a historical foundation for all knowledge and insisted on the primacy of historical method for all social sciences. As Stephen Toulmin and June Goodfield observed: "Whether we consider geology, zoology, political philosophy or the study of ancient civilizations, the nineteenth century was in every case the Century of History—a period marked by the growth of a new, dynamic world-picture."[55]

The great historicist systems had perhaps made their case too well. They showed how individuals or social forms had evolved out of their antecedents and were destined to recapitulate what had gone before. The present thus seemed predetermined and smothered by the past. As we have seen, many artists and intellectuals were sharply critical of this overbearing historicism and shared a fear that the dominion of the past would impoverish response to the present and dry up resources for the future. In this context the outbursts of the Futurists are not so absurd, for Italy was especially mired in the past, haunted by relics and monuments of the glory of a dead civilization. Hayden White has surveyed the negative response to the "burden of history" that began with Nietzsche's polemics in the 1870s and was continuously elaborated through a number of literary characters: Mr. Casaubon in George Eliot's *Middlemarch* (1871–72), Tesman in *Hedda Gabler*, Hanno in Thomas Mann's *Buddenbrooks* (1901), and Michel in Gide's *The Immoralist* (1902). Critical reaction to a "feverish rummaging of the past" intensified in the decade before the First World War, when, as White concluded, "hostility towards the historical consciousness and the historian gained wide currency among intellectuals in every country of Western Europe."[56]

Although there were exceptions to this generalization (Marxists and liberals continued to believe in history and progress), it was true for a number of thinkers and artists who rebelled against the sweeping, and at times blind, faith that the nineteenth century had had in the value of the historical approach to all living processes, especially human. And precisely as the historical past began to lose its authority as *the* theoretical framework, the personal past began to attract a variety of prominent thinkers who scrutinized it with unprecedented care and insisted that an understanding of it was essential to a healthy and authentic life. New theories about memory and forgetting and new studies in child and developmental psychology appeared and were synthesized in psychoanalysis, which popularized as never before the notion that the individual past remains active and continues to shape adult behavior.[57] Freud insisted that access to that past was essential for mental health. Bergson analyzed how the present is constantly being reformed by extrusions from the immediate past, and he insisted that only a life fully open to the fluid movement of *durée* had access to an essential source of individual freedom. For Proust the past surfaced in flashes of involuntary memory, which is the only true source of joy, beauty, and artistic inspiration. And while he made history a nightmare for Stephen Dedalus, Joyce exploited the possibilities of the new literary technique of direct interior monologue to reconstruct from a variety of perspectives the personal past of all his main characters.

Ibsen and Nietzsche also found more value in the individual's appropriation of a personal rather than an historical past. Nietzsche's essay of 1874 was far more critical of the potential paralysis from too much history than he was in his analysis of the personal past in *Thus Spoke Zarathustra.* There he acknowledged that the personal past is a necessary component of consciousness. One does not *choose* to look to the personal past as a model for action as one might do with the historical past, because the "it was" is inherent in the structure of consciousness. And since Zarathustra announces an essentially hopeful, positive philosophy—the way to the overman—he insists on the possibility and the necessity of the will's transformation of the "it was" into a "thus I willed it" or even "thus shall I will it." Nietzsche valued the essentially constructive, if challenging, function of the personal past. In *The Genealogy of Morals* he warned of the self-loathing that comes from a cumulative inheritance of generations of guilt—that poisonous residue of the historical past. "It was" is the will's loneliest melancholy, but the personal past constitutes a necessary obstacle in human consciousness that forces the will to make something of itself in the face of all the pernicious influences of habit and tradition.

Ibsen also considered both pasts and gave a far more positive and dramatically penetrating account of the force of the personal past. In *Hedda Gabler* Tesman is portrayed as a cartoon character, as flat as the history he is writing. His history is a museum of the dead and drains the energy out of his life and marriage. But in other Ibsen plays that we have surveyed the personal past is revealed with great effectiveness and gives depth to the characters' lives that is conspicuously absent in Tesman. Clearly Ibsen was far more interested in the personal than the historical past, which is collective and impersonal. By having a personal past erupt into the lives of his characters Ibsen realizes greater dramatic impact than he could have done with the introduction of anything out of the historical past. His plays achieve in the theater what Freud achieved in the clinic—a reconstruction of the life in time that breaks through and then breaks down self-mystifications and defenses. In *The Wild Duck* Ibsen deferred to the need for self-deception by defending the "life lie," but his main purpose was to suggest that one must come to terms with one's own past or go under. The past was an essential part of the structure of his drama, as it was for Nietzsche an essential part of the structure of consciousness.

This shift in attention from the historical past to the personal past was part of a broad effort to shake off the burden of history. By focusing on the immediate past of individuals these thinkers and artists sharpened the analyses of their philosophical studies, increased the effectiveness of their psychiatric interventions, and intensified the dramatic impact of their literary works. The historical past was the source of social forces over which they had little control; it created institutions that had lasted for centuries; and it limited their sense of autonomy. The overbearing deterministic formal systems of nineteenth-century historicism produced broad, general laws of history, whereas these thinkers wanted to understand the unique responses of individuals to particular circumstances. Freud was an exception, because he tried to work out laws of mental life; however, they were not laws of collective historical processes but of individuals, and they encompassed only their personal past. With the Ringstrasse looming as a monument of artistic capitulation to the historical past, modern artists affirmed the independence of their work. They did not want to imitate the art of the past, and they did not want their lives to be regulated by social conventions that were conceived in the distant past and over which they had no control. Above all they wanted freedom. They focused their attention on the personal past, because they believed it to be a richer source of subject matter than the remote and impersonal historical record. The personal past was something over which they might gain some control. One is not responsible for history in the way one is responsible for one's past, even one's childhood. And if one is more responsible for the personal past, then one can hope to understand it, perhaps even refashion it, as indeed Nietzsche, Ibsen, Freud, Bergson, Gide, Proust, Joyce, and the Futurists, each in different ways, insisted that we must.[58]

The philosophers of the Enlightenment looked to antiquity for values they lost in their struggle with Christianity, and they found in the ancients spiritual sustenance and models for action. The Romantics longed for the historical past of the preindustrial world as a retreat from a vulgar present. In the mid-nineteenth century the mystique of the historical past began to lose some of its attraction as the Realists sought subject matter in the contemporary world for their formal studies, science, and art. While they eschewed romanticization of the past, they embraced historicism as the foundation of formal thinking. But around the turn of the century, artists and intellectuals turned from the glorification of the historical past and from the method of historicism and began to consider the personal past, thereby generating an unprecedented concentration of interest in the way the personal past works on the present. These thinkers did not discover this past, but they broadened and deepened understanding of the ways it persists in germ cells and muscle tissue, dreams and neuroses, retentions and involuntary memories, guilt and ghosts.

This focus on the personal past over the historical past also lines up with the shift of focus from the homogeneous public time to the varieties of private time that we observed in the first chapter. For the personal past is private, and it varies from one individual to the next, while the historical past is collective and tends to be more homogeneous, although individuals are free to interpret it in different ways. Thus the most distinctive general development about the nature of time—elaborated as heterogeneous, fluid, and reversible—accords with these arguments on behalf of the personal past. To the massive, collective force of uniform public time we may add the sweeping force of history—making a composite temporal structure against which, or perhaps we should say out of which, the leading thinkers of this generation affirmed the reality of private time and sought to root themselves in a unique personal past.

3
THE PRESENT

On the night of April 14, 1912, the largest moving structure ever built, the *Titanic*, steamed at a recklessly high speed into an ice field in the North Atlantic. The first officer recalled that the sea was especially calm and so that night there were no "ice blinks"—flashes of light given off when waves splash against icebergs and illuminate their crystallized surfaces. Visibility was further reduced by fog. At 11:40 P.M. a lookout suddenly spotted an iceberg dead ahead. The ship turned sharply and, as it scraped by, was opened up like a tin can with a gash below the water line three hundred feet long. The captain determined that they were going to sink fast and at 12:15 A.M. ordered his wireless operator to send the distress call. Within a few minutes the airwaves were rippling with signals as over a dozen ships became aware of the disaster. This was simultaneous drama on the high seas, driven by steam power and choreographed by the magic of wireless telegraphy.

Ten ships heard the call from over a hundred miles away and remained in contact but were too distant to help, as were also the *Hellig Olav* at 90 miles and the *Niagara* at 75 miles. The *Mount Temple* was 50 miles away but had to move slowly through ice fields. The *Carpathia* at 58 miles was the first to arrive, but not until almost two hours after the *Titanic* went down with 1,522 passengers. Another ship, close enough to have saved all the passengers, was not in wireless contact. The *Californian* was approximately 19 miles away, but its wireless operator had hung up his earphones for the night about ten minutes before the *Titanic* sent out its first CQD. Two watchmen on the deck of the *Californian* saw the rockets that the *Titanic* fired but could not figure out what they meant or convince their captain to pull anchor and find out. What the eyes and ears of man could not perceive the wireless could receive over vast distances and through darkness and fog.

The operator on the *Carpathia* got the call for help when he put on his earphones to verify a "time rush" (an exchange of time signals with a neighboring ship to see if their clocks agree). At 1:06 A.M. he heard the *Titanic* tell another ship coming to help, "Get your boats ready; going down fast on the head." The world began to get news of the disaster at 1:20 A.M., when a wireless station in Newfoundland picked up the message that the *Titanic* was sinking and was putting women off in boats. Shortly after that hundreds of wireless instruments along the Atlantic coast began to transmit and the airways became jumbled in confusion. The *Titanic*'s wireless had a range of only 1,500 miles, so signals to Europe had to go first to New York and then across the ocean by cable; still, by early morning the entire world was privy to news of the disaster.[1]

To one of the survivors in a life boat it seemed as if the stars above saw the ship in distress and "had awakened to flash messages across the black dome of the sky to each other."[2] The communication that he imagined between stars was accomplished on a lesser scale between the ships at sea by wireless. On April 21, the *New York Times* commented on its magical power.

> Night and day all the year round the millions upon the earth and the thousands upon the sea now reach out and grasp the thin air and use it as a thing more potent for human aid than any strand of wire or cable that was ever spun or woven. Last week 745 [*sic*] human lives were saved from perishing by the wireless. But for the almost magic use of the air the *Titanic* tragedy would have been shrouded in the secrecy that not so long ago was the power of the sea . . . Few New Yorkers realize that all through the roar of the big city there are constantly speeding messages between people separated by vast distances, and that over housetops and even through the walls of buildings and in the very air one breathes are words written by electricity.

An editorial in the *London Times* of April 16 noted the expanded range of experience made possible by the wireless. "The wounded monster's distress sounded through the latitudes and longitudes of the Atlantic, and from all sides her sisters great and small hastened to her succor . . . We recognize with a sense near to awe that we have been almost witness of a great ship in her death agonies." An officer of the American Telephone and Telegraph Company praised the communication that made it possible to follow the rescue. The telephone and wireless, he wrote, "enabled the peoples of many lands to stand together in sympathetic union, to share a common grief." William Alden Smith, the Michigan senator who chaired an exhaustive inquiry into the sinking, as part of his summary of those hearings before the United States Senate on May 18, 1912, referred to the new sense of world unity that required worldwide safety regulations. "When the world weeps together over a common loss," he said, "when nature moves in the same directions in all spheres, why should not the nations clear the sea of its conflicting idioms and wisely regulate this new servant of humanity?"[3] Although the wireless had been used before to save lives at sea, this rescue effort was particularly highlighted because so many were aware of the tragedy: the survivors watching from life boats, the wireless operators in distant places, and the frustrated seamen in the rescue ships.

∞

The ability to experience many distant events at the same time, made possible by the wireless and dramatized by the sinking of the *Titanic*, was part of a major change in the experience of the present. Thinking on the subject was divided over two basic issues: whether the present is a sequence of single local events or a simultaneity of multiple distant events, and whether the present is an infinitesimal slice of time between past and future or of more extended duration. The latter debate was limited largely to philosophers, but the issue of sequence versus simultaneity was expressed by numerous artists, poets, and novelists and was concretely manifested in some new technology in addition to the wireless—the telephone, the high-speed rotary press, and the cinema.

Already in 1889 Lord Salisbury commented on the simultaneity of experience made possible by the telegraph, which had "combined together almost at one moment . . . the opinions of the whole intelligent world with respect to everything that is passing at that time upon the face of the globe."[4] The telegraph had been in operation since the 1830s, but its use was limited to trained operators and confined to transmitting stations. The wireless proliferated source points of electronic communication, and the telephone brought it to the masses.

The history of wireless telegraphy begins with a paper by James Clerk Maxwell in 1864, which argued that electromagnetic waves must exist and should be able to be propagated through space. In 1887 Heinrich Hertz produced those waves in a laboratory, and in 1894 Guglielmo Marconi devised an apparatus to transmit and receive them. In 1897 Marconi went to England and established the first coast station on the Isle of Wight for communication with ships at sea. In 1901 a message was sent across the Atlantic from a special high-power transmitter in England, and two years later King Edward VII and President Theodore Roosevelt exchanged messages over it. As wireless instruments proliferated, an International Congress on Wireless Telegraphy was held at Berlin in 1903 to regulate their use. The Marconi Company established the first wireless news service in 1904 with nightly transmissions from Cornwall and Cape Cod. The first distress signal from a ship at sea was sent in 1899, and in 1909, following a collision between two ships, a wireless call saved 1700 lives. The technology got some sensational publicity in 1910 when a wireless message led to the arrest of an American physician in London, who murdered and buried his wife and attempted to escape aboard a ship with his secretary dressed as a boy. The captain became suspicious of the two, wired Scotland Yard, and arranged to have a detective arrest the couple at sea before they arrived in port.

By 1912 the wireless was an essential part of international communication linking land stations and ships at sea in an instantaneous, worldwide network.[5]

The telephone had an even broader impact and made it possible, in a sense, to be in two places at the same time. It allowed people to talk to one another across great distances, to think about what others were feeling and to respond at once without the time to reflect afforded by written communication. Business and personal exchanges suddenly became instantaneous instead of protracted and sequential. Party lines created another kind of simultaneous experience, because in the early systems bells rang along the entire line and everyone who was interested could listen in. One imaginative journalist envisioned the simultaneity of telephone communication as a fabric made from the fibers of telephone lines, switchboard cables, and speech: "Before the great switchboard the girls seem like weavers at some gigantic loom, the numerous cords crossing and recrossing as if in the execution of some wondrous fabric. Indeed, a wondrous fabric of speech is here woven into the record of each day."[6]

Within a few years of its invention in 1876 the telephone was used for public "broadcasts." In 1879 sermons were broadcast over telephone lines in the United States, and in 1880 a concert in Zurich was sent over telephone lines fifty miles to Basel. The following year an opera in Berlin and a string quartet in Manchester were transmitted to neighboring cities. The Belgians began such transmissions in 1884: the telephone company of Charleroi gave a concert which could be heard by all of the subscribers, an opera in Monnaie was heard 250 kilometers away at the royal palace at Ostend, and the North Railroad Station in Brussels piped in music from the Vaux-Hall in what was perhaps the first experiment with muzak.[7]

Jules Verne envisioned "telephonic journalism" in a science-fiction story of 1888.[8] Five years later it became a reality when a Hungarian engineer started such a news service in Budapest and expanded it into a comprehensive entertainment service with outlets in the homes of its 6000 subscribers, each of whom had a timetable of programs including concerts, lectures, dramatic readings, newspaper reviews, stock market reports, and direct transmissions of speeches by members of Parliament. It focused the attention of the inhabitants of an entire city on a single experience, regulated their lives according to the program schedules, and invaded their privacy with an emergency signal that enabled the station to ring every subscriber when special news broke. An English journalist imagined that this service, if introduced in England, would "democratize" many luxuries of the rich as the "humblest cottage would be in immediate contact with the city, and the 'private wire' would make all classes kin."[9] At the same time it would diminish the isolation of individuals in cities and make it possible for one voice to be heard simultaneously by the six million people of London. In the United States in 1896, telephones were used to report presidential election returns, and, according to a contemporary report, "thousands sat with their ear glued to the receiver the whole night long, hypnotized by the possibilities unfolded to them for the first time."[10]

There was diverse critical response to the simultaneity of experience created by modern journalism. Already in 1892 the indefatigable alarmist Max Nordau complained that the simplest village inhabitant has a wider geographical horizon than the prime minister of a century ago. If the villager reads a paper he "interests himself simultaneously in the issue of a revolution in Chile, a bush-war in East Africa, a massacre in North China, a famine in Russia."[11] Nordau anticipated that it would take a century for people to be able "to read a dozen square yards of newspapers daily, to be constantly called to the telephone, to be thinking simultaneously of the five continents of the world" without injury to the nerves. Paul Claudel reacted more positively in 1904 when he wrote that the morning newspaper gives us a sense of "the present in its totality,"[12] and an editorial in *Paris-Midi* of February 23, 1914, characterized the headlines of one daily paper as "simultaneous poetry."

The discovery of the cinema between 1893 and 1896 portended its international scope and the simultaneous experience that it was able to suggest. Within a few years Edison in the United States, Robert W. Paul in England, Max and Emil Sklandowski in Germany, and Louis and Auguste Lumière in France invented instruments to project a continuous picture through moving rolls of celluloid film onto a screen. By 1910 there were 10,000 nickelodeons in the United States alone, creating the need for about two hundred one-reel films a week. From the outset the film industry was international, with a mass appeal. Film expanded the sense of the present either by filling it with several noncontiguous events or showing one event from a variety of perspectives. Three techniques were used: double exposure, the montage balloon, and parallel editing.

Photographic montage had been used by still photographers for decades when Méliès introduced it in one of his science-fiction fantasies of 1898 to create phantoms on the screen by actual multiple exposures. But that cluttered the picture. In *The Life of an American Fireman* Porter used the montage balloon. He showed a fireman sitting at his desk and, in a bubble in the upper right corner, his wife putting their infant to bed. This dramatized the fireman's later heroics in saving another child from a burning building. These two techniques were cumbersome, and directors soon discovered a more versatile way to suggest simultaneity—contrast editing or intercutting.

Porter first used contrast editing in *The Ex-Convict* (1905). A wealthy manufacturer refuses to hire an ex-convict, and to dramatize the encounter between them separate scenes show the luxury and squalor of their respective homes. In *The Lonely Villa* (1909) Griffith intercut to show a man racing home to save his wife and children who are being attacked by robbers. The suspense is intensified by showing the robbers, the man, and his family in progressively shorter sequences that build to a climax when they all come together. Griffith used it more effectively in *The Birth of a Nation* (1914), where the camera cuts ever more rapidly among simultaneous scenes of converging action: some men under attack in a cabin, gunmen closing in on them, and the Ku Klux Klan galloping to the rescue. His most ambitious effort to keep several lines of action going at the same time was in *Intolerance* (1916), which intercut four stories from different periods dealing with intolerance: the invasion of Babylon by the Persians, the conflict between Jesus and the Pharisees, the St. Bartholomew's Day Massacre, and a modern story of a man convicted of a murder he did not commit. The stories are portrayed in alternating sequences, and the film concludes with a last-minute rescue of the contemporary man as he is about to be executed. Thus was Griffith able to splice open a moment and insert a number of simultaneous activities.[13]

While many early cinema viewers complained that the stories were technically "jerky," they soon adapted their visual response and learned to sustain the continuity between lapsed sequences and integrate them in a powerful climax as all converged in a single dramatic moment. Hugo Münsterberg noted that the cinema could appear to take the viewer from one place to another instantly and achieve the effect of his being "simultaneously here and there." In films "we see the man speaking into the telephone in New York and at the same time the woman who received his message in Washington."[14] In 1916 the Futurists hailed the ability of the cinema to "give the intelligence a prodigious sense of simultaneity and omnipresence." They found the sequence in books oppressive and preferred motion pictures because they offered a "fleeting synthesis of life in the world."[15] For artists in love with machines, it was a romance made in heaven.

Henri-Martin Barzun, Blaise Cendrars, and Guillaume Apollinaire wrote several kinds of simultaneous poetry and even became involved in a public argument over who originated the new art form.[16] Most insistent about his priority was Barzun, who, in 1912, founded a journal to present his theory of simultaneity and publish works that conformed to it. One of his poems was about the unification of the world by wireless: "I radiate, invisible, from the summit of the Tower / Fluid carrying the hope of ships in distress / Enveloping the earth with my waves / Proclaiming the Word, the Time of the world."[17] Barzun also proclaimed an aesthetic. Aviation has transformed distance; the whole of humanity is involved with catastrophes around the globe; international alliances have increased the "federative" nature of the world. It is an age of democracy—of crowds and public assemblies—and only a simultaneous poetry can capture it. Song must give up its monodic character and become polyphonic; a "multiple lyricism must render the multiplicity of modern life."[18] In *Voix, rythmes et chants simultanés* (1913), he noted that past poets expressed the voices of a successive universe; the contemporary poet ought to express them all at once as they are perceived by the senses and magnified by technology. City life provides "proof of the existence of simultaneous realities" that can only be expressed by the chant of simultaneous voices. Poets must adopt an orchestrated verbalism of simultaneity, of which Barzun supplied numerous examples, consisting of parallel lines to be read at the same time by different voices and recorded on phonographs.[19]

In 1912 Cendrars began to frequent the home of Robert Delaunay where he met Sonia Delaunay, who had used her husband's technique of simultaneous art in paintings and book bindings. In February 1913 Cendrars published *La Prose du Transsibérien et de la petite Jehanne de France,* which was announced as the "First Simultaneous Book" (Figure 1). Printed on a sheet two meters long, it was meant to be seen all at once so that the spatial limitations of one page after another would not chop up its wholeness. It was illustrated down the left side with Sonia Delaunay's *couleurs simultanées.* The poem describes Cendrars's journey from Moscow to Harbin on the Trans-Siberian railway in 1904, and above the text he reproduced a map to show the route at a glance. Thus the reader saw "simultaneous"

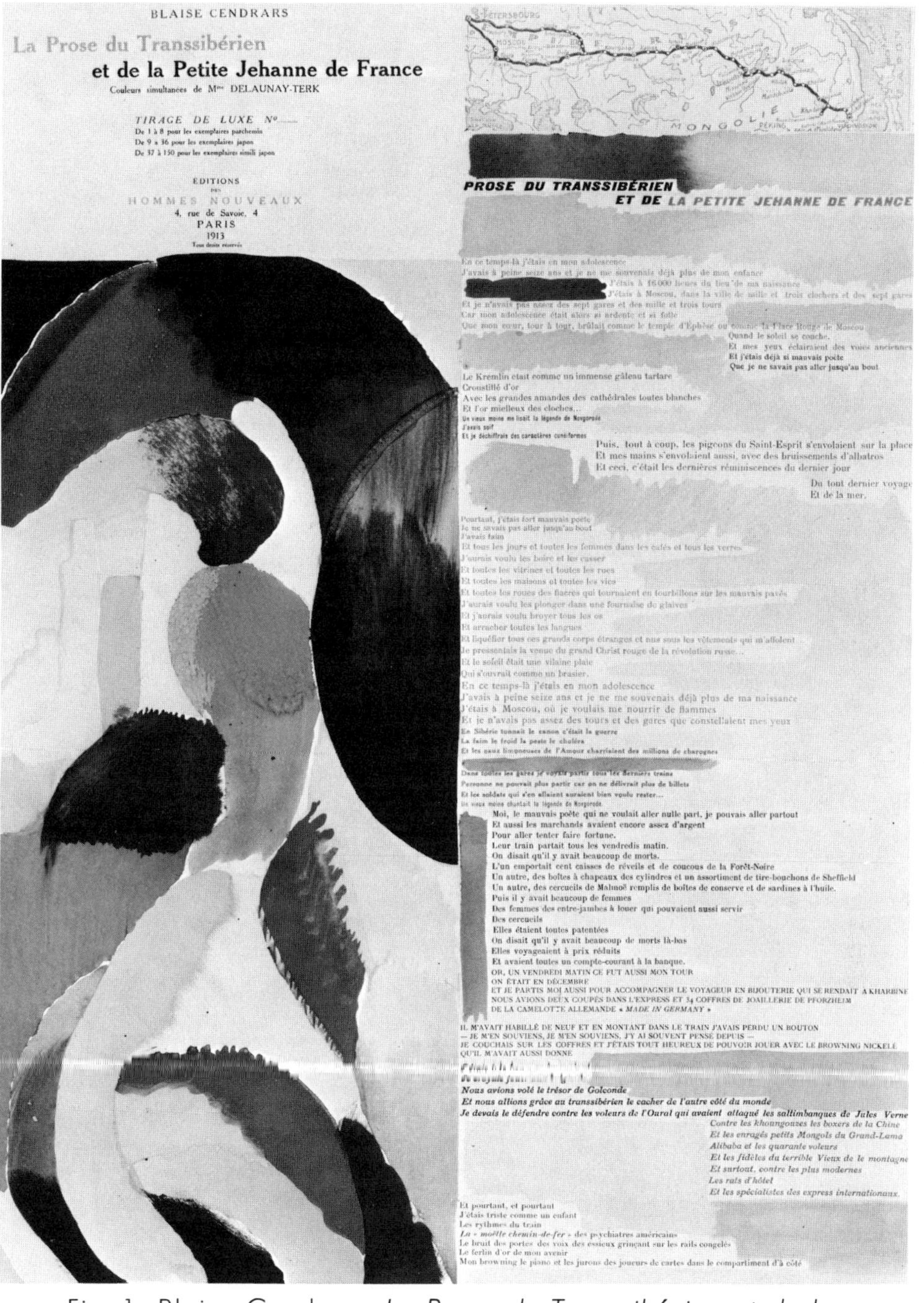

Fig. 1. Blaise Cendrars, *La Prose du Transsibérien et de la petite Jehanne de France,* 1913 (the top quarter of the poem).

colors, a map of the trip, and a poem about it at the same time. The poem was also intended to recreate an impression of the journey as a whole, the way it came to Cendrars when he began to write. Its metaphors span distance and time to suggest the sense he had of being able to experience every moment of the journey and the world beyond it simultaneously. The poem toys with chronology by uniting remote ages: "I spent my childhood in the hanging gardens of Babylon / . . . The prehistoric ancestor will be afraid of my motor." It unites the diverse timepieces of the world: "The big clapper of Notre Dame / The shrill ringing of the Louvre announcing Saint Bartholomew / The rusted bells of Bruges-la-morte / The electric bells of the New York Public Library / The city bells of Venice / and the bells of Moscow." While the trains run forward according to these, "the world, like the clock in the Jewish quarter in Prague, turns desperately counterclockwise." Time is compressed and reversed to break down the divisiveness of sequence, and space is ignored to undo the divisiveness of distance and bring together separate places in a single vision of his train racing across Russia as his mind raced around the world. Verbal montages unite what is distant as if they were quick-cut camera directions: "Now I've made all the trains run after me / Basel-Timbuktu / I've also played the horses at Auteuil and at Longchamps / Paris-New York / Now I've made all the trains run alongside my life / Madrid-Stockholm." The poet explores the new dimensions of a world shaped by railroad and wireless (indeed one critic called him *le Sans Fil*[20]). Even speed is useless in this world that "stretches lengthens and retracts like an accordion / tormented by a sadistic player." One image includes both temporal and spatial simultaneity: "All the days and all the women in the cafes and all the glasses / I should have liked to drink and break them."[21] This voraciousness echoes that of another *simultanéiste,* Arthur Craven, who, also in 1913, updated the Faustian spirit: "I would like to be at Vienna and at Calcutta / To take all trains and all ships / to fornicate with all women / and gourmandize all dishes."[22] Electronic communication and rapid transportation filled the present with diverse sensations and quickened the appetite of Cendrars and Cravens, who sought to embrace all the experiences as technology embraced places and times.

Apollinaire also indulged in fantasies of sexual ubiquity. For his story *Le Roi-Lune* he invented a belt that made its wearer able to make love to all the women of all centuries. A specially adapted piano could produce the sounds of different countries. Apollinaire also concocted the Baron d'Ormesan, whose *toucher à distance* enabled him to appear simultaneously in diverse places around the world. The Baron was a film director, and this ability of his embodied the montage techniques that directors used to suggest ubiquity and simultaneity. He died in 820 places simultaneously after having created *l'amphionie*—a tourism adapted to the speed of modern times, which provided a "complete" tour of Paris in thirty minutes.[23]

Apollinaire's poetry included several aspects of simultaneity. *Zone* (1912) sought to knit remote places and times into a single fabric of present experience—a zone between past and future, between near and remote. In *Ondes* (1913) he described the Eiffel Tower, whose electronic waves carried time signals that made it possible to determine the simultaneous occurrence of distant events. *Liens* surveyed various ties between distant places. In addition to radio waves there were the sounds of clocks ringing across Europe, rails that bound nations, cables that united continents, and rays of light that linked the earth with distant stars. The desire to give his reader everything at once led Apollinaire to create *calligrammes* with words arranged to depict a poem's content. One poem about time was arranged in the form of a pocket watch. For a collection published in *Les soirées de Paris* in July 1914, he placed at the top of the page a drawing of a telegraph pole crossed by wires like a musical time signature for the poems printed below.

The model for simultaneous art and poetry was music. In counterpoint different melodies worked simultaneously, and in opera two or more voices might sing different words at the same time. Wagner explained that he intentionally had Tristan and Isolde say important things at the same time to intensify the urgency of their encounter. But the modern composers went beyond these forms and improvised more challenging simultaneities involving music in different tonalities and rhythms. Richard Strauss combined two keys at the same time in *Also Sprach Zarathustra* (1896) as did Debussy in parts of *Pelléas et Mélisande* (1902). In the first of *Fourteen Bagatelles* of 1908, Bela Bartók produced what one historian has called the "earliest thorough-going example of two simultaneously sounding melodic parts written in different keys."[24] Prokofiev had such bitonal passages in *Sarcasmes* (1911), and Stravinsky made extensive use of tritone harmonies in *Le Sacre du printemps* (1913). While there were also precedents to polyrhythm and polymeter in counterpoint and classical symphonic music, there was a striking concentration of them in the twentieth century beginning with Charles Ives's combination of two marches in different tempi in *Three Places in New England* (1904).

In literature a famous early example of simultaneous action

occurs in *Madame Bovary* (1857), where Rodolphe's tiresome romancing of Emma is interspersed with the announcement of prizes at an agricultural fair that takes place in the background of their meeting. The juxtaposition of Rodolphe's avowal that he will never stop loving Emma with the announcement that Monsieur Bizet won a prize for the best manures accents the vulgarity and insincerity of Rodolphe's love and the banality of rural life. Frank Norris cut between paragraphs in *The Octopus* (1901) to contrast the fate of a destitute farmer's wife and child wandering the streets of San Francisco and the lavish dinner of some railroad moguls in a mansion nearby. The paragraphs become ever shorter and climax in "cinematic shots" of the woman dying as the meal is completed. In *The Death of a Nobody* (1908) Jules Romains used the technique to dramatize his philosophy of *unanimisme*—that everybody is linked by bonds of fellowship of which they might not be entirely aware. The death of a lonely railway worker brings together a number of people and taps forces of mutual affection, a common spiritual bond they discover in the course of preparing for his funeral. Simultaneous action is going on at four places: the corpse decaying on the fourth floor of an apartment in Paris, some neighbors on the second floor trying to buy a wreath for the funeral, the dead man's father journeying to Paris, and the building's concierge moving about the apartment. Andrey Biely used quick cuts in *St. Petersburg* (1913) with great dramatic effect. The story traces the lives of a conservative Russian official and his son, who becomes involved with a radical political group and is charged with setting off the explosion that killed his father. The ticking of the time bomb marks simultaneity as time runs out in ever more rapid cuts between the two. The Futurists wrote several plays in which simultaneous action takes place on a divided stage. In a manifesto of March 1914, Marinetti proclaimed the birth of a new beauty characterized by "the simultaneity that derives from tourism, business, and journalism."[25] The following year he published *Simultaneità*, in which two different places and their inhabitants interact: a cocotte penetrates a bourgeois family as her dressing table occupies part of the space of their living room. In *The Communicating Vases* (1916) Marinetti undertook three simultaneous actions. In the end the characters break down the two partitions on stage and enter one another's world in a climax of simultaneous activity and interpenetration.

The highpoint of simultaneous literature was *Ulysses*. Joyce was deeply impressed with cinematic montage, and in 1909 he was instrumental in introducing the first motion picture theater in Dublin.[26] In *Ulysses* he improvised montage techniques to show the simultaneous activity of Dublin as a whole, not a history of the city but a slice of it out of time, spatially extended and embodying its entire past in a vast expanded present.[27] In this respect he was realizing Bergson's view that the knowledge we have by intuition is analogous to that we gain by walking around a city and living in it. Joyce hoped his readers would go back to the book many times, continually building up the network of cross-references scattered throughout until Dublin came to life.

A vivid example of Joyce's technique is the "Wandering Rocks" episode, a montage of nineteen sections, each of which shows a different aspect of Dublin. Joyce used five devices to recreate the unity of the city and its simultaneous activity: multiple accounts of a character from different perspectives; repetition of action in at least one other episode; a narration that begins over again and again; multiple appearance of an object (a handbill seen three times as it floats down the Liffey, uniting Dublin spatially and providing a symbol of the passage of time); and a final recapitulation with a cavalcade that travels about the city and links the characters and places it passes. The journey of the handbill and the movement of the cavalcade are linear, but they suggest the spatial interrelatedness of the city and provide points of juncture for all that was happening. Although the cavalcade moves sequentially, the glimpses that some of the characters had of it in earlier sections anticipate the final summation of the simultaneity of its movements.

As "Wandering Rocks" was inspired by the cinema, so "Sirens" was inspired by music and constructed with counterpoint and polyphony. By identifying characters and themes with short verbal passages and then "sounding" them on the page in quick succession Joyce sought to overcome the necessarily sequential time of literature and achieve an effect similar to the simultaneous sounding of different notes in musical harmony. He truncated and augmented words and phrases, had them interrupt one another, and reversed their direction contrapuntally to suggest the overlapping of musical subjects in a fugue. Consider the opening lines:

> Bronze by gold heard the hoofirons, steelyrining imperthnthn thnthnthn.
> Chips, picking chips off rocky thumbnail, chips. Horrid!
> And gold flushed more.
> A husky fifenote blew.
> Blew. Blue bloom is on the

> Gold pinnacled hair.
> A jumping rose on satiny breasts of satin, rose of Castille.
> Trilling, trilling: Idolores.
> Peep! Who's in the . . . peepofgold?
> Tink cried to bronze in pity.
> And a call, pure, long and throbbing. Longindying call.
> Decoy. Soft word. But look! The bright stars fade. O rose!
> Notes chirruping answer. Castille. The morn is breaking.
> Jingle jingle jaunted jingling.

Several readings of the whole chapter and some scholarly investigation were needed to identify the characters and themes represented by these phrases.[28] "Bronze" and "gold" are the two waitresses in the bar where the action takes place, and "imperthnthn thnthnthn" is the stuttering of someone trying to say "impertinent." The "hoofirons" they hear are those of the cavalcade outside. "Blue bloom is on the", "rose of Castille", and "Idolores" are bits of popular songs, and the variations on "blew" and "bloom" sound the approach of Bloom like a distant horn. The "longindying call" is a tuning fork, but also Bloom's loneliness and his lament for the death of his passion for Molly. "The light stars fade" and "the morn is breaking" are from the song that sounds throughout the chapter and signifies its musical technique. "Jingle" announces Blazes Boylan, who is traveling in a jingly car to the bar before his assignation with Molly in her jingly bed. Just as the novel was intended to be reread until the cross-references revealed the artfulness of the whole, so must this chapter be reread until the opening lines can be appreciated as the verbal equivalent of the statement of themes in a fugue. The harmonies of this chapter approximate music when past readings have made the phrases intelligible for integration into the rapid mix of sequential, verbal harmonies.

In "Nausicaa" there is an interweaving of simultaneous action at three places: Gerty MacDowell on the rocks at the shore with two girlfriends watching over twin boys and an infant, Bloom on a rock nearby, and some distance away in a church the men's temperance retreat performing a benediction and praying to the Virgin Mary. Joyce unites them with concrete action, language, interior monologues, and ironic juxtaposition.

The opening paragraph pictures the last rays of sun that touch the sea, the rocks on the strand, and the church, uniting the scenes of action that will take place. It concludes with another image of ubiquity as, the sun having set, a bat begins to fly—"Here. There. Here." The episode begins on the rocks and deliberately moves to the other locales: one of the twins kicks a ball that rolls to Bloom, Gerty hears the prayers that enable her to imagine the scene in the church, and Cissy Caffrey walks over to Bloom to ask the time. (Bloom has been oblivious to public time since 4:30 when his watch stopped.) The traditional dividers of sequence and distance collapse into a unified whole which the reader must envision after several readings, like a circling bat surveying it all from a darkened sky.

Joyce also unites action at separate places by stringing it together in run-ons, as descriptions of Gerty and her friends, the men in church, and Bloom flow into one another with nothing separating them but simple conjunctions.

> . . . then Father Conroy handed the thurible to Canon O'Hanlon and he put in the incense and censed the Blessed Sacrament and Cissy Caffrey caught the two twins and she was itching to give them a ringing good clip on the ear but she didn't because she thought he [Bloom] might be watching but she never made a bigger mistake in all her life because Gerty could see without looking that he never took his eyes off her and then Canon O'Hanlon handed the thurible back to Father Conroy and knelt down looking up at the Blessed Sacrament and the choir began to sing *Tantum ergo* and she just swung her foot in and out in time as the music rose and fell . . .

The first part of the chapter is structured around Gerty's thoughts as she grows increasingly aware of Bloom, who becomes ever more interested in watching her. She realizes that he is masturbating and raises her skirt to help him along. The second part is from Bloom's perspective, and interior monologues reveal his itinerant thoughts during and after climax. As "Sirens" mixed simultaneous activity with sound, "Nausicaa" linked them visually. The bodily organ of the chapter is the eye, and there is an emphasis on its unifying function—Gerty and Bloom gazing at each other, everybody else watching the fireworks, the clergymen eyeing the host, and, negatively, the bat that cannot see.

Actions in different places are also juxtaposed with images that suggest their similarity. There is the parallel between the swinging thurible and Gerty's foot. Father Conroy knelt down and looked up at the blessed sacrament as Bloom sat down and looked up Gerty's skirt. For good measure Joyce explains that Bloom was "literally worshipping at her shrine." Everybody on the strand exclaimed when they saw the fireworks go off as Bloom exclaimed when his went off. The infant puked on his bib and Gerty said "that that was the benediction because just then the bell rang out from the steeple over the quiet seashore because Canon O'Hanlon was up on the altar with the veil that Father Conroy put round him round his shoulders giving the benediction with the blessed Sacrament in his hands."[29] The chapter concludes with an elaborate image of simultaneity.

> A bat flew. Here. There. Here. Far in the grey a bell chimed. Mr Bloom with open mouth, his left boot sanded sideways, leaned, breathed. Just for a few.
>
> *Cuckoo*
> *Cuckoo*
> *Cuckoo*
>
> The clock on the mantelpiece in the priest's house cooed where Cannon O'Hanlon and Father Conroy and the reverend John Hughes S.J. were taking tea and sodabread and butter and fried mutton chops with catsup and talking about
>
> *Cuckoo*
> *Cuckoo*
> *Cuckoo*
>
> Because it was a little canarybird bird that came out of its little house to tell the time that Gerty MacDowell noticed the time she was there because she was as quick as anything about a thing like that, was Gerty MacDowell, and she noticed at once that that foreign gentleman that was sitting on the rocks looking was
>
> *Cuckoo*
> *Cuckoo*
> *Cuckoo*

The flitting of the bat and the call of the cuckoo bird signify visual and auditory ubiquity and unite the characters and settings. The mechanical bird's nine o'clock chime is also a comment on them all. Gerty acted crazily, the clergymen talked nonsense, and Bloom appeared to Gerty to be cuckoo. Public time reappears to remind us of Bloom's cuckoldry just as his watch had stopped and public time disappeared precisely at 4:30, when Molly was being unfaithful to him.

The dialectic of thought takes unexpected pathways. At the same time that so many artists were celebrating simultaneity as a distinctive experience of the age, Einstein was arguing that there could be no such thing in a universe with moving parts.

The great symbol of simultaneity was the Eiffel Tower, which Robert Delaunay painted,[30] the poets eulogized, and the Futurists worshipped. It was used for sending out time signals that were thought by most people to travel instantaneously and make possible the calculation of simultaneous events. Actually their velocity was finite, as Olaf Römer had discovered in 1675, and so was the velocity of light, which experiments in the nineteenth century calculated to be around 186,000 miles per second. To artists and critics this incredibly high speed seemed to insure simultaneity—but not to the physicists. Ever since Mach began poking at absolute space and time, the absolute simultaneity of distant events became problematical, and with Einstein it became untenable for events observed from moving reference frames.

In the special theory Einstein concluded that spatial and temporal coordinates vary with relative motion, that no exact determination of the simultaneity of distant events is possible for an observer in motion with respect to those events, and that therefore one cannot attach any *absolute* status to the concept of simultaneity: "Two events which, viewed from a system of coordinates, are simultaneous, can no longer be looked upon as simultaneous when envisaged from a system which is in motion relatively to that system."[31] Of course almost nobody knew of, or could properly understand, these ideas, especially in the prewar years. The vast majority remained awed by electronic communication and believed that the wireless and telephone had "annihilated" space and time. Einstein's theory showed that no precise meaning could be given to the idea of absolute simultaneity; however, it applied to subatomic or cosmic events involving enormously high relative velocities and could not affect the everyday experience of anybody. So in spite of his isolated counterargument, the big news of the age was that the present moment could be filled with many distant events. As one historian concluded boldly, because of these changes, on the eve of the war "succession gave way to simultaneity."[32]

∞

While simultaneity extended the present spatially, other attempts were made to expand the traditional sharp-edged present temporally to include part of the immediate past and future. Experimental psychologists tried to measure the present; studies of after-images that made the continuous movement of cinema possible showed that we experience the present visually as a quantum of time involving the immediate past. There was also evidence of a protracted present during moments of great emotion. A Swiss geologist commented on the dilation of the present during sudden falls in mountain climbing accidents; two French psychiatrists analyzed the compression of long sequences into short dream episodes and the dramatic expansion of the lived present among psychotics.[33] Artists tried to portray duration, and novelists used various techniques to fan out their sequential narratives into a "continuous" or "prolonged" present.

In the early 1880s Wilhelm Wundt conducted some experiments to determine the duration of the present—that interval of time that can be experienced as an uninterrupted whole. He concluded that its maximum limit was about 5 seconds, and one of his students set it at 12. Another student found that the shortest interval between separate clicks that the ear could discern was 1/500 of a second and that the eye could not distinguish sparks less than .044 seconds apart, because it retained an image of an object after it had disappeared.[34] Projecting pictures at 16 frames per second created the illusion of a continuous moving picture. All these experiments provided new information for several philosophers who debated about the nature and duration of what we experience as "now."

Since the ancient Greeks there had been a controversy about the structure of time. Some thinkers argued that it was composed of discrete parts—infinitesimal instants that constituted longer durations the way points made up a line. The most radical statement of that view was made by Hume, who said that time "consists of different parts" out of which longer durations are formed.[35] This interpretation was rejected by William James, Josiah Royce, and Husserl, who argued for a "thickened" present.

In an article of 1884 James observed that it is impossible to experience the present because it is past before we can properly comprehend it. "It is only by entering into the living and moving organization of a much wider tract of time that the strict present is apprehended at all." James credited Shadworth Hodgson and E. R. Clay with anticipating his philosophy. In 1878 Hodgson distinguished between the strict present—a divider between past and future too brief to be experienced—and the practical present that is extended to include several seconds or even minutes. Clay argued that the present we experience is part of the recent past as the path of a meteor seems to be "contained in the present," and he defined this interval as the "specious present." James adopted the concept and illustrated it with one of his unforgettable images: "the practically cognized present is no knife-edge, but a saddle-back, with a certain breadth of its own on which we sit perched, and from which we look in two directions into time." He also accepted the calculations of its length from Wundt's laboratory and concluded that the present has a "vaguely vanishing backward and forward fringe; but at its nucleus is probably the dozen seconds or less that have just elapsed."[36]

Royce also accepted the term "specious present" and agreed that its length varies among individuals. He invoked this variable extended present to argue by analogy that God's infinite range enables him to experience all at once and eternally the events that we experience as a series of surprises as we make our way into the future.[37] Although James and Royce had different emphases, they agreed that the present is far thicker than Hume and the associationist psychologists would allow and that the thickness varies in different circumstances. They even shared an uncertainty that this notion was valid by using the word "specious" to describe a present that clouded over the traditional instantaneous one.

Husserl did not like the term "specious present," but he incorporated many of its features to explain how we can experience at once events that occur at different times, as in listening to a melody. Do we apprehend discrete notes and *then* combine them by some mental operation, as the associationist psychologists believed, or do we take in an extended whole all at once? Husserl believed that duration is experienced directly as a whole and is "constituted" in perception as something inherently temporal. We have already noted his explanation that past notes fade but remain present as "retentions." The present has a nucleus of a "now-apprehension," a "comet's tail" of retentions that cling to it, and a horizon of the anticipated future. The enduring tone "is constituted in an act-continuum which in part is memory, in the smallest punctual part is perception, and in a more extensive part expectation." The future components are "protentions," which fulfill the intentional nature of consciousness. Moreover our experience of the past is future-oriented: "every act of memory contains intentions or expectations whose fulfillment leads to the present." A bird flies. In every position "the echo of earlier appearances clings to it," but those past positions also point to the present and future, otherwise we would at every moment expect

Fig. 2. Giacomo Balla, *Dynamism of a Dog on a Leash,* 1912.

it to fall out of the sky.[38] We do not experience the future directly, but since retentions, recollections, and now-apprehensions point ahead, the basic direction of all consciousness is toward the future. Thus the present is a continuous unrolling field of consciousness, thickened with retentions and protentions.

The Futurists painted some comets' tails of their own. Giacomo Balla pictured a temporally extended moment in two works of 1912. *Rhythms of a Bow* shows a player's hands in successive positions, violin strings swollen with vibrations, and the air around them quivering as if the sound waves had made it visible. In *Dynamism of a Dog on a Leash* (Figure 2) Balla depicted successive stages of a dog trotting alongside a pair of feet. The undulation of the leash is represented in four fixed positions as if stopped with a modern stroboscopic light, and the intervals between are continuous lines of light reflected off the swinging links of chain. Gino Severini insisted on the historical necessity of portraying successive memories: "In this epoch of dynamism and simultaneity, one cannot separate any event or object from the memories . . . which its expansive action calls up simultaneously in us."[39] Luigi Russolo depicted *Memories of a Night* with past images scattered about the picture surface as they are in reality scattered achronologically about the mind.

These works conformed with Apollinaire's pronouncement that the painter must "contemplate his own divinity" and "encompass in one glance the past, the present, and the future."[40] Some novelists also tried to recreate moments when they felt divine, swollen with past and future. Proust's *moments bienheureux* and Virginia Woolf's "rings of light" compressed successive experiences into a single heightened moment. Joyce defined his "epiphanies" as sudden spiritual manifestations, "the most delicate and evanescent of moments."[41] Gertrude Stein developed two techniques for rendering a temporally expanded present: beginning again and the continuous present tense. In an essay of 1926 her maddeningly repetitive narration explained as well as illustrated her message:

> Beginning again and again is a natural thing even when there is a series.
>
> Beginning again and again and again explaining composition and time is a natural thing.

The technique is evident from the first pages of *Melanctha* (1909) where she tells us, in very nearly the same words, twice that Rose Johnson was a real black negress, three times that her baby died, and twice that she laughed when she was happy. The effect of this stuttering prose is to flatten out temporal distinctions and create an impression that all action occurred in a continuous present. To emphasize that present she stretched action with a verb form called the "continuous present." So we read that Melanctha was "always losing what she had," "always being left when she was not leaving others," and "always seeking quiet and rest." Melanctha was always the same, even though much happened to her; always beginning again and again, even though the story is full of sequential action; always living in the continuous present, although we learn of her full past. Since the author knows everything at the time of composition, she can tell it in any order whatever. For Stein it was only an artificial convention that the story begin at the beginning and end at the end; both are present in the author's mind at the moment of conception. The fact that it takes time to write the story or that an author makes discoveries about his subject in the course of writing does not contradict her theory. If I begin a story about a friend, I may not know where it will go, but I do know what had transpired between us. The cumulative effect of that experience is what inspired me to begin. Stein sought to do justice to that original impulse by sustaining a continuous present to show the way it embodies from the outset the entire past and the intention to write the story.[42]

In his affirmation of the now, Joyce used various techniques to suggest the experience of simultaneity, and he also compressed memories and expectations into a temporally thickened present. In 1912 he formally praised William Blake, who "by minimizing space and time and denying the existence of memory and the senses . . . tried to paint his works on the void of the divine bosom."[43] That void is God's time—the ultimate expanded present wherein what appears to us as sequence is an unchanging whole. Art tries to approximate that. It is inspired in instants, has as its temporal locus only the now, but will endure only if it is universal and eternal. The tactical problem was how to follow Blake and minimize time while writing about people who live quite palpably in it. The solution was the expanded present of *Ulysses,* where characters' moments are thickened with direct interior monologues, and their lives and the history of the universe compressed into a single day. Thus did Joyce "hold to the now." While his rendering of simultaneity united distant events in a spatially expanded present, the direct interior monologue linked past and future in a temporally thickened one. Both techniques underlined his affirmation of the present as the only real location of experience.[44]

Joyce shared this repudiation of the past with several others, among whom Nietzsche stood out. His contempt for the "abuse" of history was matched by an equally intense admiration for those who love their lives and are able to respond joyously to the theory of eternal recurrence, which he developed in the early 1880s after learning about some cosmological speculations that the present configuration of forces in the universe must repeat itself again and again eternally. He was never entirely convinced that this doctrine was true, but reflecting on the possibility that it might be, he developed a concept central to his philosophy—one of the most emphatic affirmations of the present in the history of thought. In *The Gay Science* (1882) he introduced the idea of eternal recurrence as a thought experiment, a test for evaluating one's acceptance of the present.

> How, if some day or night, a demon were to sneak after you into your loneliness and say to you: "This life, as you now live it and have lived it, you will have to live once more and innu-

> merable times more; and there will be nothing new in it, but every pain and every joy . . . must return to you—all in the same succession and sequence—even this spider and this moonlight between the trees, even this moment and I myself" . . . Would you not throw yourself down and gnash your teeth and curse the demon who spoke thus? Or have you once experienced a tremendous moment when you would have answered him: "you are a god and never did I hear anything more godlike!" . . . The question in each and everything, "do you want this once more and innumerable times more?" would weigh upon your actions as the greatest stress. Or how well disposed you would have to become to yourself and to life to *crave nothing more fervently* than this ultimate eternal confirmation. . . ?[45]

Only the overman, by means of a terrifying transvaluation of all values, can utter the great "Yes" to his own existence and view the demon's message as godlike. He does not dote upon the past as a source of guilt or long for heavenly reward. He craves "nothing more fervently" than eternal recurrence, because he has made of his life something creative in defiance of the crushing burden of the past and the seductive hope for future paradise. He alone can accept Nietzsche's motto, *amor fati.*

As cinematic montage combines distant scenes to create a unified whole, so have I drawn together pieces of the cultural record using the principle of conceptual distance and an expository technique of juxtaposition. My method involves the presentation of diverse sources that are far enough apart to justify broad generalizations about the age without being too far apart to exceed the limits of plausibility. Thus, a parallel presentation of the response to two sinking ships would not tell us as much as an identification of the thematic similarity between the reaction to a sinking ship and the musings of a philosopher. It is a long way, conceptually, from the *Titanic* to Nietzsche, and that is precisely what makes the identification of a common denominator so fruitful. It would be outrageous to link the *Titanic* and Nietzsche directly, but by following the shorter, intermediate links we see a coherent matrix of thought emerging. The juxtapositions from the *Titanic* to the wireless and telephone, to simultaneity and the spatially expanded present, to the temporally thickened "specious present," and finally to the positive evaluation of the present in Nietzsche and others outline the distinctive experience of the present in this period. The individual "shots" come from various sources relating to the two focal issues and conclude with a picture of Nietzsche's overman overjoyed at the prospect of eternal recurrence, happily affirming his fate in the here-and-now.

There was no actual confrontation over these two focal issues, because simultaneity and the thickened present attracted all of the innovative thinkers and artists. Simultaneity was the more directly influenced by technology, because electronic communication made it possible for the first time to be in a sense in two places at once, while temporal thickening derived from a theory of experience that could have been articulated in any age. Simultaneity also had the broader cultural impact. One response was a growing sense of unity among people formerly isolated by distance and lack of communication. This was not, however, unambiguous, because proximity also generated anxiety—apprehension that the neighbors were seen as getting a bit too close. Perhaps the most far-reaching impact of the new simultaneity was due to the cinema, which was able to bring together an unprecedented variety of visual images and arrange them coherently in a unified whole. German audiences moved visually between Munich and the wild American West; French audiences traveled to the North Pole and the moon. The cinema also thickened the present. Any moment could be pried open and expanded at will, giving the audience seemingly at once a vision of the motives for an action, its appearance from any number of perspectives, and a multitude of responses. A man is shot in an instant, but moviegoers saw the event prolonged and analyzed like a detailed case history. The present was thus thickened by directors who spliced time as they cut their film.

The new aesthetic and ethic joined in affirming the reality of a present that embraced the entire globe and included halos of the past and future which made it perceptible in the flux of time as atomic particles are made visible in their path through a cloud chamber. The new technology changed the dimensions of experience so rapidly that the future seemed to rush toward the present at a tempo as hurried and as irregular as Stravinsky's music. In the prewar years there was still a time to be born and a time to die, but the protracted sequence of events, each in its own time, was becoming ever more hurried and compressed. The world was racing into the future like the *Titanic* into the North Atlantic, and those who looked ahead foresaw both shipwreck and the wonders of time travel.

4
THE FUTURE

Shortly after the armistice in 1918 Eugène Minkowski began a work entitled "How We Live the Future (and Not What We Know of It)." He never published it but applied the ideas in his clinical practice in the postwar years and then incorporated them in *Lived Time,* where he distinguished two modes of experiencing the immediate future—activity and expectation. The essential difference is the orientation of the subject in time: in the mode of activity the individual goes toward the future, driving into the surroundings in control of events;

in the mode of expectation the future comes toward the individual, who contracts against an overpowering environment. Every individual is a mixture of both modes, which makes it possible for him to act in the world and maintain an identity amidst a barrage of threatening external forces. The war sharpened the contrast between the two modes. The dominant one for the soldier was expectation, as the war limited his activity and sense of control over the future. Minkowski's description of expectation reads like a phenomenology of life in the trenches. "It englobes the whole living being, suspends his activity, and fixes him, anguished in expectation. It contains a factor of brutal arrest and renders the individual breathless. One might say that the whole of becoming, concentrated outside the individual, swoops down on him in a powerful and hostile mass, attempting to annihilate him." Another image conjured up the sinking of the *Titanic.* "It is like an iceberg surging abruptly in front of the prow of a ship, which in an instant will smash fatally against it. Expectation penetrates the individual to his core, fills him with terror before this unknown and unexpected mass, which will engulf him in an instant."[1] While expectation dominated the war experience, activity dominated the prewar period, and the two modes constitute basic polarities of this generation—how they lived the future (*and* what they knew about it).

The future, to be sure, is not experienced as vividly as the present and is dependent on the past for its content of images reassembled and projected ahead. Nevertheless it is an essential component of the personality, as the organization of those projections provides a sense of direction and makes novelty, purpose, and hope possible. Although the historical data on how people viewed the future are more limited than those on the past or present, it is possible to identify this generation's distinctive experience. The new technology provided a source of power over the environment and suggested ways to control the future. The Futurists identified their movement with the promise of that technology and the new world that it offered. There was a burst of science-fiction literature that sought to appropriate the future imaginatively. Philosophers argued that the possibility of freedom required that there be an unknown future, and one political tactician considered the importance of a myth of the future for revolutionary movements. These examples cluster on the side of an active future, that expansive and creative embodiment of the *élan vital* that Minkowski, following Bergson, believed essential to mental health. The war put a swift halt to this exuberance, but even in the prewar period some thinkers envisioned the future in the mode of expectation. The entire discussion of degeneration pointed to a future in which mankind waited to be overpowered by the forces of nature and society, leading to a decline of cultures and an ultimate extinction of the species.

The effect of the telephone on the past and present was recognized at once—it eliminated the preservation of the past in letters and expanded the spatial range of the present. But there was little recognition of the impact of the telephone on the experience of the future. The historian Herbert Casson, writing in 1910, touched on the subject. He noted that "with the use of the telephone has come a new habit of mind. The slow and sluggish mood has been sloughed off . . . life has become more tense, alert, vivid. The brain has been relieved of the suspense of waiting for an answer . . . It receives its reply at once and is set free to consider other matters."[2] Actually it had a far more complex effect. In comparison with written communication or face-to-face visits the telephone increased the imminence and importance of the immediate future and accentuated both its active and expectant modes, depending whether one was placing or receiving a call. A call is not only more immediate than a letter but more unpredictable, for the telephone may ring at any time. It is a surprise and therefore more disruptive, demanding immediate attention. The active mode is heightened for the caller who can make things happen immediately without enduring the delay of written communication, while the intrusive effect of the ringing augments the expectant mode for the person called by compelling him to stop whatever he is doing and answer. He is thrust into a passive role because the caller can prepare for the conversation and control it at the outset.

Even though interpretation reveals an intensification of both modes, the general impact of the telephone was its ability to manipulate the immediate future, because the telephone was conceived largely through the experience of the caller. (Casson did not even consider the magnification of expectation for the person waiting for a call.) Evaluations of the telephone divided sharply between optimists and pessimists, and those who viewed it favorably usually had the caller in mind. The pessimists pictured the recipient of the call first suspended in waiting and then disturbed by the intrusion. Indeed, waiting for the telephone to ring became a symbol of loneliness and helplessness in the expectant mode. It is more tormenting than waiting for a letter, because the call may or may not come at any time, while the letter either does or does not arrive in the daily mail. One may thus prepare for a letter in a way that it is impossible to prepare for a telephone call.

A similar division between active and passive modes was created by the introduction of the assembly line at Ford's Highland Park factory in Detroit in 1913. While products made individually involved the worker in the manufacturing process, the conveyor belt and continuous operation of the assembly line eliminated challenges and surprises as the product moved along with every step worked out beforehand. Once uncertainty about the future was eliminated by the assembly line it became possible to streamline the productive process further by observing every stage, determining the minimum movements necessary to complete all tasks, and then instructing the workers to make them. This was the achievement of Frederick Taylor's time and motion studies,[3] which accelerated production by increasing the predictability of workers' movements and depriving workers of the opportunity to select the sequence of actions to complete an operation. The assembly line and Taylorism diminished the factory worker's active control over the immediate future in the productive process and relegated him to an expectant mode, waiting for the future to come along the line, at the same time increasing the manufacturer's control. Although the impact of the new technology on the future fluctuated between these two roles, as was the case with the telephone, the larger and more decisive historical impact was a magnification of the active mode.

Another concrete manifestation of the active mode of the future was imperialism and the prospect of European ascendancy throughout the world in years to come. Annexation of the space of others, outward movement of people and goods, and the expansive ideology of imperialism were spatial expressions of the active appropriation of the future. In a famous address before the Colonial Institute in 1893, the Liberal-Imperialist Foreign Minister Lord Rosebery interpreted British motives for the colonization of Africa in terms of the future:

> It is said that our Empire is already large enough, and does not need extension. That would be true enough if the world were elastic, but unfortunately it is not elastic, and we are engaged at the present moment, in the language of mining, "in pegging out claims for the future." We have to consider not what we want now, but what we shall want in the future. We have to consider

> what countries must be developed either by ourselves or some other nation, and we have to remember that it is part of our responsibility and heritage to take care that the world, so far as it can be moulded by us, shall receive an English-speaking complexion, and not that of other nations . . . We have to look forward beyond the chatter of platforms and the passions of party to the future of the race of which we are at present the trustees.[4]

A personal reaction to the two modes of experiencing the future, so altered by new sources of energy and forms of technology, was recorded in Henry Adams' autobiography of 1907. The nineteenth century had measured its progress by carloads of coal produced. It was regulated by Newton's laws and accepted the law of contradiction as a basis for reasoning. But this coherence began to break up in the 1890s. Adams wrote that thinking was "caught and whirled in a vortex of infinite forces," men were flung about "as though [they] had hold of a live wire or a runaway automobile," and he was forced to learn to think in contradictions.

In 1892, when Adams was over fifty, he "solemnly and painfully learned to ride a bicycle." This was an active, if somewhat creaky, appropriation of the future, but the new technology also threatened to overwhelm him. At the Chicago Exhibition of 1893 he was awed by the mechanical forces of the dynamo, the creator of a new phase of history. By the Exhibition of 1900 his fascination had turned to devotion, and he saw the dynamo as a symbol as powerful in its way as the image of the Virgin. The achievement of science and the power of technology—radium and x-rays, "frozen air" and electric furnaces, automobiles and telephones—surrounded him as he looked up at the dynamo. All mocked the slow-paced, regular accounting that had shaped his historical thinking and shattered his neat categories of history. "Satisfied that the sequence of men led to nothing and that the sequence of their society could lead no further, while the mere sequence of time was artificial, and the sequence of thought was chaos, he turned at last to the sequence of force; and thus it happened that, after ten years' pursuit, he found himself lying in the Gallery of Machines at the Great Exhibition of 1900, his historical neck broken by the sudden irruption of forces totally new."[5] Henry Adams has left us with a dual image of his response to technology—a courageous man learning to ride a bicycle and an elderly scholar lying on the ground with his historical neck broken. Here are the extremes of activity and expectation coming together in the life of a pioneer in the history of technology.

Although the world seemed to be rushing ahead at an ever faster clip, for some that was not fast enough. Science-fiction writers reached out for the future as if it were a piece of overripe fruit. Their stories came into vogue on a grand scale, indicating that the future was becoming as real to this generation as the past had been for readers of the Gothic novel and historical romance. There had been utopian writings before, but they generally meant to identify current problems rather than delineate a world to come and the processes by which it would evolve. From the 1860s on Jules Verne's *voyages extraordinaires* popularized the genre with projections of future developments from current science and technology, and in the 1890s H. G. Wells became even more fanciful with his "tales of space and time."

Wells interpreted this particular inclination of his generation in a lecture of 1902, "The Discovery of the Future."[6] In a manner remarkably similar to Minkowski's he distinguished two types of mind by their attitude toward time and "the relative amount of thought they give to the future of things." One type is retrospective, a "legal or submissive" mind that looks for precedents to decide how to deal with the future. The other is the "legislative, creative, organising or masterful type" that attacks the established order: "It is in the active mood of thought while the former is in the passive." Most people still cling to tradition: they travel on roads that are too narrow; they live in space-wasting houses out of a love of familiar shapes; their clothing, speech, politics, and religion all testify to the binding power of the past. But the modern age has turned away from a dogged adherence to tradition and has "discovered" the future as a source of values and a guide for action. While three hundred years ago people drew their rules of conduct "absolutely and unreservedly from the past," now they are more inclined to look ahead and consider the consequences of any action and modify the rules if the consequences merit it. Even modern wars are conceived and justified in terms of the future: "a comparison of the wars of the nineteenth century with the wars of the Middle Ages will show . . . in this field also there has been a discovery of the future, an increasing disposition to shift the reference and values from things accomplished to things to come." The spirit of modern science, the flood of technological discoveries, and geology, archaeology, and history have drawn attention to the flexibility of our life in time. As larger vistas of the past have been opened up and have shattered conventional ideas about its duration and effect on the present, so a new knowledge of the future is becoming possible. Gravitational astronomy is able to predict stellar movements, medical science continually improves its ability to diagnose, meteorology predicts the weather, and chemists forecast elements before they are discovered, as Clerk Maxwell announced the existence of rays before Marconi put them to use.

Until 1902 Wells's vision of the future was full of catastrophes and degeneration; later he began to foresee progress. His lecture included both. It concluded with the hope that the creative energies of life will overcome the catastrophes, but the lasting impression was an expectation of disaster: some poison from industry or outer space, an uncontrollable killer disease or predator, evolutionary degeneration, war, collision with a heavenly body, and if nothing happens earlier, the certainty that the sun will cool and its planets rotate ever more sluggishly "until some day this earth of ours, tideless and slow moving, will be dead and frozen."

Wells explored this last dismal prospect in his classic of 1895, *The Time Machine.* Its hero, the Time Traveller, invents a machine in which he is able to slip like a vapor through the interstices of intervening substances and travel into the future. He stops in the year 802,701 and discovers the Eloi, a beautiful people living on fruit and playing all day long, seemingly without a care in the world. But they do fear the dark and the Morlock, a "bleached, obscene, noctural Thing" that lives underground and supports the Eloi only to harvest them for food in raids on moonless nights. The Time Traveller concludes that the opposition between capitalist and laborer had led to this radical differentiation between the Eloi and Morlocks, who had evolved physically into different species, occupied different living spaces, acquired different character traits, and lived in perpetual fear of one another though they were mutually interdependent for survival. It was, he reflects, a "working to a logical conclusion [of] the industrial system of today."

For Wells the most disturbing thought about the future was that man is not the end of all things, and the most fascinating speculation was about what is to come after. He ventured an answer in a chapter called "The Further Vision." Fleeing an attack by the Morlocks, the hero traveled into the future and stopped at the edge of a sea. But there were no waves. The work of tidal drag was done and the earth had ceased to rotate. The sun hung motionless on the horizon, swollen and red because the earth had drawn closer. The only vegetation was a "poisonous-looking" cover like forest moss that lived in perpetual twilight, and the only animals were enormous crabs smeared with algae. When one attacked him, the Traveller sped on to his last stop, thirty million years hence, where he was horrified by an eclipse of the sun as one of the inner planets passed near the earth. There was a slight rippling from the sea but beyond that an uncanny silence, and when the eclipse was complete, it grew cold and black. That desolate scene sated his curiosity and he returned to his own time.

The story is a compendium of nineteenth-century theory projected into the future. Marx's vision of the growing stratification of classes is magnified in the conflict between the Eloi and the Morlocks. Eugenics is represented by the breeding of the Eloi. The ideal of preventive medicine is achieved since all disease is eradicated; the erosion of the family that many feared in Wells's time is complete; and the sexes have grown to look alike. The *fin-de-siècle* preoccupation with the decadence of mankind, summarized in Max Nordau's *Degeneration* (English translation, 1895), is vividly represented by the helpless, effete, and self-indulgent Eloi and the physically degenerate and cannibalistic Morlocks. Charles Darwin's theory is there, but in reverse—a devolution of the species from human beings back to giant crabs and then to a creature so elementary that Wells did not bother with his usual detailed description—merely "a round thing" with tentacles trailing behind it. George Darwin's prediction of the cessation of the earth's rotation from tidal drag and Kelvin's prediction of a cooling of the sun have come to pass. Wells utilized current speculation about the fourth dimension for an explanation of the way the Time Traveller slipped through the interstices of matter, and the time machine itself is a symbol of the hope of all technology to accelerate the processes of change.

Wells looked ahead again and again. In *When the Sleeper Wakes* (1899) the hero emerges from a cataleptic trance of 203 years and discovers an amazing technology in the service of big government that tyrannizes its subjects. Collective life has swallowed up all privacy and cities have become prisons. The story laments the passage of the character traits and social institutions that Wells valued and saw on the decline in his own day—individuality and privacy, the rivalries and jealousies of the middle classes, and the "strong barbaric pride" of the lower classes. The hero's reflections point to a moral: "It seemed to him the most amazing thing of all that in his thirty years of life he had never tried to shape a picture of these coming times. 'We were making the future,' he said, 'and hardly any of us troubled to think what future we were making.' "[7] He who does not contemplate the future is destined to be overwhelmed by it.

In *Anticipations* (1901), an ambitious essay in prophecy, Wells promised to follow a scientific method of forecasting and speculate from the trend of present forces. The reader will be a "prospective shareholder" in this sketch of the future that begins with some probable developments in land locomotion. As the railroad dominated the nineteenth century, the "explosive engine" will dominate the twentieth. There will be paved roads and "conspicuous advertisements" by the roadside; there will be traffic jams as motor vehicles replace pedestrians in the towns. By the year 2000 London will extend to Wales, and in the United States there will be a continuous city from Washington to Albany. Improvements in telephone and postal service will make possible a diffusion of talents to the suburbs. "The businessman may sit at home in his library and bargain, discuss, promise, hint, threaten, tell such lies as he dare not write, and, in fact, do everything that once demanded a personal encounter." The future will alter the "method and proportions" of human undertakings and the "grouping and character" of society. Three new classes will emerge: unskilled workers displaced by machines, technically trained people who can work them, and shareholders who do nothing.

Some of his predictions about future wars were memorably in error, notably that the submarine would do little more than suffocate its crew and founder at sea, or that the airplane would not seriously modify transport or communication. Although he got the vehicle wrong, he was right on the strategic impact of aerial warfare, which he thought would be conducted from balloons. "Stalked eyes," equipped with telephonic nerves, would hang above the front lines, observe enemy troop movements, direct artillery fire, drop explosives, and demoralize the enemy. He predicted the future of land warfare as though he had journeyed in his time machine and witnessed the battle of the Somme. He forecast the rifle with cross-thread telescopic sights and a machine-gun breech that will enable it to fire a spray of "almost simultaneous bullets." Wells's most famous prediction was the tank, called a "land ironclad," that could move fire power through no-man's-land, protect men from machine gun bullets, and tear apart barbed wire. Machines will also be used to dig miles of trenches. There will no longer be a sharply focused battlefield or a "Great General" observing from the field. Instead somewhere in the rear a "central organizer" will direct operations along a vast front from a telephone center. At times Wells wrote as though he could smell the battle and feel the percussion of exploding shells. "For eight miles on either side of the firing lines—whose fire will probably never altogether die away while the war lasts—men will live and eat and sleep under the imminence of unanticipated death."[8]

The impulse to look ahead is universal, but the quantity of science fiction in this period and its success in the market place suggest that this generation was especially eager to do it. In America, Edward Bellamy's *Looking Backward,* a vision of the future in spite of its misleading title, was an immediate success. It sold 213,000 copies within two years of its publication in 1888 and initiated what one historian has called an "outburst of literary utopianism."[9] Some authors saw the future as a nightmare—dystopias with destructive volcanos, killer diseases, and maniacal rulers who held people captive with fantastic new contraptions.[10] Others looked forward to happier utopias with less drudgery, cheaper goods, and clean, safe cities. Still others saw mixtures of progress and degeneration, islands of carefree pleasure and oppressive technocracies.

The Futurists were not troubled by any ambivalence. They created a kind of science fiction of their own out of the latest of everything in artistic works that squeaked from newness. Marinetti's "Founding Manifesto" of 1909 traced the birth of the movement. After a night of frenzied scribbling and brooding over their ennui, he and his friends were drawn outside by the sounds of the city rising. The creaking of the bones of "sickly palaces" was interrupted by the roar of automobiles, and they set off to shake the gates of life. Their rush into the unknown led first into a ditch. But some fishermen rigged a derrick to pull them out, and as their automobile revved up again Marinetti proclaimed their objective: "We intend to sing the love of danger, the habit of energy and fearlessness." Here is fixation on change. "We stand on the last promontory of the centuries! . . . Why should we look back, when what we want is to break down the mysterious doors of the Impossible? Time and Space died yesterday. We already live in the absolute, because we have created eternal, omnipresent speed."[11] They will surge into the future at full throttle—innovating, challenging, and occasionally going smash. In a manifesto of 1910 they linked the progress of science and their orientation toward the future. "Comrades, we tell you now that the triumphant progress of science makes profound changes in humanity inevitable, changes which are hacking an abyss between those docile slaves of past tradition and us free moderns, who are confident in the radiant splendour of our future."[12]

The Futurists strained the limits of traditional genres to create new forms. Enrico Prampolini defined "a new state of perception" among human beings—chromophony—the colors of sounds. Carlo Carrà announced a new painting of sounds and smells. Luigi Russolo called for a "music of noises" composed from backfiring motors, squealing electric trams, and the howl of mechanical saws sounding to such diverse rhythms as tapping valves and the irregular noises of city life. Sculptors were to fabricate wild shapes and integrate empty space in compositions out of ever new materials. Futurist theater jumped out at the audience and drew it into the action. Futurist paintings showed the new dynamics and technology of daily life. Traditional activities—running, swimming, descending a staircase—are "futurized" by depicting moving objects and the currents of water and air streaming off them. In Boccioni's *Dynamism of a Cyclist* (1931) man, cycle, and air interpenetrate in a composition of abstract volumes and lines of force, pumping limbs, and swirling eddies of light and air. However, the technology in these works is the current model—no time machines and, in spite of their praise of war—no ray guns. Bragaglia's multiple-exposure photograph of *The Typist* (1912) is very much of this world, and the Futurist content in Boccioni's *Train in Motion* is not a supercharged monorail but his innovative technique of showing movement.

The most explicit picture of a future world was drawn by the Futurist architect Antonio Sant'Elia. His manifesto of 1914 began with an attack on contemporary architecture and its "hilarious salads" of Egyptian pilasters, Gothic arches, Renaissance cherubs, and rococo scrolls.[13] The new construction should use modern materials and be responsive to the needs of contemporary life and the aesthetics of modern technology. Instead of wood, stone, and brick, architecture will exploit steel, glass, cardboard, reinforced concrete, and textile fibers. The Futurist house must be like a gigantic machine, the city like a dynamic shipyard. Streets must no longer lie dormant at ground level but plunge into the earth to hold traffic and link up with moving pavements. Roofs and underground spaces must be utilized and walkways flung high above ground. Elevators must no longer be hidden like tapeworms in the bowels of buildings but be accessible and visible on the outside of façades. The purely decorative must be abolished. "Fussy moldings, finicky capitals, and flimsy doorways" must give way to bold groupings of masses with bare or violently colored surfaces. The Futurists aim at an abandonment of the heavy and static for the light, practical, and swift. Whenever possible, emotive elliptical and oblique lines will replace rigid horizontals and perpendiculars; the "artificial" aesthetic of the mechanical world will replace the "natural" aesthetic of the past.

To identify the distinctive thought of any age, the cultural historian is on the lookout for ideas that are entirely new, like the one proposed in the final paragraph of Sant'Elia's manifesto. "From an architecture conceived in this way no formal or linear habit can grow, since the fundamental characteristics of Futurist architecture will be its impermanence and transience. Things will endure less than us. Every generation must build its own city." In earlier versions this was missing, and most likely it was added by Marinetti to bring Futurist architecture in line with the Futurists' commitment to a continually evolving, ephemeral art that would never become like the museum pieces they excoriated. According to Carrà, Sant'Elia disapproved of this statement but allowed it to remain in conformity with the larger Futurist program.[14] Imagine the pressure that would lead an architect to put his name to the first formal commitment to build buildings that would fall apart. Sant'Elia's compliance evinces the Futurist addiction to change, born in an age in which change had become routine and the future seemed more within the active control of mankind than ever before. The recommendation that every generation build its own city shows that someone thinks it can. Sant'Elia's drawings for his city provided a blueprint, but no buildings were ever built and he was killed in 1916. The Futurist architectural program that every generation would have to rebuild itself was more true of its thinkers than of anything they ever built.

The philosophy of the future of this period was an emphatic repudiation of a body of deterministic thought that had been building for a century from its foundation in the naturalistic determinism of Pierre Laplace. With a spectacular show of ambitiousness, at the beginning of a century that was spectacularly ambitious about the possibilities of reason and science, Laplace speculated that the future is determined in the present state of matter in the universe. "An intellect which at a given instant knew all the forces acting in nature, and the position of all things of which the world consists—supposing the said intellect were vast enough to subject these data to analysis—would embrace in the same formula the motions of the greatest bodies in the universe and those of the slightest atoms; nothing would be uncertain for it and the future, like the past, would be present to its eyes."[15] Throughout the nineteenth century this was the goal, if not the achievement, of science. Bergson charged that it denied time and freedom by rolling up the future in the present the way the end of a film is already determined at the start of the reel. He conceded that the isolation of phenomena in closed systems for purposes of analysis is not entirely artificial, because matter has a tendency to constitute isolable units, such as the solar system, which, to a degree, conforms to regular laws. But it is only a tendency. Gravitational forces attract the solar system to the rest of the universe and draw it into a future of endlessly new orbits and configurations. And to whatever limited extent inorganic matter may be suited to such analytical reduction, organic matter is less so. Scientists think they can measure lived time and then compare measured intervals to derive laws of change. But they are wrong, like those people who believe that their life could be unfurled like a fan, open to view at a single glance. In reality it unfolds in time very differently, as Bergson put it in the opening pages of *Creative Evolution.* "If I want to prepare a glass of sugared water, try as I may, I must wait until the sugar melts. This little fact is of great significance." The time I have to wait through is not the same as the interval that can be measured mathematically, because that interval is completed before the measurement is made and therefore different from what I live through. Time as I live it "coincides with my impatience." That waiting constitutes its essence and ensures my freedom. Without it the future unfolds as something already known and we are locked in determinism. Science seeks to discover laws and predict the future, but human experience is an uncertain chain of events in time.

Bergson was joined in his insistence on the importance of that uncertainty by the eminent French physicist Emile Meyerson, who considered the problem in a famous chapter provocatively titled "The Elimination of Time."[16] Meyerson indicted the tendency of modern science to eliminate time by the identification of cause and effect symbolized in the equal sign of an equation. This operation is based on the principle of conservation of matter and energy—that in any phenomenon nothing is created, nothing lost—and the postulate of reversibility—that in any causal action "the integral effect may reproduce the entire cause or its equal." Natural phenomena such as aging or burning wood are irreversible. Chemical reactions are also irreversible, but "chemical equations are the expression of the tendency to identify things in time; one can say 'to eliminate' time." If science succeeded in describing everything with an equation, in identifying antecedent and consequent, nothing would change, time would be refined out of science, and the future would become a necessary consequence instead of a promise of surprise. It would be "the confusion of past, present, and future—a universe eternally immutable." He conceded that this complete identification of everything in an equation is impossible, but it is a goal. Modern science has not entirely eliminated time but cannot stop trying.

The French philosopher Jean Guyau made another argument on behalf of an active sense of the future by deriving our sense of time itself from it. To make this argument Guyau reversed Kant's theory that our sense of time is an *a priori* form of perception that makes all experience possible. Instead, Guyau derived the sense of time out of activity and the future orientation of experience. In *The Genesis of the Idea of Time* (1890) Guyau held that our idea of time is a product of evolution and the psychological development of the individual. His theory is anchored in human physiology. The child experiences hunger and reaches out for the nurse—that is the germ in our idea of the future. Bodily needs generate desire, the memory of former satisfactions generates a conception of the possibility of future satisfaction, and the individual prepares to gratify the desire with intentional activity oriented ahead of itself in space and toward the future in time. Thus out of desire and activity the idea of the future and our whole sense of time originates. This is a philosophy of the future in the active mode: "The future does not come toward us, but it is that toward which we move."[17]

Guyau and Bergson have left vivid images of the active and passive modes of the future—reaching out for a nurse and waiting for the sugar to melt. However, both saw the future as a combination of the two. Guyau insisted on a "passive form of time," a substratum of continuity against which change can be observed. This is not just a passive orientation toward the future; it suggests that the entire experience of time is an integrant of passivity and activity, permanence and change. Bergson's impatience is waiting with an active edge, like a sprinter in a starting block. Bergson understood that the experience of the future is a mixture of the active and passive modes, but the emphasis is on the link between freedom and action: we become freer the more we feel "our whole personality concentrate itself in a point, or rather a sharp edge, pressed against the future and cutting into it unceasingly."[18] The two shared the central idea that an open future is the source of human freedom and with Meyerson defended it against naturalistic determinism and the ubiquitous equal sign of modern science.

In spite of all their utopian tracts and projects for future change, social and political thinkers in the nineteenth century did not explore the social or historical basis of the experience of the future as such. The revolutionary movements of the nineteenth century had always held the promise of a better world as justification for the destruction of the present; the great problem was how to get people to act. For decades socialist leaders fought over tactics, while the rank-and-file members languished in chronic inaction. By the early twentieth century it was clear that analyses of the evils of capitalism, calculations of the benefits to accrue from socialism, and incantations of the rightness and inevitability of revolution would not budge them off the rock of the status quo. It took the starvation, killing, and general madness of World War I to get rid of the Romanovs, after all, and without such disruption to shake the stability of old regimes, revolutionary movements were stymied. Socialist revolutionaries shared the Marxian notion of the future as a triumph of socialism. They had a vision of the future but no concept of it, no explicit analysis of its motivating power independent of its content. Only one radical theoretician before the war made such an analysis. Faced with working class inaction, the French syndicalist Georges Sorel developed a tactic of action-for-action's-sake that relied on the creation of an inspiring vision of, and dynamic movement toward, the future.

A pioneer of social psychology, Sorel conceived of political action as theater and believed in the necessity of creating a sense of urgency, a movement toward climax, that would give the workers a profound and lasting impression of revolution. He drew from Bergson's theory that intuitive knowledge was superior to analytical knowledge and worked out a plan by which workers would intuit socialism as a whole, instantaneously, in the drama of a general strike. To get them to act leaders must create an anticipation of the future in the form of a myth embodying their hopes. Sorel theorized: "Without leaving the present, without reasoning about the future . . . we should be unable to act at all. Experience shows that the *framing of a future, in some indeterminate time,* may, when it is done in a certain way, be very effective."[19] The idea of framing a future for mass manipulation of workers ran counter to orthodox Marxism. For Marx workers embodied the future: action was to come from class consciousness generated out of struggle with the present. For Sorel it was to come from a deception of workers with a myth about the future. Sorel's modification was a lone, but distinctive, voice on behalf of an active appropriation of the future in politics. Nothing was inevitable: everything was up for grabs, and effective political action required a vivid sense of the future, whatever the cost to the integrity of the movement.

∞

The new technology, the science fiction, Futurist art, and revolutionary politics looked at the future like a predator eyeing its prey. It was an age for planners and go-getters: for the great tomorrow of the Carnegies and the Rockefellers, anarchist terrorists and Bolshevik revolutionaries, the German Navy and the new Russian Army. But in contrast to all this active mobilization for the future some people voiced passivity and fatalism, focusing their thoughts on the concept of degeneration. Its spokesmen anticipated deterioration of the quality of urban living, breakdown of health, decline of Western civilization, extinction of life on the planet, and ultimately depletion of energy in the universe. Although the imminence of these catastrophes varied considerably, they tended to group in a single dreadful vision. Although they derived from the past, they were projected ahead into a threatening future.

The bad news in physics broke in 1852 with William Thomson Kelvin's essay "On a Universal Tendency in Nature to the Dissipation of Mechanical Energy," which predicted the death of the earth from heat loss as a result of the second law of thermodynamics—that the amount of energy available in the universe for useful work is always decreasing as entropy (randomness or disorder) increases. "Within a finite period of time past the earth must have been, and within a finite period of time to come the earth must again be, unfit for the habitation of man as at present constituted, unless operations have been, or are to be performed, which are impossible under the laws to which the known operations going on at present in the material world are subject."[20] The discovery of radioactivity in the 1890s forced Kelvin to revise his estimate of the age of the earth, but the implications of the second law of thermodynamics for the future were unchanged. Although the earth would not become unfit for human habitation until the far distant future, this prediction became the nucleus of a number of gloomy biological, social, and historical theories of contemporary degeneration: the blood of the race was becoming progressively polluted by an accumulation of diabetes, tuberculosis, syphilis, and alcohol; the intimate organic communities of the good old days were deteriorating into mechanistic societies, impersonal big cities of crime, suicide, and insanity; and civilization was heading toward spiritual collapse. Brooks Adams envisaged degeneration from the coming domination of capital, which he predicted in *Law of Civilization and Decay* (1895), and Oswald Spengler chronicled the crisis of the soul of the modern era in *The Decline of the West* in 1918.

Spengler's work is a sprawling history of the life and death of cultures, each interpreted under a unifying principle or "destiny-idea." Thus the classical world was Euclidean—spatially extended, atemporal, centered in the polis, and visibly symbolized by monumental architecture. The modern era is characterized by the restless striving of the Faustian soul and is inherently temporal. It began with the discovery of the mechanical clock and eventually produced the pocket watch that accompanies the individual to remind him constantly of his temporal existence. The drama of Spengler's message is prepared by his emphasis on the importance of a sense of the future in the modern world. While the classical world bowed in "submission to the moment," the modern world has an "unsurpassably intense Will to the Future." Western culture glorifies hard work as "an affirmation of Time and the future," and with its meaning embodied in the future, it is particularly sensitive to the pessimistic vision that Spengler sketches.

The modern age is suffering from the consequences of the rule of money allied to political democracy, but this alliance will not hold against the coming of Caesarism. Western culture labors under the tyranny of reason and the cult of science but has not produced any genius since Gauss and Helmholtz. In physics it is experiencing "the *decrescendo* of brilliant gleaners who arrange, collect and finish off." After the impressionists Spengler can find no painters; after Wagner, no musicians. But the main cause for alarm comes from nature itself. The discovery of the law of entropy in the 1850s and of atomic disintegration in the 1890s has given the life-sustaining energy of our world a time limit. Inorganic matter has acquired a perishability previously reserved for living matter, and it is heading for a period of steady decline. "What the myth of *Götterdämmerung* signified of old, the irreligious form of it, the theory of Entropy, signifies today—*world's end as completion of an inwardly necessary evolution.*"[21]

The timing of the publication of this book in Germany in the aftermath of military collapse accounts for a large measure of its impact. The war seemed to show that Western civilization was worn out, and the book captured the sense of powerlessness, of passivity, that many experienced at that time. The dynamics of thought and emotion cluster in opposites. In an age of energy, while many sensed the great promise of things to come, others dreaded it and felt helpless. For all who thought that the future was theirs to control, there were those like Spengler, who braced for catastrophe, and the characters of Thomas Mann's *The Magic Mountain,* who spent the years from 1907 to 1914 waiting to die.

In 1912 Mann visited his wife at a sanatorium in the Swiss Alps where she was being treated for tuberculosis. He developed a cold and was advised by the doctors to stay, but left and began writing a story about his experience that swelled into an immense novel completed only twelve years later. *The Magic Mountain* is about Hans Castorp, who visits his cousin Joachim in a tuberculosis sanatorium, intending to stay for three weeks, and winds up staying seven years. Mann draws us into this community as Hans was drawn into its seductive monotony. Against the austere backdrop of Alpine sky the patients pace about, their coughs cutting the silence of the thin air. We follow Hans through the corridors, eavesdropping on discussions about a myriad of subjects including the nature of time, the past, and the present. The novel thus recapitulates the ideas surveyed in the first three chapters of this book and offers a vision of the future in the passive mode for the patients who helplessly awaited the progress of their disease as they did the daily routine of measuring temperature and sipping soup.

Both *The Decline of the West* and *The Magic Mountain* were conceived before the war, worked on during it, and published after. They spanned the period and sought to identify what it signified. Spengler's characterization, mired in cultural pessimism, was of a twilight of the Faustian soul. Mann ingeniously reconstructed the diplomatic community of Europe in the fictional community of the Berghof: explosive, feverish, constantly taking its temperature, struggling from one crisis to the next, with patients separated along national lines at their dining tables. And, as we would expect in a retrospective view of an age leading up to war, they were portrayed waiting for it to happen. Difficult as it is for contemporary historians to keep in mind that the "prewar" period did not become prewar until after war broke out, it was impossible for Mann or Spengler to conceive of it in any other way in the immediate "postwar" period when they completed their works.

Although Mann's narrative moves ahead with surprises for the reader, the characters anticipate only more of the same, and the dominant mode of their future is passive expectation. The patients at the Berghof curled up in their lounge chairs and awaited the onslaught of disease as, a few years later, front line soldiers would curl up in their fox holes and await the burst of artillery shells. When Hans saw an x-ray of his own hand, he had a gloomy vision of a future of endless waiting. For the first time in his life he understood that he would die, and all that remained in the time ahead was to "measure, eat, lie down, wait, and drink tea."[22] The waiting, Mann explained, actually accelerates the passage of time: it consumes large chunks of it like a greedy man whose digestive tract processes great quantities of food without absorbing its nutritional value. Undigested food does not make him any stronger; time spent waiting makes him no wiser. The patients just waited and grew old. Some died and some recovered, but the end brought no resolution. The thunderclap of war shook Hans off the mountain, but he disappeared into the front lines, lost amid the shelling and killing as formerly he was lost amid the coughing and dying. Europe was finally choking to death after years of waiting.

The novel, like Spengler's history, contrasts with and sets off the active mode of the future that dominated the thinking of the age. But for all the age's hopeful action and aggressive, prospective thinking, there was also passivity and caution. The dialectic of thought and experience presented a mixture of contrasts. The telephone and assembly line accentuated both active and passive modes of the future. Henry Adams felt exhilarated by the dynamo, but it also broke his neck. Wells's Time Traveller sailed ahead assuredly with the aid of the new technology but found a world of stagnation and degeneration. Among these contrasts there are no simple syntheses, but we can identify the terms of discourse and get a sense of what people thought and why they acted. This generation had a strong, confident sense of the future, tempered by the concern that things were rushing much too fast. The *Titanic* symbolized both. It is appropriate that Hans was studying engineering before he came to the Berghof and spent his time reading a book called *Ocean Steamships* during his first months there. One of the patients, Settembrini, compared the lives of the patients with the voyage of an ocean liner, and, considering Mann's symbolic intention, the comment also applied to Europe before the war. The comfort, the luxury, the hubris of tempting fate and controlling the wild elements was a triumph of the human spirit, a "victory of civilization over chaos," but envious gods may take swift revenge and wreck the luxury liner. And, he asked Hans, "are you not afraid of the hurricane which is the second circle of the Inferno" that whips and whirls those who sacrifice reason to desire? Settembrini concluded his argument with a suggestive image of Hans, like a small boat, "flapping about in the gale, head over heels"[23] (The *Titanic* went down in a calm sea, but her stern did flip straight up in the air before the final plunge.) The age had its doubts and hesitations, but it was essentially characterized by hubris that ignored the warning messages and pushed the throttle full speed ahead.

5

SPEED

In 1897 Germany embarked on a policy of *Weltpolitik* and began to build a battle fleet to challenge British control of the seas. That same year the German passenger steamer *Kaiser Wilhelm der Grosse* took away from the British Cunard Line the Blue Riband for the fastest Altantic crossing. In 1903, with its national prestige at stake, the British government subsidized the construction of a vessel capable of reaching 25 knots and beating the German record. The Cunard yards produced the *Mauretania,* which regained the Blue Riband in 1907 and retained it for twenty-two years.[1] The White Star Line that designed the *Titanic* hoped to surpass all rivals in speed and luxury. As several expert mariners testified at the British inquiry into her sinking, the pressure to keep to schedule obliged many captains to steam at recklessly high speeds through fog and ice.[2] One surivor commented that the public demanded more speed every year and refused to patronize the slower lines.[3] A bishop in Chicago condemned the "insane desire" for excessive speed on both land and sea. Another critic observed a "mania for speed and smashing records."[4] In a letter to the London *Daily News and Leader* George Bernard Shaw criticized the captain of the *Titanic* for deliberately steaming into an icefield at full throttle, and Joseph Conrad wrote an angry article in the *English Review,* predicting more irresponsibility in the future when steamships could plow across the ocean in all weather at forty knots.

The arrogance of large ocean liners, their pursuit of speed records at risk of life, was the subject of *Futility* (1898), a novel of uncanny prevision. It is a story about the largest craft afloat, a symbol of modernity that incorporated the knowledge of "every science, profession, and trade known to civilization." The designers had discovered how to close compartments automatically in case of collision, and the ship was advertised as "practically unsinkable" and carried as few life boats as law would permit. The owners announced that it would steam at full speed in all weather. The first night out it cut another ship in two, and the man on watch insisted on reporting all he knew in the hope of ending "this wanton destruction of life and property for the sake of speed."[5] The name of the ship was *Titan.*

The sinking of the *Titanic* was but the most tragic consequence of speed made possible by a broad technological revolution that also affected how people traveled to work and how fast they worked when they got there, how they met each other and what they did together, the way they danced and walked and even, some said, the way they thought. There was no question that the pace of life was greatly accelerated, but there was sharp debate about the meaning and value of speed.

The German historian Karl Lamprecht observed that in the last decades of the nineteenth century there was a sharp rise in the domestic production and importation of pocket watches (he estimated 12 million imported watches for a German population of about 52 million). At the same time people began paying specific new attention to short intervals of time—"five-minute interviews, minute-long telephone conversations, and five-second exchanges on bicycles."[6] The new profusion of watches was a response to, as well as a cause of, a heightened sense of punctuality in this period, especially in urban centers. In an essay on "The Metropolis and Mental Life" (1900), Georg Simmel commented on the impact of the "universal diffusion of pocket watches" in accelerating modern life and instilling a sense of punctuality, calculability, and exactness in business transactions as well as human relations.[7]

The bicycle was about four times faster than walking and warnings were issued about getting "bicycle face" by moving against the wind at such high speeds.[8] Its design made the bike difficult to ride, but this became easier after 1886 when the wheels were made of equal size, and more comfortable in 1890 with pneumatic tires. In America Sylvester Baxter observed that the bicycle "quickened the perceptive faculties of young people and made them more alert."[9] A French critic attributed the excitement of cycling to the sheer pleasure of movement, enhanced by a sense of mastery over the environment.[10] The popular French writer Paul Adam wrote that it created a "cult of speed" for a generation that wanted "to conquer time and space."[11]

A penetrating evaluation of its impact on human sensibilities and social relations was made in 1898 by Maurice Leblanc in a novel about cycling, *Voici des ailes!* On its title page is a drawing of a bare-breasted woman with an unbuttoned chemise trailing down over her belt, hair streaming in the wind, strings flying free from her wrists, pedaling a winged bicycle, all of which suggests the sexual, social, and spatial liberation that the two married couples of the book experience during a bicycle tour (Figure 3). The first day out Pascal observes to his friend Guillaume that nothing evokes the idea of speed more than the humming spokes of a bicycle. On the road the couples feel a new rhythm of movement, a unique sense of penetrating the surrounding world as their senses open to new parts of the terrain. They experience a new sense of time, as if they were moving through a dream rather than the French countryside. Social restrictions loosen when they address each other by first names. Sartorial and sexual liberation begins when Pascal's wife unbuttons her blouse and bathes her neck and shoulders in a public fountain. The next day both women appear without corsets. Later they strip off their blouses and cycle bare-breasted, and eventually the bonds of marriage break down as the couples exchange spouses and finish their tour re-paired.

Fig. 3. Title page from Maurice Leblanc, *Voici des ailes!* 1898.

Pascal comments on the dimensions of experience opened up by the bicycle. Steam and electricity only serve man, but the bicycle alters his body with a faster pair of legs. "This is not two different things like man and horse. There is not a man and a machine. There is a faster man." Speeding along he finally declares his love for Guillaume's wife and shouts "we have wings"—to escape the narrow spatial framework of their former city lives, the constricted social world of their ill-suited marriages, the physical confinement of corsets and tight clothing, and the emotional restrictions of their sexual morality.[12]

The automobile captured the imagination in the 1890s and became a major means of transportation in the first years of the twentieth century. In France there were about 3000 automobiles in 1900 and about 100,000 by 1913. Between 1896 and 1900 at least ten journals about "automobilism" appeared, all attentive to the ever breaking speed records, which by 1906 had exceeded 200 kilometers per hour. Commenting on its impact the French novelist Octave Mirbeau mixed metaphors as rapidly as the movement of his subject—the mind of modern man. Under the impact of the automobile it has become an "endless race track." "His thoughts, feelings, and loves are a whirlwind. Everywhere life is rushing insanely like a cavalry charge, and it vanishes cinematographically like trees and silhouettes along a road. Everything around man jumps, dances, gallops in a movement out of phase with his own."[13]

In England the Highways and Locomotives Act of 1878 required that any vehicle using public roads be preceded by a man on foot and not exceed a speed of four miles per hour. This law was abolished by another of 1896 that opened public roads to the faster "light locomotives," but as the number of traffic accidents rose opposition mounted. In 1903 the *Daily Telegraph* campaigned for a new speed limit, about which C. S. Rolls protested: "Our hereditary instincts are shocked at seeing anything on the road faster than a horse, but as our senses become educated we shall recognize the fact that speed of itself is not dangerous but the inability to stop is dangerous."[14] Parliament was not fooled by such doubletalk and in 1904 imposed a limit of 20 miles per hour on public highways, 10 miles per hour if required by local authorities. During the next year 1,500 motorists were charged with reckless driving. The number of traffic fatalities in London increased from 769 in the period 1892–1896 to 1,692 in the period 1907–1911.[15] In April 1914, when a child was killed by the chauffeur-driven car of Hildebrand Harmsworth, son of the famous newspaper magnate Lord Northcliffe, public outrage peaked. Annoyances and inconveniences of car travel were substantial. The dust that trailed behind autos engulfed pedestrians and cyclists and ruined the crop of lettuce farmers. And since the dust *was* the road, there were complaints from taxpayers. In *The Condition of England* of 1909, C. F. G. Masterman hissed his protest about automobiles that "scramble and smash and shriek all along the rural ways."

Nothing moved faster than the electricity that raced through conduits, powering motors and accelerating a variety of activities. The first electric tram was put into operation by Werner Siemens in Berlin in 1879; the first one in America ran between Baltimore and Hampden in 1885.[16] They sashayed about the inner city like those that marked the regular pace of public time in Bloom's Dublin. The electrified London underground was completed in 1890, and in the following decade there was a proliferation of electric rails everywhere. In the United States the 1,261 miles in 1890 increased to 21,290 miles by 1902.[17] Visitors to the 1900 World Exhibition in Paris were impressed by the new Otis escalator and a moving pavement designed by the French that portended faster pedestrian traffic. The telephone accelerated business transactions and enabled Wall Street to become a truly national financial center by increasing the liquidity of securities and the speed of fund raising. J. P. Morgan averted a financial panic in 1907 when, over the telephone, he extended $25 million credit to several major banks threatened with excessive withdrawals.[18] The great generating station that opened at Niagara Falls in 1895 converted the rush of water into an even faster rush of electrical current that transformed the pace of life and, some speculated, the very processes of life. An article in *The Fortnightly Review* proposed that electricity might accelerate the growth of crops and increase agricultural yield.[19] This theory was elaborated upon by the Belgian chemist Ernest Solvay in a lecture at the opening of his Institute of Physiology in Brussels.[20] Enthusiasm for this theory peaked with the work of the Nobel Prize winning chemist Svante Arrhenius, who tested the effect of electrical stimulation on the growth of children. He placed one group in schoolrooms with wires carrying high-frequency alternating current. After six months the "electrically charged children" had grown twenty millimeters more than those in the control group. The "magnetised teachers" reported that "their faculties were quickened."[21] While some researchers tried to use electricity to accelerate the processes of life, others used it to speed up death; in 1888 New York passed a law substituting death by electricity for hanging. In 1890 the New York prison authorities first used the "electric chair" to execute a convicted murderer, although it proved to be far less speedy than expected. The first charge of current failed to kill the man, and after some delay a second charge was given. When it was all over eight minutes had elapsed, the victim was covered with blood from cuts sustained at the points of contact of the circuitry, the District Attorney was in tears, one witness had fainted, everybody was horrified, and a reporter for the *New York Times* wrote that it had been a "revolting spectacle," "far worse than hanging."[22]

The technology of speed affected newspaper reporting and modified the language of journalistic communication. On February 12, 1887, a reporter for the *Boston Globe* used a telephone for the first time to report a speech made by Graham Bell in Salem, Massachusetts; and in 1880 the London *Times* installed a direct telephone line to the House of Commons to gain 45 minutes in the reporting of late night debates for their morning edition. Robert Lincoln O'Brien noted in an essay of 1904 on "Machinery and English Style" that the telegraph came into ever greater use as the need for fast reporting increased. Because economy of expression produced monetary savings, reporters were inclined to write their stories with the fewest possible words. The telegraph also encouraged the use of unambiguous words to avoid any confusion, and the language of journalism came to be more uniform as certain words came into more frequent use. Adverbial phrases at the beginning of a sentence were especially "dangerous," because they might be confused with the preceding sentence, and writers used the simplest syntax. Information tended to be written with a minimum of punctuation. "The delicacy, intricacy, nuance of language is endangered by the wires," O'Brien concluded, as the need for speed, clarity, and simplicity shaped a new "telegraphic" style.[23] No doubt Hemingway's simplification of the English language was in part a consequence of his experience as a foreign correspondent, obliged to prepare his articles for transmission over the Atlantic cable.

Factory work was accelerated by applying Frederick W. Taylor's "scientific management," which he first conceived in 1883.[24] Taylor observed skilled workers and determined the exact series of elementary operations that make up their job, selected the quickest series, timed each elementary operation with a stop-watch to establish minimum "unit times," and reconstructed jobs with composite times as a standard. Although there was nothing new about cracking the whip, scientific management was, as the name implied, scientific, or at least systematic, and avoided the caprice of a foreman's shifting moods. Wages were raised as workers approached their maximum efficiency rate, and those who fell short of a minimum rate were discharged. One of Taylor's reports shows the kind of psychological harassment caused by this systematic speedup: "it was found necessary to measure the output of each girl as often as once every hour and send a teacher to each individual who was falling behind to find what was wrong, straighten her out, and encourage her and help her to catch up."[25] He began to publicize his methods in 1895, stressing that workers complete jobs in the shortest possible time.[26] The following year a Massachusetts builder, Sanford Thompson, devised a "watch book" with stop-watches concealed in the cover, so that they could be operated without the worker's knowledge. Taylor disapproved of "spying" because it undermined the mutual commitment to speed and efficiency between worker and management that he thought essential, but he conceded that some workers object to being timed, and for them concealment might be necessary.[27]

Taylor's disciple, Frank B. Gilbreth, applied the methods of scientific management to work in space. A "motion study" of bricklaying in 1909 enabled him to devise an adjustable scaffold for piling up bricks that tripled worker output. He conducted research by means of "cyclegraphs" produced by attaching small electric lights to the body and making photographic time exposures of motions that appeared as continuous white lines. These made it possible to see the path of a motion and reconstruct it in three dimensions with a stereoscopic light. For more precision he adapted a motion picture camera to take "chronocyclegraphs," which would show "the paths of each of several motions made by various parts of the body and their exact distances, exact times, relative times, exact speeds, relative speeds, and directions."[28] In an article on scientific management of households Gilbreth boasted that with chronocyclegraphy "we can now for the first time record the time and path of individual motions to the thousandth of a minute."[29] His wife, Lillian, who collaborated with him, conceived of a new managerial position—the "speed boss"—whose job was to demonstrate to a worker how a task is to be done in the specified time.[30] But not all was rush, work, and profit. The Gilbreths also sought to reduce worker fatigue, and their book on that subject stressed the need to offset the dreariness of factory routine by providing a certain number of "Happiness Minutes" for the workers: they concluded with the uplifting thought that "the good in your life consists of the quantity of 'Happiness Minutes' that you have created or caused."[31]

Scientific management, the motion studies of Muybridge and Marey, early cinematography, Cubism, and Futurism reflect aspects of each other across the cultural spectrum like images in a house of mirrors. As the Cubists broke up and recreated bottles and guitars, Gilbreth broke down and reconstructed work processes. He made wire models of workers' movements from cyclegraphs similar to the wire-and-plaster models of birds in flight that Marey made from chronophotographs. Gilbreth's use of successive photographs to analyze motion derived from Muybridge's serial photographs of a galloping horse. Muybridge later used the technique to capture the grace of a woman stooping to pick up a basket; Gilbreth applied it to improve the speed of workers picking up bricks. Cinema was the technological link: Muybridge and Marey were searching for a way to make moving pictures; Gilbreth used the motion picture camera to make chronocyclegraphs; the term for a film's composition—"montage"—is the French word for the assembly of a product from component parts; around 1912 the Cubists began to experiment with "Cubist Cinema"[32]; and the Futurists were inspired by its suggestion of new possibilities for a kinetic visual art. Marcel Duchamp observed that "the whole idea of movement, of speed, was in the air," and acknowledged that his *Nude Descending a Staircase* was inspired by chronophotographs and motion pictures.[33] The cinema reproduced the mechanization, jerkiness, and rush of modern times.[34]

The very name of the new medium identified its effect—moving pictures. The turning projector supplied movement of images on the screen. In 1896 one of Lumière's cameramen, M. A. Promio, hit upon the idea of taking pictures from a moving boat along the Grand Canal of Venice.[35] With creative editing action could move as fast as it did in Griffith's last-minute rescues or at a more leisurely pace in cuts between widely separate places. The story could change settings as rapidly as the interval between frames, and since in the early movies the picture was taken at 16 frames per second and projected at 24, the actors themselves seemed to hurry across the flickering screen. The cinematograph so exaggerated the quickness of movement that some actors moved more slowly than they would in real life in order to give the final result a normal tempo.[36] One critic explained that the decline in the popularity of the theatrical melodrama, which relied on a fast pace to sustain interest, was caused by competition with the cinematograph, which could intensify action and present it much more rapidly than was possible on the stage. "The swiftness develops the breathlessness and excitement [that] the melodrama proper fails to evoke."[37] Some film makers intentionally accelerated motion for special effects: flowers boiled out of buds in seconds, and the metamorphosis of a caterpillar into a butterfly could be compressed from weeks into minutes.[38] Cinematic news coverage was greatly accelerated in 1911 when a special express train outfitted with a dark room was used to develop and transport a film of the investiture of the Prince of Wales at Carnarvon at four o'clock in the afternoon and have it ready for public viewing in London at ten o'clock that night.[39]

These "rushes" dazzled audiences. Erwin Panofsky concluded that the basis for enjoyment of moving pictures was not the subject matter "but the sheer delight in the fact that things seemed to move."[40] The early viewers were fascinated by any simple moving subject: Niagara Falls, horses jumping hurdles, workers emerging from a factory, a train pulling into a station. Some inexperienced viewers would duck in their seats to avoid an approaching train. Already in 1899 the Kinetoscope made its way into a novel, as Frank Norris's McTeague was "awestruck" at an approaching cable car on the screen.[41]

The French Cubist painter Fernand Léger identified the effect of the cinema and of technology in general on the aesthetic sensibilities of artists and the viewing public. In 1913 he observed that life was "more fragmented and faster-moving than in previous periods" and that people sought a dynamic art to depict it. Cinema and color photography have made it unnecessary to paint representational and popular subject matter. "The few working class people who used to be seen in museums, gaping in front of a cavalry charge by M. Detaille or a historical scene by M. J. P. Laurens, are no longer to be seen; they are at the cinema."[42] Evolution of the means of locomotion has affected the way people see and the art they like: "A modern man registers a hundred times more sensory impressions than an eighteenth-century artist."[43] The view through the door of a moving railroad car or the windshield of an automobile is fragmented, although at high speeds it becomes continuous the way continuity is created out of a series of stills by the cinema. Léger responded to these new dynamics with paintings that incorporated machine-like elements in figure studies and landscapes—one can almost hear the clanging of machinery in his art.

In 1915 Luigi Pirandello created a character who might have stepped out of one of Léger's paintings. The narrator of his novel *Shoot: The Notebooks of Serafino Gubbio, Cinematograph Operator,* has internalized the characteristics of the "clamorous and dizzying" world in which he lives and of the motion picture camera that he operates. "Already my eyes and my ears too, from force of habit, are beginning to see and hear everything in the guise of this rapid, quivering, ticking mechanical reproduction." The identification with his occupation becomes so complete that Gubbio finally loses his identity to the camera: "I cease to exist. *It* walks now, upon my legs. From head to foot, I belong to it: I form part of its equipment." This fantasy of self-abnegation culminates with the outburst, "My head is here, inside the machine, and I carry it in my hand."[44]

With metaphor and fantasy, artists sought to portray the impact of technology on human experience. Leblanc envisioned the union of man and bicycle winging along the open road, Léger fused humans and machines in sleek metallic forms, and Pirandello created a character whose head got lost in a camera. The Futurists also lost their heads over the new technology and proclaimed a "new aesthetic of speed," first announced by Marinetti. "We say that the world's magnificence has been enriched by a new beauty; the beauty of speed. A racing car whose hood is adorned with great pipes that seems to ride on grapeshot is more beautiful than the *Victory of Samothrace* . . . We cooperate with mechanics in destroying the old poetry of distance and wild solitudes, the exquisite nostalgia of parting, for which we substitute the tragic lyricism of ubiquity and omnipresent speed."[45] Unfortunately Marinetti continued to exalt speed during the war and lost most of his audience. When in 1916 he wrote that "the new religion-morality of speed is born this Futurist year from our great liberating war," the public turned to thoughts about the breakdown of religion, the corruption of morals, and the killing pace of machine-gun fire. He hovered between hyperbole and madness with his fantasy that the acceleration of life would cut down the arabesque of valleys and straighten the meandering of rivers, that someday the Danube would run in a straight line at 300 kilometers an hour.[46]

Although Marinetti's bombast at times exceeded the ambition of many Futurists, his principles provided the inspiration and theoretical framework for their art. In 1912 Balla began to paint movement.

His first subject was the dachshund in *Dynamism of a Dog on a Leash,* scurrying along next to its mistress (see Figure 2 above). With *Rhythm of a Violinist* he depicted several different movements simultaneously—vibrating strings, gliding bow, left hand grasping the neck, and sound vibrations pulsating through air. *Girl Running on the Balcony* did not picture speed better than the action of the dog or the violinist, but with it he began to shift from concrete to abstract movement. There is a suggestion of a swirling skirt and running feet, but the girl's motion is generalized and her successive forms are of equal value, rendered alike in size, shape, composition, and color. In 1913 Balla produced a series of paintings of the flight of swifts in successive stages, wing overlapping wing as in a Marey chronophotograph, strung like links on a chain of continuous flight. With *The Swifts: Paths of Movement + Dynamic Sequences* (1913) he approached abstract movement. The schematized birds flutter all over the canvas without any specific direction and follow oscillating, luminescent lines that both channel and break up their patterns. Balla then turned to painting automobiles, but with barely recognizable forms. The speeding windows flash like facets of a turning gem, and spinning wheels spiral into lines of force. The title of one listed the themes: *Speed of an Automobile + Light + Noise.* Toward the end of 1913 he entirely abandoned concrete subject matter and rendered simply *Abstract Speed.* The force lines that formerly eddied about birds and autos now coil out of artistic forms alone. The arcs of force lines bend from movement itself; light reflects along lines of unidentifiable objects, energized by unknown sources.

While Balla was pursuing the image of abstract speed, Boccioni sought to create continuous movement, and at the end of 1913 produced his masterpiece, *Unique Forms of Continuity in Space* (Figure 4). He worked up to it over several years with statements of purpose and partial solutions of the artistic problems in drawings and sculpture. In a manifesto of 1910 he proclaimed the Futurist intention "to express our whirling life of steel, of pride, of fever and of speed." The artist will render not a fixed moment but the dynamic sensation of movement itself.[47] Boccioni was intrigued by Bergson's distinction between relative motion (that we know from outside) and absolute motion (that we intuit from within), but challenged Bergson by insisting that an artist could synthesize both in a single image. The title of a manifesto of 1914 expressed the argument as an equation: "Absolute Motion + Relative Motion = Dynamism."[48] This dynamism avoided two bogus methods of rendering movement—chronophotog-

Fig. 4. Umberto Boccioni, *Unique Forms of Continuity in Space*, 1913.

raphy and photodynamism. He agreed with Bergson that no series of still photographs, no set of fixed images, could ever properly reconstruct motion. He also rejected the visual slurs of Bragaglia's photodynamism as a facile time exposure of a moving object that lacked artistic merit. Boccioni believed that the artist could find a single form of continuous movement that would suggest the immediate past and future of the action and the interpenetration of object and environment that is generated by it.

In Boccioni's paintings of *Dynamism of a Human Body* and *Dynamism of a Cyclist* moving limbs and spinning wheels are lost in abstract patterns of the "force-forms" and "plastic dynamism" he was defining in manifestos. In a sculpture called *Synthesis of Human Dynamism* (1912), he tried to realize two requirements of plastic dynamism: to create a sense of motion and bring surroundings into the form. But both efforts were only partially successful. Another striding figure, *Speeding Muscles* (1913), offered a sharper depiction of movement. The head is a convoluted geometrical form, face distorted as if it had been racing against the wind. The torso is twisted and arched forward like a fullback bracing to charge. Movement is most forcefully suggested by the flux of lower limbs, rendered by continuous swirls. The penultimate sculpture in this series was *Spiral Expansion of Speeding Muscles* (1913). This figure is upright, its head a hybrid of man and machine. The torso is armless and streamlined, and the striding legs, now separate, spiral up from the base. But the bulky forward leg is caught in a cluster of spiral forms like a pile of wood shavings, and interferes with the continuous forward movement.

In a successful resolution of the earlier artistic problems, *Unique Forms of Continuity in Space* is a mixture of man, energy, and machine—a fulfillment of the Futurist goal to create a new beauty of speed. The head is a montage of skull, helmet, and machine parts with a sword hilt for a face. The forward thrust of the figure is balanced by calves that are shaped like exhaust flames, suggesting propulsive energy and speed of movement. Its thigh muscles are contoured for strength and aerodynamic efficiency. The torso is armless, but the shoulders, fanned out like budding wings, suggest another source of continuous movement. The chest is shaped to withstand air pressure and must have been the inspiration for Marinetti's 1915 vision of the superhuman man of the future, who will be "built to withstand an omnipresent speed . . . He will be endowed with unexpected organs adapted to the exigencies of continuous shocks . . .

[There will be] a prow-like development of the projections of the breastbone which will increase in size as the future man becomes a better flyer."[49] The body has the pliability of a wing and the hardness of steel; it is driven by muscle, machinery, and fire. Boccioni avoided the complete abstraction of speed of Balla and the excessive concretization of it as in Marinetti's racing car. He rejected the stuttering motion of chronophotography and the sloppiness of photodynamism. He attempted to reconcile Bergson's relative and absolute motion and created an image of modern man that transcended traditional shapes and proportions as Nietzsche's overman transvaluated all values. With this sculpture the culture of speed made its most eloquent statement.

Historians of music are right to be wary of making simplistic connections between the pace of life and the tempo of music, between jazz and modernity, but in this period many composers consciously wrote music to reflect a changing world.[50] The new rhythms were not simply faster; indeed some innovations delayed or even stopped the beat unexpectedly, but the mixture of syncopation, irregularity, and new percussive textures gave an overall impression of the hurry and unpredictability of contemporary life.[51]

The driving pulse of the new ragtime music that began to be heard from the Mississippi, Missouri, and Ohio river valleys around 1890 expressed the playful, hopeful side of the American blacks, shifting between oppression and bursts of emotional release, between work routines and wild celebrations. The first ragtime composition was published in 1897, and its popularity quickly spread in America and Europe. The name may have come from the ragged appearance of the early performers, but most likely it came from the irregular movement of syncopated rhythm and its effect on traditional time—literally, time in tatters. The tempo was steady but there were fluid progressions of rhythmic variations. Most distinctive was the heavy syncopation that stressed the weak beat and created the oompah accents, the sudden "break" in which the rhythm pattern of the bass line stopped to accentuate the treble, and the more dramatic "stoptime" that disrupted the rhythm completely with gasping silences. In the ragtime classic, *Maple Leaf Rag* (1899), Scott Joplin generated subtle tensions with frequent shifts in rhythm all within the steady movement of a four-beat or syncopated march tempo.[52] And how it does move—deliberately with the beat, haltingly with delays and unexpected accents, and hurriedly with animated accelerations as if the fingers could not wait for the next beat. Contemporary music critics suggested possible connections between ragtime and life style. In an article of 1915, one critic wrote: "Our children dance, our people sing, our soldiers march to rag-time." And in an essay on the "current unrest" in American society, published in 1914, Walter Lippmann observed: "We make love to ragtime and we die to it."[53]

Around 1900 in New Orleans another new music was created by blacks. Unlike ragtime, which remained locked within a steady tempo, jazz made constant inventions and variations in the tempo and allowed a free *rubato* style. Its new orchestral timbre enforced the rhythmic irregularity. Wild squawks of a saxophone and squealing cries of a muted horn accentuated the strangeness of unfamiliar cross-rhythms, polyrhythms, or other unidentifiable rhythms. While jazz had its slow parts, the early Dixieland bands especially seemed to keep to the quick step of modern life. One of many speculations about the origin of its name was that "jazz" was a slang term for speed.[54]

In concert music the climax of the breakup of traditional metres was the rhythmic pyrotechnics of Stravinsky's *Le Sacre du printemps.* The 1913 audience was shocked on opening night. They interrupted the first dance with laughter, then began to shout, and the noise became so loud that the dancers could not hear the music. Stravinsky recalled the scene: "During the whole performance I was at Nijinsky's side in the wings. He was standing on a chair, screaming 'sixteen, seventeen, eighteen'—they had their own method of counting time."[55] Indeed. Even without the din from the audience, the complex rhythms were extremely difficult to execute. Throughout the composition there are frequent metre changes, and in the first thirty-four bars of the climactic *Danse sacrale* it changes twenty-eight times. In that finale the entire orchestra turns into a percussion section with blaring horns, pizzicato strings, and hooting woodwinds dominated by timpani, bass drum, and cymbals all beating a savage rhythm for the sacrificial dancer who leaps and spins to her death.

∞

The barrage of new speeds brought out the dark side of modernity in mournful jeremiads, snap judgments, and threatening prognoses. In

1881 George M. Beard, who introduced the diagnostic category of neurasthenia (nervous exhaustion) into psychiatric nomenclature, published his *American Nervousness,* which set the tone for literature on the increasing tempo of life and its nefarious consequences. Beard argued that the telegraph, railroads, and steam power have enabled businessmen to make "a hundred times" more transactions in a given period than had been possible in the eighteenth century; they intensified competition and tempo, causing an increase in the incidence of a host of problems including neurasthenia, neuralgia, nervous dyspepsia, early tooth decay, and even premature baldness.[56] An article on old age by Sir James Crichton-Browne in 1892 attracted a good deal of attention with comparative death statistics for the periods 1859–1863 and 1884–1888. He found that heart disease in England killed 92,181 in the former period and 224,102 in the latter. Deaths from cancer and kidney disease revealed a similar increase, which he explained by the tension, excitement, and incessant mobility of modern life.[57] Max Nordau added these statistics to similar figures on the rise of crime, madness, and suicide to fuel his impassioned lamentation about the degeneration of man. Never before, he argued, did inventions "penetrate so deeply, so tyranically, into the life of every individual," and the result has been a drain on the nervous system, a wearing down of body tissue. "Every line we read or write, every human face we see, every conversation we carry on, every scene we perceive through the window of the flying express, sets in activity our sensory nerves and our brain centers. Even the little shocks of railway travelling, not perceived by consciousness, the perpetual noises and the various sights in the streets of a large town, our suspense pending the sequel of progressing events, the constant expectation of the newspaper, of the postman, of visitors, cost our brains wear and tear." In spite of his cultural hypochondria, Nordau did not follow Beard in the assumption that man is capable of just so many sensory impressions per unit of time. He believed that people can respond to most demands made upon them if there is time for gradual adaptation. But the onset of modernity came too fast. "No time was left to our fathers. Between one day and the next, with murderous suddenness, they were obliged to change the comfortable creeping gait of their former existence for the stormy stride of modern life, and their heart and lungs could not bear it."[58]

The turn of the century brought no letup from the fear of progressive degeneration. The title of John Girdner's book of 1901, *Newyorkitis,* identified a new disease—a special kind of inflammation that results from living in the big city and includes, among its numerous symptoms, "rapidity and nervousness and lack of deliberation in all movements."[59] In *L'Energie française* (1902) Gabriel Hanotaux inventoried sources of national power and evaluated the new technology and the mobility it created: bicycles magnify the locomotive energy of the feet, automobiles liberate travelers from the constraints of railroad timetables, and thought moves with the speed of lightning. Like a contemporary conservationist he warned that the enormous increase in coal consumption is rapidly using up the accumulated reserves of antediluvian forests, the energy of millennia—"We are burning our way during our stay in order to travel through more rapidly."[60] The German writer Willy Hellpach catalogued these worries in a popular medical tract, *Nervosität und Kultur* (1902). Following Beard he set the beginning of the age of nervousness in 1880 and explained its onset with the standard list of causes including a speedup in transportation and communication that created an "overwhelming increase of normal mental processes."[61]

Cultural rejection of speed mounted as philosopher and novelist joined physician and psychiatrist.[62] In 1907 Henry Adams wrote that power has outgrown its servitude and that the unprecedented speed of life has made people "irritable, nervous, querulous, unreasonable, and afraid."[63] William Dean Howells agreed, in a sketch of life in New York:

> People are born and married, and live and die in the midst of an uproar so frantic that you would think they would go mad of it; and I believe the physicians really attribute something of the growing prevalence of neurotic disorders to the wear and tear of the nerves from the rush of the trains passing almost momently, and the perpetual jarring of the earth and air from their swift transit . . . Imagine . . . a wife bending over the pillow of her husband to catch the last faint whisper of farewell, as a train of five or six cars goes roaring by the open window! What horror! What profanation![64]

Robert Musil wrote of the rush of traffic and its profanation of a death in *The Man Without Qualities.* Although written in the 1920s, the opening of the novel is set precisely in August 1913 in downtown Vienna. Musil recreated the fabric of the city's rush: the "loose-woven hurrying" of pedestrians crossed the "stronger lines of speed" of motor-cars that came shooting out of narrow streets into the squares. A man was run over by a truck, and people walked by unconcerned as if it were part of the natural order of things. " 'According to American statistics,' the gentleman observed, 'there are over a hundred and ninety thousand people killed on the roads annually over there, and four hundred and fifty thousand injured'."[65] Musil introduced the man without qualities, standing at a window with a watch in hand, counting the cars and pedestrians, estimating "the speed, the angle, the dynamic force of masses being propelled past, which drew the eye after them swift as lightning, holding it, letting go, forcing the attention—for an infinitesimal instant of time—to resist them, to snap off, and then to jump to the next and rush after that." But for all the hurrying, the imperial city is going nowhere, the empire is without a future, and people dream of living elsewhere.

> For some time now such a social *idée fixe* has been a kind of super-American city where everyone rushes about, or stands still, with a stop-watch in his hand . . . Overhead-trains, overground-trains, underground-trains, pneumatic express-mails carrying consignments of human beings, chains of motor-vehicles all racing along horizontally, express lifts vertically pumping crowds from one traffic-level to another . . . At the junction one leaps from one means of transport to another, is instantly sucked in and snatched away by the rhythm of it, which makes a syncope, a pause, a little gap of twenty seconds between two roaring outbursts of speed, and in these intervals in the general rhythm one hastily exchanges a few words with others. Questions and answers click into each other like cogs of a machine . . . One eats while in motion.[66]

Within a year there would scarce be time to grab a napkin. This is a caricature of Europe speeding out of control, heading toward war.

Stefan Zweig recalled the slow-paced and secure world of his childhood in Austria before the introduction of the new technology. "It was a world with definite classes and calm transitions, a world without haste." The adults walked slowly and spoke with measured accents; many were corpulent at an early age. He could not remember his father ever having rushed up the stairs or done anything in a visibly hasty manner. "Speed was not only thought to be unrefined, but indeed was considered unnecessary, for in that stabilized bourgeois world with its countless little securities, well palisaded on all sides, nothing unexpected ever occurred . . . The rhythm of the new speed had not yet carried over from the machines, the automobile, the telephone, the radio, and the airplane, to mankind; time and age had another measure."[67]

Many writers, however, welcomed the collapse of old palisades and viewed the new speed favorably as a symbol of vitality, a magnification of the possibilities of experience, or an antidote to provincialism. Some, like the Futurists, became so giddy with the thrill of it that their one-sidedly positive assessment lacked nuance. In a more sober vein, the French psychiatrist Charles Féré challenged the vogue of deploring the hurried pace of life by arguing that active and challenged minds became more resistant to nervous breakdown and better able to cope with diverse stimuli, precisely as they become more complex. He presented evidence that many breakdowns occur after long years of hard work when one is suddenly idle and argued that the mind deteriorates more from lack of use than from overuse, more from ignorance than from a surfeit of culture.[68] He also pointed out that improvements in transportation and public safety had made it possible for a contemporary woman to take a long voyage by herself with less strain and less anxiety and in far less time than would have been possible a century earlier for a prudent man armed to the teeth.

Yet there was often a touch of regret for the end of an era even among those admiring the new technology of speed. Octave Uzanne's *La Locomotion à travers le temps, les moeurs et l'espace* (1912) is a good example. Uzanne will miss the slow rhythmic clopping of horses along the road, their heavy breathing pulling up hills, but he is still carried away by the "fever of speed." The automobile, he explained, has broken down class barriers and reduced sectionalism. "Magnificent" long railway lines such as the Berlin-Bagdad and the Trans-Siberian have promoted international understanding. His enthusiasm is expressed with rapturous praise and immoderate metaphor. "The citizen is a mole with his undergrounds; he is an antelope, thunderbolt, cannon ball with his automobiles; he is an eagle, sparrow, albatros with his airplanes." Modern life is undergoing a "stupefying transfiguration" and "the rapid movement which sweeps us in space and piles up a variety of impressions and images in a short time gives life a plenitude and a unique intensity." Here the torrent of new stimuli that Beard thought inherently pathogenic, and Nordau, too fast to assimilate, Uzanne sees as a liberation from the impoverished routines and wearisome repetition of daily life.[69]

Among the many responses to the new technology those of the alarmists appear more impassioned and more numerous than those of the defenders of speed. But protests, however moving, cannot negate the fact that the world opted for speed time and again. People complain about the intrusion of a telephone but rarely do without one and organize their lives with as many time-saving devices as they can. Despite all the mixed feelings, however, it can be said without qualification that the new speed had a profound impact on civilization.

It is precisely this consensus that invites further interpretation, because in the dialectic of experience opposites are linked, and whenever one dynamic is so markedly pronounced we must look for unconscious countercurrents. If a man travels to work on a horse for twenty years and then an automobile is invented and he travels in it, the effect is both an acceleration and a slowing. In an unmistakable way the new journey is faster, and the man's sense of it is as such. But that very acceleration transforms his former means of traveling into something it had never been—slow—whereas before it was the fastest way to go. Suddenly his old horse has become obsolete. Thus for Zweig the way his father walked up stairs never used to seem particularly slow or relaxed—it was the way things were. But years later the course of history transformed his memories, and his father's gait became a symbol of "the Golden Age of Security." So, in the larger world, the impact of the automobile and of all the accelerating technology was at least twofold—it speeded up the tempo of current existence and transformed the memory of years past, the stuff of everybody's identity, into something slow.

Memories have the potential for becoming nostalgic only after changes have made comparisons possible and the past seems irretrievably lost. As steamships monopolized ocean travel, sailing vessels suddenly appeared to be majestic and graceful, instead of unreliable and cramped. Just as contemporary reaction to airplane crashes momentarily obscures the fact that air travel is safer, mile for mile, than any other means of transportation, so the sinking of the *Titanic* raised questions about the value of speed and brought to mind the virtues of slower travel. The anger coughed up in the dust of speeding autos muted complaints about the slow pace of traveling by foot or by carriage. Modern workers looked back fondly on the good old days of "inefficient" production precisely because they suffered the drawbacks of scientific efficiency. For every speed lover like Marinetti there were thousands who preferred the way rivers wandered and the way barges drifted on them. The Danube never seemed so deliciously slow until he suggested speeding it up. And of all the technology that affected the pace of life, the early cinema most heightened public consciousness of differential speeds. Since many early projectors were hand cranked, no two showings ever went at the exact same speed. They varied from scene to scene with inspired nudges from the cinematograph operator, and there were more irregularities from the interplay between the organ player and the film operator. To the delight of audiences they would suddenly shift the tempo, foil and tease each other with unpredictable lapses and rushes.

On the surface there was agreement: Taylorism and Futurism, the new technology, the new music, and the cinema had set the world rushing. But beneath there ran countercurrents. As quickly as people responded to the new technology, the pace of their former lives seemed like slow motion. The tension between a speeding reality and a slower past generated sentimental elegies about the good old days before the rush. It was an age of speed but, like the cinema, not always uniformly accelerated. The pace was unpredictable, and the world, like the early audiences, was alternately overwhelmed and inspired, horrified and enchanted.

6

THE NATURE OF SPACE

In an autobiographical sketch Einstein recalled two incidents from his childhood that filled him with wonder about the physical world. When he was five years old his father showed him a compass. The way the needle always pointed in one direction suggested that there was "something deeply hidden" in nature. Then at twelve he discovered a book on Euclidean geometry with propositions which seemed to be about a universal and homogeneous space.[1] These early memories embodied two opposing views about the nature of space. The traditional view was that there was one and only one space that was continuous and uniform with properties described by Euclid's axioms and postulates. Newton defined this "absolute space" as at rest, "always similar and immutable," but the action of the compass suggested that space might be mutable, with orientations that varied according to its contents. The quivering needle pointed to the north pole and to a revolution in physics.

New ideas about the nature of space in this period challenged the popular notion that it was homogeneous and argued for its heterogeneity. Biologists explored the space perceptions of different animals, and sociologists, the spatial organizations of different cultures. Artists dismantled the uniform perspectival space that had governed painting since the Renaissance and reconstructed objects as seen from several perspectives. Novelists used multiple perspectives with the versatility of the new cinema. Nietzsche and José Ortega y Gasset developed a philosophy of "perspectivism" which implied that there are as many different spaces as there are points of view. The most serious challenge to conventional space came from physical science itself, with the development in the early nineteenth century of non-Euclidean geometries.

Geometry is the branch of mathematics most directly concerned with the nature of space and with the properties of points, lines, planes, and objects in it. Euclid stated without proof certain axioms and postulates that seemed self-evident and from them derived other theorems by deductive logic. His geometry was of two and three dimensions, and for over two millennia it was considered to be the only true geometry of real space. Kant assumed that its propositions were necessarily true and about the world, hence synthetic judgments *a priori.* At the beginning of the nineteenth century it lay at the heart of classical physics and Kantian epistemology. But in the course of that century other geometries challenged the idea that Euclid's was the only valid one. Crucial to it was the Fifth Postulate: that through a point in a plane it is possible to draw only one straight line parallel to a given straight line in the same plane. The non-Euclidean geometries replaced the postulate with others and modified the rest accordingly. Around 1830 the Russian mathematician Nicholai Lobatchewsky announced a two-dimensional geometry in which an infinite number of lines could be drawn through any point parallel to another line in the same plane. In his geometry the sum of the angles of a triangle is less than 180 degrees. In 1854 the German mathematician Bernhard Riemann devised another two-dimensional geometry in which all triangles had angle sums greater than 180 degrees. Riemann's space was elliptical; that of Lobatchewsky was hyperbolic. These alternative surface spaces contrasted with the flat planar surface of Euclid's two-dimensional geometry in which the angle sum of a triangle is exactly 180 degrees. By the end of the century other mathematicians had developed geometries for all kinds of spaces—a doughnut, the inside of a tunnel, even a space like a venetian blind.[2]

The parallel postulate was a weak point in Euclid. As early as 1621 Sir Henry Savile identified it as a blemish in the system, and to many mathematicians thereafter it did not seem sufficiently self-evident to warrant acceptance without proof. It is therefore ironic that Lawrence Beesley, in his account of the sinking of the *Titanic*, referred to the law of parallels as if it were a symbol of order in the natural world. From a lifeboat he described the beauty of the ship at night, marred by the "awful angle" made by the level of the sea with the rows of porthole lights. "There was nothing else to indicate she was injured; nothing but this apparent violation of simple geometrical law—that parallel lines should 'never meet if produced ever so far both ways.' "[3]

If the spaces of non-Euclidean geometry were not bewildering enough, there were other new spaces that could not be accounted for by any geometry. In 1901 Henri Poincaré identified visual, tactile, and motor spaces, each defined by different parts of the sensory apparatus. While geometrical space is three-dimensional, homogeneous, and infinite, visual space is two-dimensional, heterogeneous, and limited to the visual field. Objects in geometrical space can be moved without deformation, but objects in visual space seem to expand and contract in size when moved different distances from the viewer. Motor space varies according to whatever muscle is registering it and hence has "as many dimensions as we have muscles."[4] In a similar manner Mach defined visual, auditory, and tactile spaces that varied according to the sensitivity and reaction times of different parts of the sensory system. These spaces constituted the physiological foundation for the "natural" development of geometrical space. Symmetry has a bodily source, and the positive and negative coordinates of Cartesian geometry derive from the right and left orientation of our body. Our notion of surface comes from the experience of our own skin. "The space of the skin," Mach wrote, "is the analog of a two-dimensional, finite, unbounded and closed Riemannian space." Terms for basic units of measurement such as "foot" and "pace" reveal anatomical origins, and thus "notions of space are rooted in our physiological organism."[5]

Speculation that there are two- and three-dimensional spaces other than the one described by Euclid and that our experience of space is subjective and a function of our unique physiology was disturbing to the popular mind. Perhaps the most famous critic of these notions was V. I. Lenin, who, in *Materialism and Empirio-Criticism* of 1908, cried "enough" to the proliferation of spaces, to the "Kantian" notion that space is a form of understanding and not an objective reality, and to "reactionary" philosophies such as those of Mach and Poincaré. Like a man trying to hold down a tent in a wind, Lenin raced about defending the objective, material world in absolute space and time that he believed to be the foundation of Marxism and which, he feared, was threatened by recent developments in mathematics and physics. It is an embarrassing performance by a man straining in a field beyond his expertise, but it gives a sense of the concrete implications and political overtones of this seemingly abstract thought.

Lenin began the chapter on "Space and Time" with a statement of the materialist position: there is an objective reality in which matter moves in space and time independently of the human mind. This is in contrast with the Kantian view that time and space are not objective realities but forms of understanding. He conceded that human conceptions of space and time are "relative," but this relativity moves toward the "absolute truth" of objective reality. Mach's statement that space and time are "systems of series of sensations" was "palpable idealist nonsense." He labeled "absurd" Mach's speculation that physicists might seek an explanation for electricity in a space which is not three-dimensional, and he reaffirmed the orthodox position: "Science does not doubt that the substance it is investigating exists in three-dimensional space." He tossed off Poincaré's famous anticipation of the relativity of time and space and then criticized that "scrupulous foe of materialism" Karl Pearson, who had written that time and space are "modes under which we perceive things apart." The kind of thinking that denies the objective reality of time and space is "rotten" and "hypocritical."[6]

Lenin engaged in this polemic because he believed that the reputation and political effectiveness of the Bolshevik party were at stake. When an article appeared in *Die Neue Zeit* (1907) about certain Bolsheviks who had embraced a Machist philosophy and compromised orthodox Marxism, Lenin decided to attack publicly to define the Bolshevik position and show that Machism was simply an aberration of certain individuals in his party, one manifestation of a general disease of doubting material reality that was infecting modern society as a whole and that could break out in any political party.[7] In the concluding paragraphs Lenin singled out the prominent Bolshevik philosopher A. Bogdanov, who had argued for the social relativity of all categories of experience in *Empirio-monism* (1904–1906). Bogdanov had written that time, like space, is "a form of social co-ordination of the experiences of different people." Such relativistic idealism undermined materialism and the belief that there is one and only one real framework of time and space in which the events of all cultures take place. According to Bogdanov, Lenin charged, "various forms of space and time adapt themselves to man's experience and his perceptive faculty."[8] This formulation contradicted Lenin's materialism in two respects. The reference to a plurality of spaces challenged the universality of a single space, and the suggestion that these various forms of space and time "adapt" to man's experience identified Bogdanov with the genetic epistemology of both Mach and Poincaré.

While Lenin was combating the social relativism of Bogdanov, a far more important theory of relativity was being developed by Einstein. Efforts by physicists to fit the negative findings of the Michelson-Morley experiment into the body of classical physics were like those of a squirrel trying to bury a nut in a tile floor. Lorentz hypothesized a dilation of time for the beam of light traveling in the direction of the "ether current" just enough to reconcile the experiment with absolute time. George Fitzgerald suggested a similar compromise to hold on to absolute space. He hypothesized that the arms of the apparatus in the experiment actually contracted in length in the direction of the ether flow just enough to compensate for the longer time that the light should take to travel with and against the current as compared with the beam of light that traversed the same distance across and back. Einstein scrapped the Fitzgerald contraction together with the Lorentz dilation and proposed relativity instead. In the special theory of 1905 space was redefined as a quasi-perspectival distortion. The contraction was not a real change in the molecular construction of the apparatus but a distortion created by the act of observing from a moving reference system. This perspectival effect differed from ordinary perspective because it was not due to optics and would occur no matter how far the object observed in motion was from the observer. The relative velocity of the object and viewer was the crucial factor, not the distance between them. With Einstein's explanation no absolute meaning could be given to the concept of the actual length of the apparatus or of the space it occupies. Length is not in anything; it is a consequence of the act of measuring. Thus absolute space has no meaning. In 1916 Einstein explained: "We entirely shun the vague word 'space,' of which, we must honestly acknowledge, we cannot form the slightest conception and we replace it by 'motion relative to a practically rigid body of reference.' "[9] With the general theory of relativity the number of spaces increased beyond calculation to equal the number of moving reference systems of all the gravitational fields generated by all of the matter in the universe. In 1920 Einstein summed up boldly: "there is an infinite number of spaces, which are in motion with respect to each other."[10] Fortunately Lenin was too busy making a revolution to take notice.

While physical scientists were trying to come to terms with the heterogeneity of abstract space, natural scientists began to investigate the relation between the structure of living organisms and their spatial orientation. In 1901 the Russian physiologist Elie de Cyon published an article on the "natural" foundation of Euclidean geometry based on results of experiments that he had been conducting for over twenty years on the physiological origins of experiencing space.[11] His hypothesis was that the sense of space is rooted in the semicircular canals of the ear. Animals with two canals experience only two dimensions and those with one canal are oriented in one. Humans experience three dimensions because they have three canals set in perpendicular planes, and three-dimensional Euclidean space corresponds to the physiological space determined by the orientation of these canals. From these experiments Cyon concluded that the sense of space is not inherent and that Kant's theory that it is an *a priori* category of the mind was wrong. Only the semicircular canals are inherent, and our sense of space derives from them and remains dependent upon them. The boldness of these claims, particularly the attack on Kant, triggered a good deal of scholarly criticism,[12] but Cyon was undaunted and continued to extend his theory. In 1908 he argued that the sense of time also was dependent upon the semicircular canals.[13] The following year his results were incorporated into a classic of theoretical biology, Jacob von Uexküll's *Umwelt und Innenwelt der Tiere.*

Uexküll asks the biologist to set aside everything that he takes for granted in his own world—nature, earth, heavens, objects in space—and focus on only that part of the environment that a particular organism can actually experience. Although all animals live in the same environment, they have their own surrounding world (*Umwelt*). Each species responds to the outer world in its own way, and that response creates its special inner world (*Innenwelt*). The lower animals react to stimuli directly, and only higher animals with some organ of sight develop a proper sense of space. Their brains recognize the surrounding world not merely by direct contact but are also able to mirror objects and spatial relations in the environment. This mirror world or counterworld (*Gegenwelt*) differs with each type of nervous and muscular system. Thus the inner worlds, surrounding worlds, and counterworlds vary with the "building plans" of each animal and constitute different senses of space.

Uexküll modified and extended Cyon's theory to the entire animal kingdom and concluded that the sense of space of all animals, however rudimentary, varied with their unique physiology. Each had special dimensions, even the space sense of one-celled animals. The amoeba's space was a limited one, but he reconstructed it in great detail and characterized it as a "most lively work of art." His appreciation of the creative force generated by the needs and structural patterns of animals led him to a critique of Darwin's theory of natural selection. "It is not true, as people are accustomed to think, that nature compels the animal to adapt, but on the contrary, the animal forms its nature according to its special needs."[14] Among the throng of worlds and living spaces, he speculated, there may also be higher worlds of greater dimensions that we are unable to see, as the amoeba is unable to see the stars in our sky.

This reminder that there are complete worlds with distinctive spatial orientations scattered all along the phylogenetic scale challenged the egocentrism of man. Another challenge came from social scientists. Adventurers and scholars had long sailed about the earth and dug into its crust to find out about other societies, but they always reconstructed them in the uniform space of the modern Western world, never imagining that space itself might vary from one society to another as much as did kinship patterns and puberty rites. Durkheim's arguments for the social relativity of space and its heterogeneity were part of his general theory of the social origin of basic categories of experience.[15] In *Primitive Classification* he challenged the theory, attributed to Sir James Frazer, that social relations are based on logical relations inherent in human understanding. He argued the opposite—that logical categories derive from social categories, space being one of them. To illustrate he described the Zuñi Indians who divided space into seven regions—north, south, east, west, zenith, nadir, and center—which derived from social experience and in which all objects belonged. The wind and air belonged to the north, water and spring to the west, fire and summer to the south, earth and frost to the east. Different birds and plants belonged to specific regions as did the energies of life. The north was the region of the pelican and crane, the evergreen oak, force and destruction. He concluded that their space was "nothing else than the site of the tribe, only indefinitely extended beyond its real limits."[16] Space is heterogeneous in two senses: it varies from society to society, and within societies such as the Zùñi it has different properties in different regions.

In *The Elementary Forms of the Religious Life* Durkheim elaborated on the heterogeneous nature of space, again as part of a general theory of the social origins of the categories of thought. If space were absolutely homogeneous, he argued, it would be useless to coordinate the varied data of sensuous experience. To identify things in space it must be possible to place them differently—to put them above and below, right and left—and so in every society space is heterogeneous. But there is a collective sense of these unique spaces, shared by all member of a society, hence they must have a social origin; and there is evidence that these spatial classifications are structurally similar to social forms: "There are societies in Australia and North America where space is conceived in the form of an immense circle, because the camp has a circular form; and this spatial circle is divided up exactly like the tribal circle, and is in its image. There are as many regions distinguished as there are clans in the tribe, and it is the place occupied by the clans inside the encampment which has determined the orientation of these regions."[17] Durkheim believed that there was a multitude of such spaces about the surface of the globe, differing from each other like patterns of Oriental rugs.[18] In Germany another social scientist unearthed a plurality of spaces buried in time.

Spengler believed that different cultures had a unique sense of space (as well as time) manifested in a symbolism that embraced every aspect of life. This sense of space or extension is the "prime symbol" of a culture, inherent in political institutions, religious myths, ethical ideals, principles of science, and the forms of painting, music, and sculpture. But it is never conceptualized directly, and it is necessary to interpret many aspects of a culture to grasp its particular notion of extension. The infinitely extended space of the modern "Faustian" era is but one of several in which the great cultures of history have been staged.

The Egyptians conceived of space as a narrow path down which the individual soul moves to arrive at the end before ancestral judges. Their most distinctive constructions are not buildings but paths enclosed by masonry. Reliefs and paintings are done in rows and lead the beholder in a definite direction. In Chinese culture space is also a path that wanders through the world; but the individual is led to his ancestral tomb by nature, by "devious ways through doors, over bridges, round hills and walls," not by rows of stones like the Egyptians. Greek space was dominated by a sense of nearness and limit. The universe was a cosmos, a "well-ordered aggregate of near and completely viewable things" covered by the corporeal vault of heaven. Its government was a clearly circumscribed city-state; its temples, finite structures formed about a center, enclosed by a colonnade. Classical art had "closed" figures with sharply bounded surfaces, and the predominance of the body brought the eye from the distant to the "near and still." Its statues, like its buildings, were clearly delimited, with no suggestion of the infinite or unbounded, and it produced a geometry of regular, closed figures that were the ideal forms of the earth and heaven.[19]

Spengler's account of space in the modern era expands with an exuberance that parallels his thesis—that the prime symbol of the Faustian soul of the modern age is limitless space. Faust's restless striving, the soaring of Gothic cathedrals, and the proliferation of geometric spaces reflect this sense of infinity. Modern music such as Wagner's *Tristan* liberates the soul from material heaviness and sets it free to move towards the infinite. He concludes with a cannonade of evidence for the modern era's sense of the limitlessness of space: "The expansion of the Copernican world picture into that aspect of stellar space that we possess today; the development of Columbus's discovery into a worldwide command of the earth's surface by the West; the perspective of oil painting and of tragedy-scene; . . . the passion of our civilization for swift transit, the conquest of the air, the exploration of the Polar regions and the climbing of almost impossible mountain peaks."[20]

The proliferation of geometrical and physical spaces had a great effect on mathematics and physics but did not generally influence thinking in other areas. The exploratic [illegible] ne experience of space of the amoeba, the Zuñis, and the ancie[illegible] gyptians was important to some natural and social scientists but made little stir outside their respective disciplines. However, the multiplication of points of view in painting had an impact far beyond the world of art. It created a new way of seeing and rendering objects in space and challenged the traditional notion of its homogeneity.

The depiction of space in painting reflects the values and fundamental conceptual categories of a culture. In the Middle Ages the importance of persons and things in heaven and earth determined their size and position in space. With the introduction of perspective, objects were rendered to scale according to their actual size and were located in space to reproduce the relations of the visible world.[21] In 1435 the Florentine painter Leon Battista Alberti formulated the rules of perspective that were to govern painting for four hundred and fifty years. He intended to help painters create a unified pictorial space in which God's order, the harmony of nature, and human virtues would be visible. Samuel Edgerton has observed that this formulation of perspective was a "visual metaphor" for the entire Florentine world at that time: its politics were just coming under the authority of the Medici oligarchy; there was a growing rationality in banking and commerce that relied on mathematical orderliness and utilized the system of double-entry bookkeeping; the Tuscan hills were terrassed with neat rows of olive trees and parallel strings of grape vines, all controlled by a centralized land management; proportion and orderliness were valued in every area of culture and were expected to regulate decorum and dress.[22] Although there were occasional variations or intentional violations of the rules of perspective, they governed the rendering of space in art until the twentieth century. Then, under the impact of the Impressionists, Cézanne, and the Cubists that perspectival world broke up as if an earthquake had struck the precisely reticulated sidewalks of a Renaissance street scene.

When the Impressionists left their studios and went outside to paint, they discovered a new variety of points of view as well as shades of color and light. They broke Alberti's rule that the canvas should be placed precisely one meter from the ground, directly facing the subject, and positioned it up and down and at odd angles to create new compositions. They moved in and out of the scene, and the frame ceased to be the proscenium of a cubed section of space that it had traditionally been. Daubigny carried to an extreme their rejection of the fixed point of view when he painted from a houseboat as it rocked at anchor or actually sailed along the Seine. With these new points of view the Impressionists abandoned the scenographic conception of space.[23]

However varied the scope and angle of Impressionist space, it was essentially one space as seen from one point of view. Cézanne was the first to introduce a truly heterogeneous space in a single canvas with multiple perspectives of the same subject. In *Still Life* (1883–1887) a large vase is reconstructed from two points of view with the elliptical opening more rounded than a strict adherence to scientific perspective would allow and gaping fuller than the opening of the other vase standing next to it on the same flat surface in the same plane. In *Still Life with a Basket of Apples* (1890–1894) the corners of the table are seen from different vantage points and grafted together to create balance with the other shapes. His *Portrait of Gustave Geoffroy* (1895) combines a frontal view of the seated subject with an aerial view of the table before him on which open books are lying with almost no perspectival foreshortening. This optically impossible mixture of points of view enabled Cézanne to show all that he wanted of the man and his work and at the same time conform to the requirements of composition. Cézanne was enamored of the shape of Mont Sainte-Victoire and painted it hundreds of times. By using different perspectives for different parts of the landscape he gradually pulled it out of the distant background toward the foreground until in the later paintings it loomed large as a symbol of his lifelong fascination with form and space. His landscapes broke ground for modern art as he gouged out quarries and cleared trees to make the terrain of Aix-en-Provence conform to his artistic needs.

Cézanne's primary commitment was to the composition of forms on the flat surface of the canvas; conventions for accurately rendering volume and depth were secondary.[24] While most painters had tried to create an illusion of three-dimensional space, Cézanne accentuated the flatness of the picture surface and frequently violated the rules of perspective in deference to it. He never entirely abandoned the techniques for showing depth but compromised them when necessary. And so he broke up consistent linear perspective with multiple perspectives, he violated aerial perspective in landscapes by painting objects in the distance as bright or brighter than those in the foreground, and he occasionally chipped off a piece of pottery when overlapping would interfere with his overall design. He sought to reconcile the properties of volumes in three-dimensional space with the two-dimensionality of the picture plane, and his paintings vibrate from the tension. He also wanted to fuse perceptions and conceptions—the way we see things from a single point of view and the way we know them to be from a composite of several views. Experience tells us that the opening of a vase is circular, but when viewed from the side we see it as an ellipse. Cézanne combined the two perceptions visually with multiple perspectives.

These daring innovations were possible only for someone with a sharp sense of space. Cézanne's unique sensitivity to the effect of slight shifts in point of view is revealed in a letter to his son of September 8, 1906: "Here on the edge of the river, the motifs are plentiful, the same subject seen from a different angle gives a subject for study of the highest interest and so varied that I think I could be occupied for months without changing my place, simply bending more to the right or left."[25] Subtle differences in form and perspective that most painters would not notice occupied Cézanne—fascinated him—for months. He wrestled with them until, as Merleau-Ponty believed, he created "the impression of an emerging order, of an object in the art of appearing, organizing itself before our eyes."[26] He "realized" objects in space as they take form, as the eye darts about the visual field and hovers around things until they are identified in space and integrated into our world of experience. For Cézanne an object in space was a multitude of creations of the seeing eye that varied dramatically with the most minute shifts in point of view.

One of the great fallacies of historical reconstruction is the characterization of events as transitional. The work of Cézanne is one of the most fully realized corpuses in the history of art, and it is particularly misleading to view it as a transition to modern art. Nevertheless the important innovations he made in the rendering of space—the reduction of pictorial depth and the use of multiple perspective—were carried further by the Cubists in the early twentieth century and have therefore come to be viewed as transitional. The Cubists repeatedly expressed their debt to Cézanne and used his techniques to create even more radical treatments of space. Their use of multiple perspective also shows a strong similarity to the cinema, which broke up the homogeneity of visual space.

Like modern art, the cinema offered some new and varied spatial possibilities. Theater viewers saw action in the same frame, from a single angle, and from an unchanging distance in a space that was stationary and uniform from beginning to end. But the cinema could manipulate space in many ways. The frame could be changed by moving the camera or changing the angle of the lens. The point of view or distance from the action could be shifted with different camera positions, and the space in view could move continuously with a pan. The multiplicity of spaces produced by these camera techniques was augmented by editing, which made it possible to shift quickly between points of view and break up spatial coherence even further.[27] The cinema also showed places around the world to which the audience rarely had access. In 1898 a Viennese physician made a film of a surgically exposed pulsating heart. The camera also looked into the interior space of the human body by means of the new x-rays. An article of 1913 on "The Widening Field of the Moving-Picture" described the "Roentgencinematography" of a radiologist at Cornell Medical College who made a film from a succession of x-rays of a mixture of bismuth subcarbonate and buttermilk as it passed through the intestines.[28]

The two pioneers of Cubism, Picasso and Braque, incorporated the innovations of Cézanne and the cinema and brought about the most important revolution in the rendering of space in painting since the fifteenth century. They abandoned the homogeneous space of linear perspective and painted objects in a multiplicity of spaces from multiple perspectives with x-ray-like views of their interiors. Picasso's first Cubist work, *Les Demoiselles d'Avignon* (1907), showed two figures in frontal pose but with noses in sharp profile. The seated figure has her back to the viewer but her head is seen from the front. Delaunay's Cubist *Eiffel Tower* (1910–11; Figure 5) is assembled to suggest the ubiquity of the tower in Parisian life. Houses from different parts of the city are clustered under and about its base like gifts under a Christmas tree. Their windows peer at it from all sides, even from inside it. The lower section is shown from a corner and the ironwork of the rear is perched on the side to indicate both the airiness of the structure and that it can be seen from all directions. Part of the tower has been taken out and upper sections collapsed toward the base to suggest its height. The tower was a particularly good subject because it really could be seen from anywhere and symbolized the Cubist objective to rearrange objects as seen from multiple perspectives.

One explanation for multiple perspective was that it enabled the Cubists to transcend the temporal limitations of traditional art. In 1910 the essayist Roger Allard described the Cubist painting of Jean Metzinger as "elements of a synthesis situated in time."[29] The following year Metzinger explained that Cubists have "uprooted the

Fig. 5. Robert Delaunay, *Eiffel Tower,* 1910–11.

prejudice that commanded the painter to remain motionless in front of the object, at a fixed distance . . . They have allowed themselves to move round the object, in order to give, under the control of intelligence, a concrete representation of it, made up of several successive aspects. Formerly a picture took possession of space, now it reigns also in time."[30] In 1913 Apollinaire commented that Cubists have followed scientists beyond the third dimension and "have been led quite naturally . . . to preoccupy themselves with new possibilities of spatial measurement which, in the language of the modern studies, are designated by the term: the fourth dimension."[31] There was a popular interest in the fourth dimension in France at that time, which might have inspired the Cubists.[32]

In addition to rendering multiple points of view, the Cubists also revised the traditional concept of depth. Formerly artists conceived of painting as the representation of an object in three-dimensional space, but modern artists rejected the notion that art was supposed to represent anything. Rather it must be what it is—a composition of forms on a flat surface. In 1900 the art critic Maurice Denis announced this essential characteristic of modern art: "a picture—before being a war horse, a nude woman, or an anecdote—is essentially a flat surface covered with colors assembled in a certain order."[33] This flattening was accomplished by the Cubists in part by multiple perspective but also by multiple light sources, the reduction of aerial perspective, and the breakdown of discrete forms and consistent overlapping. All of these techniques can be seen in Braque's *Still Life with Violin and Pitcher* (1910; Figure 6). The violin is broken up and shown from several points of view. Color is limited to shades of white, black, and brown, and there is no aerial perspective. The wild overlapping suggests forms and depth, but it is impossible to determine exactly what forms in what depths. The light source is ambiguous and casts shadows in different directions, but the fold of paper at the top throws a distinct shadow to the left while the illusionistic nail casts one to the right. This contradiction further interrupts a consistent sense of depth. There is another ambivalence about two- and three-dimensional space with the molding on the wall, which indicates depth clearly at one corner but then breaks into the flatter composition of the rest. The Cubists, like Cézanne, never entirely abandoned depth but reduced it, creating tensions between the world of three dimensions that was their inspiration and the two-dimensionality of painting that was their art. The *trompe-l'oeil* nail is a symbol of this creative tension. It is the most unambiguously three-dimen-

Fig. 6. Georges Braque, *Still Life with Violin and Pitcher,* 1910.

sional object in the painting and is represented clearly with an identifiable light source, but it also contradicts the illusion of depth by proclaiming that the painting *is* flat and could be nailed to the wall like a piece of paper. It is a stake in the heart of the third dimension of painting.

The Cubists' break with the space of traditional art was the subject of an essay of 1912 by Gleizes and Metzinger. They argued that the convergence technique of perspective records only visual space, but to establish pictorial space the artist must react to the world, as does the viewer, with all of the faculties. "It is our whole personality which, contracting or expanding, transforms the plane of the picture. As it reacts, this plane reflects the personality back upon the understanding of the spectator, and thus pictorial space is defined—a sensitive passage between two subjective spaces." Modern art is no longer content with slavishness to the rules of scientific perspective. "The worth of river, foliage, and banks, despite a conscientious faithfulness to scale, is no longer measured by width, thickness, and height, nor the relations between these dimensions. Torn from natural space, they have entered a different kind of space, which does not assimilate the proportions observed." That different kind of space must no longer be confused with "pure visual space or with Euclidean space." It is the space of all of the faculties and emotions and, if it is to be linked with any geometry, it would be a non-Euclidean geometry such as Riemann's.[34]

The proliferation of perspectives and the breakup of a homogeneous three-dimensional space in art seemed to many to be a visible representation of the pluralism and confusion of the modern age. As early as 1923 Picasso tried to defend his achievement from such forced juxtapositions: "Mathematics, trigonometry, chemistry, psychoanalysis, music and whatnot, have been related to Cubism to give it an easier interpretation. All this has been pure literature, not to say nonsense, which has only succeeded in blinding people with theories."[35] This is an important reminder that Cubism came out of pressures and challenges within art. Nevertheless Cubism did influence, and was influenced by, other developments. Chronophotography and cinema no doubt had some effect, however indirect, on the way Cubists rendered space and sought to give a sense of the development of an object in time as a construction of successive points of view. X-ray must have had something to do with the Cubist rendering of the interior of solid objects. In spite of Picasso's warning, critics continued to draw parallels between Cubism and a number of other cultural developments. Fritz Novotny suggested that the "alienation of objects from reality" in Cubism was symptomatic of a culture that affirmed the "unreality of place" and that was plagued by nihilism.[36] Siegfried Giedion linked Cubism with a new sense of the many-sidedness of moral and philosophical issues.[37] Pierre Francastel saw Cubism as a reflection of the fragmented space of the modern age.[38] Max Kozloff saw a connection with the relaxation of rules of grammar where words are run together as in the writing of Joyce.[39] Wylie Sypher stressed its similarity to the shifting perspectives of the new cinema and used it as a metaphor for the modern "world without objects."[40]

Painters and novelists faced contrasting challenges in reproducing the dimensions of experience. Painters, limited to a single instant, used multiple perspective to portray objects as they came into view in time. Writers, limited to a series of single settings, used multiple perspective to depict different views of objects in space. Proust and Joyce used the technique in several ways.

While riding in a carriage Marcel was moved by the sight of the twin steeples of the church of Martinville, which continually changed position as he approached them along a winding road. His description of the shifting steeples is a literary analog of a Cubist painting.[41] His account of successive views of a sunrise seen through the windows of a speeding train made the connection with painting directly: "I was lamenting the loss of my strip of pink sky when I caught sight of it afresh, but red this time, in the opposite window which it left at a second bend in the line, so that I spent my time running from one window to the other to reassemble, to collect on a single canvas the intermittent antipodean fragments of my fine scarlet, ever-changing morning, and to obtain a comprehensive view of it and a continuous picture."[42] In addition to such multiple perspectives of objects viewed over a relatively short time, there is another proliferation of space in Proust that is produced over long stretches of time by the action of feelings on the settings of important events. After many years Marcel returned to the Bois de Boulogne to try and recapture the pleasures of his childhood. But all was changed. The carriages were replaced by motor cars; the women wore different hats. Space itself, he realized, was as malleable as the objects in it: "The places that we have known belong now only to the little world of space on which we map them for our convenience. None of them was ever more than a thin slice, held between the contiguous impressions that composed our life at that time; remembrance of a particular form is but regret for a particular moment; and houses, roads, avenues are as fugitive, alas, as the years."[43] Spaces are subject to changing perspectives, thoughts, and feelings and suffer the unceasing transformation of things in time.

We have already observed in a discussion of simultaneity how Joyce reconstructed events, such as those in the "Wandering Rocks" episode, from a number of points of view in order to give a fuller sense of them. He also envisaged a multiplicity of coexisting universes of different dimensions. Bloom reflects on the size of his universe and sees it as one of an infinite number enclosed within one another as in a set of Chinese boxes. He thinks of the star Sirius 57,000,000,000,000 miles distant, 900 times as large as the earth, and then of the nebula of Orion in which 100 of our solar systems could be contained. He then considers the infinitesimally small universes around him, "the incalculable trillions of billions of millions of imperceptible molecules contained by cohesion of molecular affinity in a single pinhead" and "the universe of human serum constellated with red and white bodies, themselves universes of void spaces constellated with other bodies." In the final account of his hero, Joyce mocks the convention of giving a precise, single location of action. Bloom is in bed next to Molly and telling her about his day: "Listener S.E. by E.; Narrator, N.W. by W.: on the 53rd parallel of latitude, N. and the 6th meridian of longitude, W.: at an angle of 45° to the terrestrial equator."[44] Here the relative position of the two lying head to foot is identified by means of this incongruous navigational jargon, which ironically brings to mind the impossibility of knowing the precise location of bodies in space. We know their exact location on earth, but where is the earth? Moreover, even if we did know that, Joyce implied, it would not reveal the crucial information about place. Odysseus's Mediterranean, Bloom's Dublin, his bed at 7 Eccles Street are not the essential settings, because the real action takes place in a plurality of spaces, in a consciousness that leaps about the universe and mixes here and there in defiance of the ordered diagraming of cartographers. Edmund Wilson has interpreted these shifting perspectives as part of a general movement in European culture. "Joyce is indeed really the great poet of a new phase of human consciousness. Like Proust's or Whitehead's or Einstein's world, Joyce's world is always changing as it is perceived by different observers and by them at different times."[45] Thus the two most innovative novelists of the period transformed the stage of modern literature from a series of fixed settings in a homogeneous space into a multitude of qualitatively different spaces that varied with the shifting moods and perspectives of human consciousness.

In geometry and physics, biology and sociology, art and literature attacks were launched on the traditional notions that there is one and only one space and that a single point of view is sufficient to understand anything. Sometimes the historical record is generous and supplies abundant evidence for a cultural change. In this period it also supplied an interpretation of that change with the philosophy of "perspectivism."

After Nietzsche left the university he began to criticize the narrowness of academic thinking—a Platonism that denied the validity of knowledge acquired through the senses, a positivism that was blind to the inherent subjectivity of knowledge. Scholars, he wrote, "knit socks for the spirit."[46] He came to life in the clear air outside the academy, and like the Impressionists who discovered a world of new colors *en plein air,* he found new philosophical topics and a fresh poetic language with which to write about them. In opposition to the positivists' belief in the truth of objective facts, he insisted that there are no such things, only points of view and interpretations, and he urged philosophers "to employ a *variety* of perspectives and affective interpretations in the service of knowledge." This philosophy was called "perspectivism," and in 1887 he proclaimed its method.

> Henceforth, my dear philosophers, let us be on guard against the dangerous old conceptual fiction that posited a "pure, will-less, painless, timeless knowing subject"; let us guard against the snares of such contradictory concepts as "pure reason," "absolute spirituality," "knowledge in itself": these always demand that we should think of an eye that is completely unthinkable, an eye turned in no particular direction, in which the active and interpreting forces, through which alone seeing becomes seeing *something,* are supposed to be lacking; these always demand of the eye an absurdity and a nonsense. There is *only* a perspective seeing, *only* a perspective "knowing"; and the *more* affects we allow to speak about one thing, the *more* eyes, different eyes, we can use to observe one thing, the more complete will be our "concept" of this thing, our "objectivity." But to eliminate the will altogether, to suspend each and every affect, supposing we were capable of this—what would that mean but to *castrate* the intellect?[47]

We must look at the world through the wrong end of the telescope as well as the right one, see things inside out and backwards, in bright and dim light. In this philosophy spaces proliferate with points of view.

In the twentieth century perspectivism was formalized by the

Spanish philosopher José Ortega y Gasset. Rationalists argue that there is one and only one truth that can be grasped by factoring out the errors that arise from viewing things from subjective points of view. Rejecting this approach, Ortega formulated his own theory of perspectivism in 1910: "this supposed immutable and unique reality . . . does not exist: there are as many realities as points of view."[48] In 1914 he made perspective into the stuff of reality: "God is perspective and hierarchy; Satan's sin was an error of perspective. Now, a perspective is perfected by the multiplication of its viewpoints."[49] The rationalist position maintained the homogeneity of space, and Ortega countered that there were as many spaces in reality as there were perspectives on it. In a manifesto for the first issue of the journal *El Espectador* (1916), he reaffirmed the validity of the individual point of view. Reality is perspective. The war itself, he suggested, was brought about by a narrow-mindedness among nations that failed to see the larger context of their actions. People must react against this "exclusivism" and develop a broad outlook that embraces a multitude of perspectives.[50]

In a lecture on the historical significance of Einstein, Ortega linked perspectivism and the general theory of relativity and maintained that the coincidence of their publication in 1916 was a sign of the time. The two doctrines signified a breakdown of the old notion that there is a single reality in a single, absolute space. "There is no absolute space because there is no absolute perspective. To be absolute, space has to cease being real—a space full of phenomena—and become an abstraction. The theory of Einstein is a marvellous proof of the harmonious multiplicity of all possible points of view. If the idea is extended to morals and aesthetics, we shall come to experience history and life in a new way."[51] He also suggested ethical and political consequences. The peace broke down in Europe because each nation was fixed in a narrow outlook. The British "white man's burden," the French "*mission civilisatrice*," and the German "*deutsche Kultur*" were but different points of view on the same landscape, but each nation viewed its own as the only true one.

Ortega once described perspectivism in terms applicable to Cubism: "The truth, the real, the universe, life . . . breaks up into innumerable facets and vertices, each of which presents a face to an individual."[52] His philosophy itself reflected many others. He was influenced by, or noted parallels to, Riemann, Lobatchewsky, Mach, Einstein, Uexküll, Proust, and Joyce and shared their restlessness with conventional notions about the sanctity of a single space or point of view. He challenged what he felt to be an arrogance deeply embedded in Western culture, an egocentrism that believed that one point of view—be it that of a mathematician, philosopher, or nation—was alone correct. Knowledge progresses and cultures advance as the diversity of concrete experience is allowed to be heard. The world is understood by the observer who localizes reality "in the current of life which flows from species to species, from people to people, from generation to generation and from individual to individual, gradually possessing itself of more and more universal reality."[53] There is danger that such a philosophy of perspective can become a runny, undisciplined pluralism, an excuse for having no point of view at all, but in this period it provided a corrective to the epistemological and aesthetic egocentrism that had dominated Western culture for so long.

Durkheim's theory of the social relativity of space gave weight to societies outside the Western world, and even Spengler was able to appreciate the broad range of achievements of cultures based on a different sense of space. Ortega's philosophy of perspectivism in its social and political implications lined up clearly on the side of pluralism and democracy against monism and monarchy. It implied that the voices of many, however untrained or chaotic, are a desirable check on the judgment of a single class, a single culture, or a single individual. Even Nietzsche, who had contempt for democracy and who railed against the leveling effect of the masses, understood that the overman must achieve transcendence through a continual struggle, and hence dialogue, with the masses. Zarathustra repeatedly returned to the masses, even though he was always misunderstood and continually threatened by contact with them. Although these various arguments on behalf of the heterogeneity of space did not always address themselves to the social and political terms of social equality versus social privilege and democracy versus monarchy, they form part of a general cultural reorientation in this period that was essentially pluralistic and democratic.

∞

A second major issue raised about the nature of space was its constituency.

The traditional view that space was an inert void in which objects existed gave way to a new view of it as active and full. A multitude of discoveries and inventions, buildings and urban plans, paintings and sculptures, novels and dramas, philosophical and psychological theories, attested to the constituent function of space. I will refer to this new conception as "positive negative space." Art critics describe the subject of a painting as positive space and the background as negative space. "Positive negative space" implies that the background itself is a positive element, of equal importance with all others. The term is somewhat unwieldy, but it is accurate and suggests the historical sense of the developments in this period, since it implies that what was formerly regarded as negative now has a positive, constitutive function.

One common effect of this transvaluation was a leveling of former distinctions between what was thought to be primary and secondary in the experience of space. It can be seen as a breakdown of absolute distinctions between the plenum of matter and the void of space in physics, between subject and background in painting, between figure and ground in perception, between the sacred and the profane space of religion. Although the nature of these changes differed in each case, this striking thematic similarity among them suggests that they add up to a transformation of the metaphysical foundations of life and thought.

From the time of Democritus scientists had believed that the stuff of the world was composed of solid bits of matter. In 1897 J. J. Thomson announced his discovery of some even more basic "corpuscles" out of which the elements were built, and developed a model of the atom with these corpuscles (eventually called electrons) orbiting around a nucleus.[54] The Thomson atom was thus largely empty space, and it wiped out the classical distinction between the plenum of matter and the void of space. By 1914 a book about atoms explained that matter had a "spongy" consistency and was "prodigiously lacunary."[55]

In 1876 William Clifford, the English translator of Riemann, formulated a theory that matter and its motion were manifestations of the varying curvature of space. He hypothesized that matter was the location of curvatures in space analogous to "little hills" on a flat surface; "that this property of being curved or distorted is continually being passed on from one portion of space to another after the manner of a wave"; and "that this variation of the curvature of space is what really happens in that phenomenon which we call the motion of matter."[56] In 1898 the American philosopher Hiram M. Stanley identified a trend among physicists of seeing all things as different states of energy. For them space was not an epistemological form but a product of the struggle for existence among the opposing forces that might displace it. Stanley concluded that space is "not full of things, but things are spaceful."[57] This adjectival form emphasized the active and constituent function, but most nineteenth-century physicists could not conceive of attributing physical functions to space, so they posited a medium called ether, pervading space, which transmitted electromagnetic phenomena like wireless waves and x-rays. A book on the wireless maintained that there is "nothing absolutely solid in nature" and that it is possible for a medium to penetrate all things.[58] Wells's Time Traveller was able to avoid collision with the solid objects that occupied the places through which his machine moved by slipping through the interstices of intervening substances. Another science-fiction writer imagined a "Y-ray" that could increase the spaces between matter to allow one solid body to pass through another.[59] Thus space constituted a large portion of matter, and the medium that was thought to pervade it played an active role in the transmission of energy.

Physical space came fully to life with Einstein's field theory. In 1873 Clerk Maxwell hypothesized that electricity and light travel in waves through fields like those around magnets. Fifteen years later Heinrich Hertz developed instruments to propagate electromagnetic waves through a vacuum, but he, like Maxwell, could not imagine how the wave could oscillate in nothing and so clung to the theory of an ether. Even after the Michelson-Morley experiment failed to detect an ether, physicists continued to spin theories to accommodate the mechanical model for the propagation of waves through a medium of ponderable matter. Einstein boldly abandoned that model. His special theory removed the idea "that the electromagnetic field is to be regarded as a state of a material carrier. The field thus becomes an irreducible element of physical description, irreducible in the same sense as the concept of matter is in the theory of Newton." In Newton's mechanics a particle of light moves through empty and static space. In Einstein's mechanics everything is in movement throughout the field at the same time, and space is full and dynamic and has the power of "partaking in physical events."[60] According to the new physics the universe is full of fields of energy in various states, and space can be thought of to be as substantial as a billiard ball or as active as a bolt of lightning.

The history of architecture is the history of the shaping of space for a variety of political, social, religious, or purely aesthetic reasons.

Greek temples and theaters, Roman basilicas and baths, Byzantine churches, Romanesque and Gothic cathedrals, Renaissance and Baroque palaces, each style had a distinctive sense of space unique to its period and self-consciously created by architects schooled in its respective artistic conventions.[61] However, around the turn of the century architects began to modify the way they conceived of space in relation to their constructions. Whereas formerly they tended to think of space as a negative element between the positive elements of floors, ceilings, and walls, in this period they began to consider space itself as a positive element, and they began to think in terms of composing with "space" rather than with differently shaped "rooms." Although this change was essentially a rethinking of the nature of architectural design, it was facilitated by three inventions that liberated architects from many structural requirements for illumination, load-bearing, and ventilation and made it possible to sculpt interior space freely.

There was an enormous increase in the use of artificial illumination even before the introduction of the electric light: between 1855 and 1895 an average household in Philadelphia increased its use of illumination twenty times.[62] But this came mostly from burning oil or gas and imposed great architectural limitations. The invention of the gas mantle in the 1880s eliminated the soot, but even so the electric light bulb quickly came to dominate the market and by the mid-nineties began to revolutionize architecture and interior design. It was cooler and cleaner than gas and could be placed almost anywhere, so architects could build with whatever natural light they desired or eliminate it completely.

In 1892 the French engineer François Hennebique increased the load-bearing strength of reinforced concrete by replacing the iron rods with steel and bending them near the supports. He used it in his own house to support a tower that cantilevered four meters out from the building. The French remained leaders in the development of concrete architecture until the First World War, their fascination with the new material culminated in the monumental Maginot Line. Reinforced concrete enabled architects to fling dramatic new forms all over Europe and America in the early twentieth century, and since it could be poured into molds, there was no end to the unusual shapes or spaces that could be created.[63]

A fully air-conditioned building controls temperature, cleanliness, humidity, and circulation of air. There is some controversy about which building was the first to have all four, but it came into being some time between 1903 and 1906. Reyner Banham identified the Royal Victoria Hospital in Belfast (1903) as "the first major building to be air-conditioned for human comfort," because the entire plan was adapted to environmental considerations. Stuart W. Cramer coined the term "air-conditioning" in lectures and patents filed in 1904–1906. But the crucial invention was the dew-point control system for humidity regulation that Willis Carrier patented in 1906.[64] Liberated from the necessity of providing structural openings for ventilation, architects could open or close spaces at will.

With the flood of industrial goods in the nineteenth century, Europeans lost their sense of the dignity of space and rooms were cluttered with knickknacks and mementos, bird cages and aquariums, ornate picture frames, moldings, drapes, and overstuffed furniture. Large interior spaces were thought to be a sign of incompleteness or poverty. As Siegfried Giedion observed, these fashionable interiors "with their gloomy light, their heavy curtains and carpets, their dark wood, and their horror of the void, breathe a peculiar warmth and disquiet."[65] Around the turn of the century, as Art Nouveau designs crawled everywhere, there was a movement among interior designers and architects to clean up the gobbledygook in rooms and the excessive ornamentation on exteriors. In an article of 1895 the British architect Charles Voysey expressed disgust with the "motley collection of forms and colors with which most rooms are crowded."[66] He criticized the clutter and eclecticism of nineteenth-century taste and appealed for flat surfaces and simple, functional structures. In Germany, Friedrich Naumann praised ships, bridges, railway stations, and market halls as the new buildings of a machine age that had "no stuck-on decoration, no frills."[67] In a famous essay of 1908, "Ornament and Crime," the Austrian designer Adolf Loos argued that erotic cave drawings, bathroom graffiti, and architectural ornamentation were manifestations of the same primitive impulse that in the contemporary world leads to degeneration and crime. He concluded that "the evolution of culture marches with the elimination of ornament from useful objects."[68]

The Dutch architect Hendrick Berlage subdued ornament on the Amsterdam Stock Exchange building that he constructed between 1890 and 1903. He articulated his aesthetics of unadorned design in 1905, and in 1908 commented on the excessive concern for ornamentation in earlier times: "The nineteenth century forgot to build from the inside out; it was an architecture of façades that sacrificed reality to appearance." Architecture must recognize its true purpose as an "art of space." The primary subjects of architecture are not so much walls and ceilings as the spatial enclosures created by them.[69] This conceptual shift was presented even more forcefully in the writings and buildings of Frank Lloyd Wright. His Larkin Soap Company building in Buffalo (1904) was essentially a single room closed to the outside. Wright himself identified its role in the history of architecture. It was "the original affirmative negation" that showed "the new sense of 'the space within' as reality." His interior spaces were carefully designed to conform to human needs and were to be the rationale for the entire structure. Space was the basic element in Wright's architectural design of Unity Temple (Oak Park, Illinois, 1906), which had a simple cubical interior that was visible on the outside of the building constructed with simple blocks of cement and an unadorned concrete slab roof. He explained that his initial conception was "to keep a noble ROOM in mind, and let the room shape the whole edifice." Although this account used the more traditional architectural terminology that conceived of space in terms of rooms, the sense of it was modern, as Wright went on to make a bold historical claim about his conception of the positive function of space: "The first conscious expression of which I know in modern architecture of the new reality—the 'space within to be lived in'—was Unity Temple in Oak Park. True harmony and economic elements of beauty were consciously planned and belong to this new sense of space-within . . . In every part of the building freedom is active. Space [is] the basic element in architectural design."[70]

This reference to a sense of freedom evoked by the space of a building echoed the aesthetic theory of the German philosopher Theodor Lipps and its application to architecture by the British architect Geoffrey Scott. In 1903 Lipps argued that our bodies unconsciously empathize with architectural forms. We feel free when there are no external constraints on our bodily movements, and buildings with large open spaces offer that freedom.[71] In 1914 Scott elaborated an "architecture of humanism" based on this theory. Architects project human feelings into a building, and it in turn impresses viewers with an immediate physical response. We feel uncomfortable in a room fifty feet square and seven feet high, because it constricts our sense of freedom. Heretofore architects have neglected the importance of space in their art. "The habits of our mind are fixed on matter. We talk of what occupies our tools and arrests our eyes. Matter is fashioned; space comes. Space is 'nothing'—a mere negation of the solid. And thus we come to overlook it." Architecture is the one art form that deals with space directly. Painting can depict space, poetry can form an image of it, music can offer an analogy, but only architecture can actually create it. "To enclose a space is the object of building; when we build we do but detach a convenient quantity of space, seclude it and protect it, and all architecture springs from that necessity. But aesthetically space is even more supreme. The architect models in space as a sculptor in clay. He designs his space as a work of art." Scott summarized the striving of a generation of architects to recognize the constituent function of space.[72]

The crowding of interior space by objects in rooms was matched by a growing crush of people, vehicles, and buildings in cities; urban planning arose to deal with the problem. All the different proposals conceived of space as a positive, constitutive factor in urban planning. Reinhard Baumeister and Joseph Stübbens oriented their designs to the needs of traffic and cut large arteries through cities to accelerate the flow.[73] Ebenezer Howard, who pioneered the modern "garden city" idea, planned cities around areas of greenery. Camillo Sitte insisted that the rhythmic distribution of spaces in pleasing and functional patterns should be the top priority. He argued that urban spaces should be enclosed to give them a definite shape. He also criticized the horror of empty space that repeatedly led planners to put statues and monuments in the center of town squares. His model was the open plazas of medieval towns that functioned as market or meeting places, and his motto was to "keep the middle free."[74] While most interior decorators, architects, and city planners felt that their principal decision was where to place solid objects, others, like Voysey, Berlage, and Sitte, reversed that priority and sought to utilize the aesthetic potential of space itself.

Changes in stage design conformed to the same lines. In the 1890s the German designer Adolphe Appia abandoned painted backdrops and created "rhythmic" spaces with sculptured architectural forms and dramatic chiaroscuro lighting.[75] In England Gordon Craig carried the ideas of Appia further toward making the stage a positive space.[76] He also eliminated the deceptive orchards and arcades on painted backdrops and recomposed the space with drapes, screens, and simple geometric forms. As painters rejected the illusionistic perspective of traditional art, Appia and Craig eliminated the illusions of depth created by traditional stage design. Accompanying these simplified stage designs were simplified costumes, stripped of excessive ornament in the manner of Loos's interiors and Berlage's façades. The stage must be adorned only with light, shadow, and nonrepresentational sets that merely accented the space in which actors moved.

National festivals in Germany in this period were staged in spaces around national monuments where masses of people could sing and dance. Earlier designers had provided a space for national worship in the form of cemeteries around monuments, but this period saw an evolution "from dead to a living space, one which was taken up not by graves but by living people acting out their national liturgy."[77] George Mosse identified the emergence of such a "living" space around the Kyffhauser monument, completed in 1896. Erected to celebrate the hundredth anniversary of the Battle of Leipzig, the Völkerschlachtdenkmal, completed in 1913, included both a cemetery to memorialize the past and a large open space on which to hold national festivals that made the monument come alive.[78]

Sculpture provided the most graphic and explicit affirmation of positive negative space. In Boccioni's *Development of a Bottle in Space* (1912) the bottle spirals out of a pool of silvered bronze into space that itself coils into the solid form (Figure 7). In a manifesto Boccioni announced that Futurists will create masses "in such a way that the sculptural block itself will contain the architectural elements of the sculptural environment in which the object exists."[79] Space is no longer a setting for the subject but a constituent element of the work that the sculptor must model.

A more dramatic use of positive negative space occurs in the sculpture of Alexander Archipenko, who created figures with concaves and voids. He reversed the traditional notion that space was a frame around the mass, that sculpture begins where material touches space, and maintained "that sculpture may begin where space is encircled by the material." *Woman Walking* (1912), in which the torso is a void enclosed by material form, was, as he recalled, his first successful creation of "space with symbolic meaning."[80] The female belly that was emphasized in the nineteenth century by tight lacing is here rendered by shapely emptiness; space has become the guts of his art. In *Woman Combing Her Hair* of 1915 (Figure 8) the arching arm frames the empty space that is her head, and its shape is repeated in the convex severed arm and the concave neck. There were precedents for the use of concavity in bas relief, intaglio, and African masks, as Archipenko himself observed, but never before in sculpture were essential elements such as a figure's face represented by completely empty space. In this work the traditional division of pos-

Fig. 7. Umberto Boccioni, *Development of a Bottle in Space*, 1912.

itive and negative space is dissolved as material and spatial forms flow together and constitute the woman with equal force.

The emergence of positive negative space in painting contrasts sharply with earlier conventions of rendering the subject with far greater emphasis than the background. For centuries the background had framed the subject as the pillow frames a head. Portrait painting of the eighteenth century, for example, was often executed by a team of the portraitist and his assistant. The best known artist to exploit this hierarchical arrangement was Sir Joshua Reynolds. The critical parts of the portrait—the overall design and the face—were executed by Reynolds, while the subject's clothing and the background were done by an assistant, the drapery painter.[81] In the modern period the background took on a positive, active function of equal importance with the subject and demanded the full attention of the artist.

The Impressionists took a first step to give space its due with their depiction of atmosphere.[82] They used coastal fog, steamy summer haze, diffused forest light, overcast winter twilight, the orange wash of a low sun, to fuse subject and background into a single composition of color and form. Monet unified the pictorial surface in his series on the Rouen Cathedral at different times of the day and seasons of the year. Space and light preempted the nominal subject, which he painted twenty times over as though it mattered less than the play of light around it. Cézanne deplored the loss of clear forms in the Impressionist atmosphere but affirmed the constituent function of space by ignoring the former distinction between the subject and the less important background and according equal significance to every portion of the canvas.[83] In portraits he gave as much attention to the shape of a space between the head and the picture frame as to the shape of the head itself. In his still life paintings the part between the edge of the table and the edge of the canvas was as crucial to the overall composition as his apples and vases. And in his late landscapes the skies are filled with interlocking faceted sections of empty space itself. There is no negative space in Cézanne's painting. All forms are of equal value, all constitute the subject of the work. A similar statement of the positive function of the background appears in the work of the Austrian Secessionist artist Gustav Klimt. As Carl E. Schorske observed, "In a series of three portraits painted between 1904 and 1908, Klimt progressively extended the dominion of the environment over the person of the subject."[84] Although in each of these the background does function as a frame for the portrait subject, it has a solid geometric structure that rivals the subject for the viewer's attention, and in the 1907 portrait of Adele Bloch-Bauer it engulfs the figure in a gold metallic brilliance.

With Cubism the emergence of space as a constituent element is complete. Braque and Picasso gave space the same colors, texture, and substantiality as material objects and made them all interpenetrate so as to be almost indistinguishable. Spatial forms became especially prominent in Braque's *Harbor in Normandy* (1909), where the lighthouses, docks, boats, and sails are rendered with the same faceted elements as the sea and sky and the spaces between the objects. In an interview Braque explained that the main attraction of Cubism was "the materialization of that new space which I sensed." He discovered a "tactile space" in nature, and he wanted to paint the sensation of moving around objects, the feeling of the terrain, the distances between things: "This is the space that attracted me, because that was what early Cubist painting was all about—research into space." The leveling of space and material object and the interpenetration of the two reached a high point in his *Violin and Pitcher* (see Figure 6 above). The neck of the violin retains its discreteness but the body is fractured into sections that open into a space rendered as substantially as the splinters of wood. It is impossible to distinguish clearly between subject and background as plaster, glass, wood, paper, and space are rendered in a fluid pattern of similar forms. Braque explained: "The fragmentation enabled me to establish the space and the movement within space, and I was unable to introduce the object until I had created the space."[85] The pitcher and violin are just different kinds of space, occupied by solid objects that can be simplified, geometrized, fragmented, and then reformed in space. In Braque's painting all spaces are qualitatively equal.[86]

The American poet William Carlos Williams was especially struck by these Cubist techniques and tried to approximate them in "Spring Strains" (1916) by giving substance to the space of the sky as well as to the objects in it.

> . . . Vibrant bowing limbs
> pull downward, sucking in the sky
> that bulges from behind, plastering itself
> against them in packed rifts, rock blue
> and dirty orange!
>
> But—
> (Hold hard, rigid jointed trees!)
> the blinding and red-edged sun-blur—
> creeping energy, concentrated
> counterforce—welds sky, buds, trees,
> rivets them in one puckering hold! . . .

The limbs suck in the sky, but the sky, rendered substantial in "packed rifts," plasters itself against them. Even the colors are the rather subdued hues of the early Cubist landscapes that contributed to the unification of the entire picture surface. And, as if welding were not strong enough, the buds and trees and sky are also riveted together in "one puckering hold."[87]

Just as in physics space was recognized as both constituent and active with atomic theory and field theory, so in art space was realized in two positive modes. Its constituent function was most explicit in the Cubist representation of the space between objects, and its active function can be seen in Van Gogh, Munch, Cézanne, and the Futurists, who depicted space energized by objects in it.

In the extraordinarily creative last two years of his life Vincent

Van Gogh created on canvas an unforgettable dynamic world. His landscapes are visual metaphors for the turbulence in his mind. Roofs undulate with the contours of the terrain, skies flow with surging mountains, and trees grow before our eyes, whipping lines of force into an atmosphere that spirals into stars, eddies around a prominent sun. In the self-portraits speckles of color explode as if the energy in his eyes had burst into the space around his head. His universe was a continuous field of energy circuiting through mind, world, and art; in his last months, when he was insane, a pervasive scream seemed to fill all space.

The Norwegian painter Edvard Munch gave visible form to such intensities in *The Cry* (1893). It shows a terrified screaming figure on a bridge, clasping its head in its hands, cut off from two people in the distance who are walking away. The emptiness of the surrounding space and the isolation of the figure contrast with the ubiquity of the scream and the feeling of intense pressure it evokes. The landscape behind the skull-like head and the space above it pulsate with the sound waves. Cézanne also energized the space around objects. In his late landscapes Mont Sainte-Victoire comes to life like a volcano, disrupting contours of the countryside and erupting into space. In a painting of it in 1904 the foreground is dematerialized and broken into vertical and horizontal brush strokes of greens, yellows, and blues.[88] The earth lunges toward a peak accented by sparks of color that hover in the space above it. The sky echoes the forms of the land as if the mountain had just pushed out of it and was still sending shock waves through the atmosphere. There is a similarity to Van Gogh's landscapes, as terrain, verdure, and sky form continuous patterns of line, color, and brush work.

While Cézanne's canvases bulked with muscular spatial forms, the Futurists depicted lines of force in space created by movement, light, and sounds. In a manifesto of 1910 Boccioni articulated their belief in an active, dynamic space: "To paint a human figure you must not paint it, you must render the whole of its surrounding atmosphere."[89] He gave visible form to his idea in *The Forces of a Street* (1911), where the clanging sounds, beaming headlight, and lurching of a streetcar take on substance and modify the colors and forms of the surrounding persons, buildings, and atmosphere. In 1909 Balla filled a canvas with the radiance of a street lamp, and in 1912 he painted another with scalloped and puff-ball formations of "atmospheric densities." In 1912 the Futurists explained that an object would be expanded by the use of "force-lines" determined by its form at rest, its continuity with surrounding space, its past and future trajectories, and the way it would be "decomposed according to the tendencies of its forces."[90] With such multiple determinants its actual depiction took many forms, but in all variations space is rendered as an active and constituent element of equal importance with the "subject." In Boccioni's study for *Dynamism of a Cyclist* (1913) force-lines spin off the speeding bicycle as if it were racing through a puddle and make continuous lines with the bicycle itself. In the finished work bicycle and rider merge with the surrounding space and give a single image of movement.

Western historians began to ponder the concept of "empty space," as their nations discovered that none was left. In America the census of 1890 declared that the frontier was closed, and by the end of the century the dominant world powers had finished taking the vast "open" spaces of Africa and Asia. Government officials considered the political impact of the closing of the world frontier, and scholars developed a new discipline to codify its significance. The great pioneer of the new "geopolitics" was the German researcher Friedrich Ratzel.[91] In an essay of 1893, "The Significance of the Frontier in American History," Frederick Jackson Turner applied geopolitical theory to explain the development of American character and institutions. The presence of an open frontier, he speculated, created a spirit of individualism. Settlers, compelled to adapt to the challenges of crossing a wilderness and rebuilding their lives again and again, sacrificed traditions and leveled religious, social, and political hierarchies. The constant expansion fragmented religious authority and led to the proliferation of rival churches scattered in the frontier towns. Continuous social dislocation made it impossible to maintain the fixed social order of the older Eastern cities, where families remained in the same place and intensified class distinctions with each passing generation. But the most important effect of the frontier was "the promotion of democracy here and in Europe." Life in the wilderness broke down complex society into a primitive organization based on the family. The need for improvising brought out new social organizations in which everybody played a role and was vital to the survival of the community. These circumstances produced "antipathy to control, and particularly to direct control."[92] No single person could monopolize power in frontier settlements where cooperation and democracy flourished. In an article of 1903, he elaborated: "Whenever social conditions tended to crystallize in the East,

Fig. 8. Alexander Archipenko, *Woman Combing Her Hair,* 1915.

whenever capital tended to press upon labor, there was this gate of escape to the free conditions of the frontier. These lands promoted individualism, economic equality, freedom to rise, democracy."[93] But when the gate closed, capital concentrated in the fundamental industries and there was commercial and political expansion overseas. These developments reconstituted hierarchies of wealth and imperial power and reversed the leveling tendencies of an open frontier. Turner's thesis and the historical circumstances that suggested it form part of a general appraisal of the constituent function of empty space. The closing of the frontier highlighted the significance of open territorial space—especially the erosion of traditional hierarchies—for the entire population, and Turner's interpretation focused attention on its social and political consequences.

A contemporary historian, Roderick Nash, pointed out that "the establishment of Yellowstone National Park on March 1, 1872, was the world's first instance of large scale wilderness preservation in the public interest." The intention was to protect the geysers, but in the 1880s and 1890s a few people began to realize that the wilderness in general had been protected, and by then a movement was under way to protect open territory for public use around the world. Such parks were open to the public, in contrast with the private preserves of kings and noblemen and the wealthy.[94] Their significance in promoting the democratic spirit, like that of the frontier, was clarified precisely as the empty spaces of the world threatened to disappear. The exploration of entirely uncharted territory also came to an end in this period. Robert Peary reached the North Pole in 1909, and two years later Roald Amundsen made it to the South Pole. Bootprints tracked over the untrodden snow and the last great frontiers of the world closed.

From the 1880s the literary bounty of stories about empire matched the psychological, political, and financial yield of the land grab of the major imperialist powers. In a survey of hundreds of novels in this genre, the literary critic Susanne Howe concluded that their characters suffered from claustrophobia at home. They became greedy for land, annoyed with boundaries, enraged by fences, and "intoxicated by space." While some were thrilled others were horrified, like the woman in Olive Schreiner's *Story of an African Farm* (1883), who, fresh from England, exclaimed at her first glimpse at the endless miles of bush: "Oh it's so terrible! There's so much of it! So much!"[95] Whether they found it an inspiration or a horror, a setting for riches or for ruin, the vast emptiness weighed upon them and shaped their lives. In one of the greatest stories about empire, Conrad's *Heart of Darkness* (1899), the empty space was overpowering: it drew Marlow and destroyed Kurtz, the man he went to find. Marlow's journey is an allegory of the history of mankind in reverse, a devolution of the species into the past, into darkness, into nothing.

As a boy Marlow used to stare at the great blank spaces on the map and dream of the glories of empire. The Congo especially continued to fascinate him. By the time he had grown up it had become filled with rivers and lakes and names. "It had ceased to be a blank space of delightful mystery—a white patch for a boy to dream gloriously over. It had become a place of darkness." He remained intrigued and got a commission to patrol the Congo River for a company that traded in ivory. At the coastal station he first heard about Kurtz, the company's man in the interior, who had sent out great quantities of ivory and who, it was feared, was in some terrible trouble. Marlow's journey thus became a quest to find and rescue him. On a trek to the central station to get his steamer his surroundings were images of negation—empty land, abandoned villages, dead carriers, and "a great silence around and above." His trip up the river, he observed, was like traveling back to the beginning of the world when vegetation rioted on the earth, but in spite of the lush flora this world seemed to be "an empty stream, a great silence, an impenetrable forest." He felt cut off from everything he had ever known as he penetrated "deeper and deeper into the heart of darkness." Strange sounds came from natives along the shore. There were outbursts of shrieking which suddenly stopped and left "appalling and excessive silence." But the stillness was not peaceful—"it was the stillness of an implacable force brooding over an inscrutable intention." The natives in the bush were cannibals, and their hunger symbolized more of the emptiness. Marlow found darkness everywhere—in the wilderness, in its people, in Kurtz, and, finally, in the condition of man.

At the inner station he saw severed heads stuck on poles surrounding Kurtz's house. They were "black, dried, sunken, with closed eyelids"—a final symbol of negation. Deep in this theater of hunger and emptiness only the prospect of speaking with Kurtz offered the hope of some illumination, some affirmation. But he discovered in Kurtz a man who had been stripped of the values of modern civilization. The wilderness had taken revenge for his invasion and whispered terrible things that echoed loudly within him, "because he was hollow at the core." Kurtz was dying and Marlow took him away. They spoke on the steamer, but, Marlow observed, "his was an impenetrable darkness." And, as though at the very end he had a sudden vision of his life, Kurtz expired with a cry: "The horror! The horror!"

Marlow concludes on a positive note as he ponders Kurtz's dying words. Kurtz's life was an adventure into the darkness where terrifying urges surfaced, but he gave his life a form, and at the end, "He had something to say." Marlow is impressed that he was able to sum it up—"The horror!" This was a judgment. "It was an affirmation, a moral victory paid for by innumerable defeats, by abominable terrors, by abominable satisfactions. But it was a victory!" At the conclusion the ivory remains. In spite of the slaughter of the elephants and the evil it inspires, it is the stuff of art, a dazzling whiteness in the heart of darkness. The emptiness itself is the subject of the novel, a force of darkness that rules the wilderness and triggers the actions of men who seek to survive in it.

This novella is a comment on the age, and Conrad took pains to make Kurtz a man of his times: "His mother was half-English, his father was half-French. All Europe contributed to the making of Kurtz." It is a catalog of literary images of the void applied in the context of imperialism. Conrad interprets the darkness as a leveling force that negates the status distinctions of class and privilege that regulated European life. In the wilderness the older class lines were obsolete. Cannibalism and head-hunting obliterated status distinctions. Marlow noticed the sharp contrast between the hierarchical society at home and the more egalitarian Congo when he returned to London and was at last able to understand the creative potential of the wilderness. In the face of danger, in the darkness, all men are pretty much alike.

A few years later another story was written about a jungle, a journey, and the void—Henry James's "The Beast in the Jungle" (1903). It is about John Marcher, who is convinced that a rare and strange fate awaits him, crouching like a beast in the jungle to leap out and slay him. He gains the affection of May Bartram, who undertakes to watch and wait with him, and over the years she comes to understand, but does not tell him, what the beast is. Marcher too is a man of his age—well-mannered, disciplined, reserved, and, except for his dependency on May, self-reliant. Everything in his life is in order—his library, his garden in the country, his feelings. When May becomes seriously ill he anticipates that the loss he will feel over her death must be the beast, but she tells him that it has already leaped and that he failed to notice it. Her explanation is confusingly negative—"your not being aware of it is the strangeness *in* the strangeness." She confounds him further by saying that she is glad "to have been able to see what it is *not.*" In the end Marcher will learn the meaning of this second negation: that the beast was not his love for her or his sense of loss. After her death he must wait alone for something that has already happened but that he does not yet understand. A year later, while making one of his dutiful visits to her grave, he notices another man deeply stricken with grief. Marcher realizes that the stranger's face shows an intensity that he had never felt. He looks back at May's grave and suddenly sees the beast. The name on the tombstone becomes "the sounded void of his life." She was what he had missed, and that was his special destiny—"he was the man of his time, *the* man, to whom nothing on earth was to have happened." He had been anesthetized by the refinements of modern civilization and could not feel deeply for her either before or after her death. The beast in the jungle was—a lack of feeling. This was not an active spirit of negation, like Mephistopheles in Goethe's *Faust,* but an inner emptiness, like the void that sounded from the silence of May's grave.

His terrifying insight, like Kurtz's dying words, are two modes of negativity. In Strindberg's *A Dream Play* (1901) there is another climactic discovery of nothing. For many years an officer attempts to get past a guard and open a door that is prominently visible in the center of the stage. Like John Marcher's obsession with the beast, the officer is obsessed with looking behind the door. "That door," he exclaims, "I can't get it out of my mind . . . What's behind it? There's got to be something behind it." In the course of the play a number of other characters come to want it opened, and when they finally succeed several university officials are clustered about and discover that there is nothing behind it. The Dean of Theology immediately interprets its significance: "Nothing. That is the key to the riddle of the world. In the beginning God created heaven and earth out of nothing." The Dean of Philosophy observes: "Out of nothing comes nothing." The Dean of Medicine makes a diagnosis as if he had just lanced a harmless boil: "Bosh! Nothing. Period." The Dean of Law suggests that the whole thing is a case of fraud. Faust had found nothing to help him affirm life from his mastery of these four fields, and Strindberg has the custodians of them struggle to explain away this reminder that the end of life is nothingness itself.

The beasts of nineteenth-century novels were generally tangible—forces of nature, vices, machines, institutions. There was prostitution, alcoholism, and gambling; there were railroads, factories, and coal mines; and there was materialism, capitalism, and the big city. As terrifying and overwhelming as these things seemed, they could at least be named. But the beasts of the twentieth century would be far less identifiable, living in the mysterious realm of negativity we find in Conrad, James, and Strindberg. For them the void supplies the focus. Their characters seek meaning outside themselves—in a jungle, in a cemetery, behind a door—and find only the horror of nothingness within.

Positive negative time is silence. The recognition of its constituent role resembles the recognition of the constituent function of empty space in its various literary forms as darkness, emptiness, nothingness, the void. While creative silences were most explicit in poetry and music, they figured prominently in some prose works.

In the opening of *Silence* (1910) by the Russian writer Leonidas Andreiyeff, Father Ignatius and his wife are discussing their daughter who has just returned pregnant from a trip to St. Petersburg that she took against their will. They are chastising her mercilessly. She refuses to speak, and, after several days of brooding silence, commits suicide. The mother has a stroke that leaves her silent too. From the day of the daughter's funeral the house was silent. "It was not stillness," Andreiyeff explains, "for stillness is merely the absence of sounds: it was silence, because it seemed that they who were silent could say something but would not." Just as empty space became the focus of Archipenko's sculpture, so silence assumed the central role in this story. The house was filled with symbols of it. The wife did not utter a sound, the daughter's portrait seemed especially mute, and, after her pet canary flew away, the cage "kept silent" as a reminder of the emptiness. Every morning Father Ignatius sat and agonized in the silence of the house, and upon his return from work he always felt as if he had been silent the entire day. It enveloped him as darkness enveloped Marlow. Father Ignatius visited his daughter's grave and pleaded for some response to fill the void. She seemed to be speaking but with the same unbroken silence. "He fancied that the entire atmosphere trembled and palpitated from a resounding silence." The absence of sound became a presence that took a substantial form: "With icy waves it rolled through his head and agitated the hair." He returned home and begged his wife to break the terrifying quiet, but she remained silent as the story ends with them staring at each other "dumb and silent" in a "dark, deserted house."[96]

The Belgian mystic writer Maurice Maeterlinck wrote like the wailing of oboes. He explored the mysterious experiences of intuitions and supernatural occurrences, the inexpressible feelings and unconscious thoughts that course beneath the surface. In an essay on "Silence" he argued that we fear the absence of sound because it betokens death and therefore spend a good deal of time making senseless sounds. Many ordinary friendships or even loves are based on a common "hatred of silence." These negative feelings illuminate the importance of silence and point on the positive side to its binding and creative powers. Far from being the mere absence of sound, silences express what no words or sounds possibly could. The most memorable moments between lovers are made up of silences, and the quality of them reveals the quality of the love. "As gold and silver are weighed in pure water, so does the soul test its weight in silence, and the words that we let fall have no meaning apart from the silence that wraps them round." This revaluation of the relative importance of silence and sound suggests a broader kind of leveling in the social sphere that Maeterlinck elaborated explicitly. "The silences of a king or a slave in the presence of death or grief or love reveal the same features."[97] On the other hand, monarchs are announced by trumpet fanfares.

The silence of a generation that came home after four years of killing and discovered that nobody spoke the same language or felt the old feelings any more was one subject of Proust's novel. Forgetting, silence, and time lost are variations on the theme of negation. The novels of the nineteenth century were as vivid as Jean Valjean's flight through the sewers of Paris, as palpable as the Count of Monte Cristo's treasure. They revolved around great noisy events—war and revolution, crime and punishment. Even Flaubert—who claimed that he wanted "to write a book about nothing, a book without any exterior support, which would sustain itself by the inner force of its style . . . a book which would be almost devoid of subject, or at least in which the subject would be almost invisible"—even he structured *Madame Bovary* around the passionate outbursts of seduction and adultery, the agonized cries of the victim of a botched surgical operation, and the sounds of a grotesque suicide. But Proust centered his novel on the forgetting and remembering of tea and cakes and the lost time that that recollection enabled him to understand. He was a great architect of silence. The ting of a spoon striking a cup was one of the most significant sounds of his novel. Silences cast lovers into despair as forcefully as the whisperings of betrayal and the shouting of insults. It was Rachel's failure to write that tormented Saint-Loup: "Thus her silence did indeed drive him mad with jealousy and remorse. Besides, more cruel than the silence of prisons, that kind of silence is in itself a prison. An immaterial enclosure, I admit, but impenetrable, this interposed slice of empty atmosphere through which, despite its emptiness, the visual rays of the abandoned lover cannot pass."[98] The words "Mademoiselle Albertine has gone!" begin a volume of the novel that dwells exhaustively on someone who is no longer present. Proust explores the initial shock of Marcel's discovery that Albertine has left him and everything that follows—regrets over the way their affair had gone, fantasies about her feelings of loss, the transformation of his happy memories into bitter ones, the blow of learning of her subsequent death and the realization that it only intensified his jealousy, and eventually the indifference and forgetfulness—all triggered by her absence and the silence that settled into his life.

While empty space and silence were used as subjects of novels and short stories, in poetry there was a formal shift in the conception of the poem from an arrangement of words to a composition of words *and* the blank spaces between them. Already in the 1880s some French symbolists began to experiment with "free verse" stretched across consciously shaped white spaces on the page.[99] This technique was most fully developed by Stéphane Mallarmé, who used the blanks between words for a kind of visual pause to establish a rhythmic movement of words and images like notes in a musical composition.[100] He also believed that poetry should be evocative, urging, in an often-quoted instruction, "Paint not the thing, but the effect it produces." Once again the subject—the thing—lost its former prominence. As Braque toppled its pictorial authority by rendering the space around it with equal substance, Mallarmé diminished its literary authority by leaving it out of poems and creating verbal compositions out of its shadows and effects. In a lecture of 1895 he explained that the new poetry dispenses with precise descriptions and employs rather evocation, allusion, and suggestion. It makes "sudden jumps and noble hesitations" that hint at things and allow the reader to respond freely with his own imagery and associations. He challenged the older aesthetic that rested on the metaphysical assumption that "only what exists exists." (This was a literary analog of Geoffrey Scott's attack on the older architectural aesthetic that concerned itself only with "what occupies our tools and arrests our eyes" and failed to appreciate the central task of creating spaces.) Mallarmé held that the most important part of the poem may be what the poet has left out.[101]

Later in his life he developed a way to make the evocative nature of his poetry visible by representing the absences with empty spaces between words and between lines. These breaks symbolized the lacunae of sequential thinking, the gaps in human communication, the silence surrounding every utterance. As he explained: "The intellectual framework of the poem conceals itself but is present—is located—in the space that separates the stanzas and in the white of the paper: a significant silence, no less beautiful to compose than the lines themselves."[102] In one essay he insisted on the historical uniqueness of this kind of composition: "We must bend our independent minds, page by page, to the blank space which begins each one; we must forget the title, for it is too resounding. Then, in the tiniest and most scattered stopping points upon the page, when the lines of chance have been vanquished word by word, the blanks unfailingly return; before, they were gratuitous; now they are essential; and now at last it is clear that nothing lies beyond; now silence is genuine and just."[103] Blanks were indeed essential in his last poem, *Un Coup de dés* (1897), a final testament to the creative force of negativity.[104] The poem is extremely difficult to understand, and his statement of purpose in a preface written in prose suggests how important it was that his method be clearly grasped. There he explained that the white spaces "even out and scatter" the words across the page, make possible a simultaneous vision of the entire page, and indicate the rhythm of the lines so that the poem may be read like a musical score.

The poem itself was first published in a journal, but Mallarmé wanted it to be a separate book. It never appeared in that form during his lifetime, but before he died he did get so far as to correct galleys for the book, and they show the importance of the white spaces. His marginal notes directed the printer, who had taken some liberties with the original spacing, to move words and lines back to create the exact "measures" he intended. The publisher printed the cover on grey paper, but Mallarmé was emphatic about keeping it on the same white paper as the rest of the poem.[105] He insisted that it be printed across two pages, with the crack in the middle as an essential part of the spacing of the poem as a book. He carefully studied the printing types available and chose eight different ones to present the various voices within the poem. The unique typography drew attention to the printed surface and contributed to the visual unity of the entire page, with all elements playing an essential constituent role.

Although Mallarmé was particularly concerned about the visual presentation of the poem, he also cared about the way it sounded and read it first aloud to Paul Valéry, who recalled the extraordinary impact of that reading. Mallarmé read in a low, even voice, to let the words and pauses give the poem its full force without the theatrical ornamentation in vogue among professional speakers. Valéry heard "embodied silences," "whispers and insinuations made visible," and saw a new language that seemed to shine out of the paper like stars. In a literary fragment Mallarmé had compared the black-on-white of printing with the "luminescent alphabet of stars" on the "dim field" of the heavens. Valéry used that same image of the interdependence of black and white, remarking that in *Un Coup de dés* Mallarmé tried "to elevate a page to the power of the starry heavens."[106] Poetry had always been evocative and concise, but with Mallarmé the blanks assumed a more active role than the incidental, background function they formerly had.

In music silence is as essential to the recognition of sound and rhythm as the white of paper is to the identification of print. Throughout the history of music there had been significant silences, but they generally occurred at the end of movements and had a separating function. In the new music of this period the pauses occurred in the middle of sections and took on a more constituent function. Several critics noted conspicuous silences in Debussy, Stravinsky, and Webern, and indeed their music does contain some novel auditory negativities.

Just as Mallarmé was inspired by music, so Claude Debussy was inspired by Mallarmé's poetry, especially his *L'Après-midi d'un faune.* In 1893 he finished *Prélude à l'après-midi d'un faune*—a musical impression of the same theme. The notes of the flute solos sound like steps of the faun pacing, stopping, and starting again. There is one mysterious pause in the sixth measure, and other, subtler suggestions of the faun moving in and out of sight amidst the foliage, as in the poem, "an animal whiteness ripples to rest." In reaction against the massive and ornate orchestration of Wagner, Debussy keeps his score simple. Notes flash and fade like the light on the skin of a moving animal in the forest. The exchange of the melody line by different instruments creates the impression of disappearance as each becomes conspicuously silent. The explicit debt to Mallarmé indicates that Debussy intended the pauses or the suggestion of them to be as essential to the overall musical effect as the blank spaces were in the poem.

Archipenko identified a similarity between silences in music and the concavities and empty spaces of his own sculpture. He explained that rhythm in music is possible only if there is some kind of alternation between sound and silence. "Silence thus speaks. In the Ninth Symphony of Beethoven, a long pause occurs twice and evokes mystery and tension. The use of silence and sound in a symphony is analogous to the use of the form of significant space and material in sculpture."[107] His example from Beethoven implies that there was nothing new in the modern music, but there was. Composers began to use silences more consciously and more conspicuously than ever before. Roger Shattuck suggested that the conclusion to Stravinsky's *The Firebird* (1910) contains silences that are unique in musical composition. "The accumulated grace notes and syncopations and sforzandos lead up to silence—twenty-two measures of rests alternating with single percussive chords. And these stretches of silence are the most moving of all, as if all the earlier pages had to be composed in order to allow a few bars of peace to emerge at the end. Many have tried, but few composers have succeeded so well in inverting our conventions of hearing so that silence has more weight than sound."[108]

The most daring composer of negativities was Anton von Webern. The extreme brevity of his compositions (whole movements less than a minute long) echoes with all that is left out, and what can be heard is laced with frequent breathtaking silences. In the first nine measures of the *Passacaglia* (Op.1) there are as many pauses as notes. As Otto Deri explained: "The function of the pause is not that of a rest as used in common musical practice; the pause in Webern's music has a functional significance in the rhythmic scheme. There is in Webern's music a new relation between *sound* and *no sound*."[109] And even when notes are sounding there are suggestions of silence, as one instrument tosses the melody to another and begins a long rest. Musicians playing such works become intensely aware of those rests as they wait to take up the melody again. Webern rejected polytonality and composed symphonies with single notes that sound as if they were ringing in outer space, surrounded by the silence of the rest of the full orchestra for which they were written. The performance of the symphonies gives a sense of the orchestra as missing and of musicians listening, counting rests, and waiting to resume playing.

There is some exaggeration in the claim that the negativities are of equal importance with the subjects of art. Braque's painting was titled *Violin and Pitcher* and not *Space between Violin and Pitcher,* and the words of *Coup de dés* intrigued the critics far more than the spaces that Mallarmé put between them. A more tempered evaluation of the relation between the subject of perception and the background that frames it was made by a group of psychologists that emerged just before World War I. The pioneers of Gestalt psychology—Max Wertheimer, Wolfgang Köhler, and Kurt Kaffka—elaborated laws explaining how the "ground" and the "figure" create each other in perception, but they also maintained that the figure was more prominent. Their theory rejected the associationist view that complex perceptions are built up out of simple, discrete elements. They argued rather that perception is an experienced whole, and the task of understanding must be of that whole and not of separate parts. In a book on the visual appearance of figures of 1915, Edgar Rubin reproduced a drawing of the so-called "Peter-Paul Goblet," which may appear either as two faces in black staring at each other or as a goblet in white framed by two black shapes. It illustrates the interdependence of figure and ground and the way they may flip back and forth with shifts in the viewer's attention. Rubin maintained that the figure is more "striking and predominant" than the ground, but the ground does play an essential role.[110] In considering the whole perceptual field, the smallest detail of a Gestalt may be as important as the more conspicuous figures in it, for all elements interact and give each other meaning.

Insistence on the unity of the perceptual field accorded with the radical empiricism of William James. In *The Principles of Psychology* he illustrated the power of negativities in a discussion of the stream of thought or, in his specific example, of sound: "what we hear when the thunder crashes is not thunder pure, but thunder-breaking-upon-silence-and-contrasting-with-it."[111] The hyphens bridge the gap between words to illustrate the continuity of experience and reverse the analytical tradition in experimental psychology. The interdependence of sound and its absence is but one example of the mutual interaction of positives and negatives that make up our mental life. James also pointed out the constituent function of negativities in his brother Henry's essay *The American Scene.* In a letter of 1907 to Henry he wrote that his style was to avoid naming something straight out "but by dint of breathing and sighing all round and round it to arouse in the reader who may have had a similar perception already . . . the illusion of a solid object, made . . . wholly out of impalpable materials, air, and the prismatic interferences of light, ingeniously focused by mirrors upon empty space . . . your account of America is largely one of its omissions, silences, vacancies. You work them up like solids."[112]

A contemporary of William James perceived the leveling effect of positive negative space in James's thought. In 1914 Horace Kallen, a philosopher at the University of Wisconsin, wrote: "Pure experience has no favorites. It admits into reality . . . evil as well as good, discontinuities as well as continuities . . . James . . . is the first democrat of metaphysics." James refused to detest the material world as did the idealists—nothing was more or less real to him than anything else. He recognized "the democratic consubstantiality of every entity in experience with every other."[113]

Few of the developments we have surveyed had a direct relation to major political, social, or religious change. But the affirmation of positive negative space, the notion that what was formerly regarded as a void now has a constituent function, had one feature in common with the progress of political democracy, the breakdown of aristocratic privilege, and the secularization of life at this time: they all leveled hierarchies. Although the link between these two clusters of developments was rarely explicit, the thematic similarity is striking and makes the connection compelling. The challenge of this generation to the notion that the subject was more important than the background spread in ever widening circles to the notions that some people were more important than others in selecting political leaders, that aristocrats were entitled to social privileges and hereditary rights, and that the sacred space of religion was more important than all other "profane" ones. Most people continued to accept the old hierarchies and defer to rank, but there was nevertheless a significant change that affected many aspects of life and thought. It came even from those who, like Nietzsche, insisted that a nobility become worthy of its status by creating new forms out of the uniform dust of the universe. There were to be no special materials anointed by priests, no special classes ennobled by kings, no special individuals enfranchised by laws. There were to be only the special creations of artists out of simple materials for anyone who could appreciate them.

Some drew an explicit connection between the new sense of space and democracy. Turner saw the open space of the frontier as a force for democracy. William James's thought was characterized by

Horace Kallen as a democratic metaphysics, while James himself implied that his brother's interpretation of the American scene with its "omissions, silences, vacancies" captured something of its uniquely democratic spirit. George Mosse suggested that the "nationalization of the masses" involved the creation of an extended "living space" around monuments used by the masses, thereby denying the hierarchy so forcefully implied by the imposing granite or bronze of the monuments themselves. The architect Louis Sullivan envisioned a new "democratic" architecture that would challenge conventional design and create new structures appropriate to the antimonarchical modern ethos.

The most profound and troubling disturbance of traditional hierarchy occurred as a consequence of the secularization of life and thought. According to the concept of divine right, legitimacy to rule comes from God. In the course of the eighteenth century the rationale for monarchy began to change from divine right to the principle of popular sovereignty. As a result the court and aura surrounding Christian monarchs lost much of its mystique and "sacred" aspect and was replaced in popular imagination by the corridors of power of parliaments and congresses. Just as the decline of a Christian metaphysical framework transformed the sense of time, so did it affect the sense of space; and the setting for significant events in history shifted from the sacred spaces of heaven, the church, and the palace to the profane spaces of the battlefield, workshop, marketplace, and home.

Throughout the nineteenth century intellectuals and artists struggled to come to terms with an ever more secularized and "profane" world. Some, including Feuerbach and Marx, rejoiced that the loosening grasp of religion had at last made it possible to begin to create a city of man in an entirely human space. The critic J. Hillis Miller has analyzed reactions to the "disappearance of God" among five English writers who spanned the century—Thomas de Quincey, Robert Browning, Emily Brontë, Matthew Arnold, and Gerard Manley Hopkins. They found the prospect of such a disappearance intolerable and struggled to create a new relation between man and God. But it was a losing battle, and the last of them, Hopkins, sank into the despair of his final days, alone and impotent—"time's eunuch," straining to build "and not breed one work that wakes."[114] In 1882 Nietzsche's "madman" announced that "God is dead." But the madman was not mad. His elaboration of the consequences included a vivid picture of the new sense of emptiness in a world without holy sanctuary. "What were we doing when we unchained this earth from its sun? Whither is it moving now? Whither are we moving? Away from all suns? Are we not plunging continually? Backward, sideward, forward, in all directions? Is there still any up or down? Are we not straying as through an infinite nothing? Do we not feel the breath of empty space?"[115] Thus did Nietzsche metaphorically link positive negative space and the profanation of religious space by suggesting that the death of God had forced man to feel "the breath of empty space."

Not only directions in space but the values of the Western world lost their former inviolability with the collapse of traditional faith. The most material consequence of the loss was a blurring of the distinction between the sacred space of the temple and the profane space outside. "Profane" means "outside the temple," and many artists and intellectuals found themselves outside, not only wondering which way was up but also faced with the realization that there was no longer a temple to return to. In a world without God all men confront nothingness, and, as Nietzsche noted, most "would rather will nothingness than not will."[116] But some people avoided nihilistic despair and learned to create their own sanctuaries. This was to be the great creative effort of the overmen, the artists and intellectuals who affirmed life and learned to love their fate in the face of the void. If there are no holy temples, any place can become sacred; if there are no consecrated materials, then ordinary sticks and stones must do, and the artist alone can make them sacred. It is no accident that the leading architects of this period displayed the simple materials of wood, stone, brick, and glass and stripped away the façades and ornaments that had adorned sacred and royal structures of the hierarchical past.

New constituent negativities appeared in a broad range of phenomena: physical fields, architectural spaces, and town squares; Archipenko's voids, Cubist interspaces, and Futurist force-lines; theories about the stage, the frontier, and national parks; Conrad's darkness, James's nothing, and Maeterlinck's silence; Proust's lost past, Mallarmé's blanks, and Webern's pauses. Although these conceptualizations were as diverse as the many areas of life and thought from which they emerged and upon which they had influence, they shared the common feature of resurrecting the neglected "empty" spaces that formerly had only a supporting role and bringing them to the center of attention on a par with the traditional subjects. If figure and ground, print and blanks, bronze and empty space are of equal value, or at least equally essential to the creation of meaning, then the traditional hierarchies are also open to revaluation. Value was henceforth to be determined by aesthetic sensibility, public utility, or scientific evidence and not by hereditary privilege, divine right, or revealed truth. The old sanctuaries of privilege, power, and holiness were assailed, if not entirely destroyed, by the affirmation of positive negative space.

7

FORM

The very title of Walter E. Houghton's *The Victorian Frame of Mind 1830–1870* suggests that there was a definite "frame" to the mind of that period. In contrast to those who viewed the mid-Victorian age as one of complete certitude, Houghton shows that leading figures in England were engaged in profound questioning about the intellectual and moral basis of life. But for all their searching they still hoped to find such a basis, and he concluded that they looked at things from a single and unchanging point of view; they tended "to divide ideas and

people and actions into tight categories of true-false, good-bad, right-wrong; and not to recognize the mixed character of human experience."[1] The conviction that order underlay experience was tersely expressed by that quintessential Victorian, Samuel Smiles—"A place for everything, and everything in its place."[2] In their homes, in their minds, in their social lives all things had a proper place as they did in their desks with separate cubbyholes for everything.

In the face of such complacency about the secure scaffoldings of life and thought, a number of leading intellectuals all over Europe reacted to the breakdown of conventional forms. In 1905 Hugo von Hofmannsthal wrote that the nature of his epoch was "multiplicity and indeterminancy" and that "what other generations believed to be firm is in fact sliding."[3] In America, in the wake of the explosive Armory Show in New York in 1913, Mabel Dodge wrote that "nearly every thinking person nowadays is in revolt against something, because the craving of the individual is for further consciousness, and because consciousness is expanding and bursting through the molds that have held it up to now."[4] Musil characterized the change as one in which "sharp borderlines everywhere became blurred, and some new indescribable capacity for entering into hitherto unheard-of relationships threw up new people and new ideas."[5] What Musil embroidered into a novel, Simmel elucidated in a formal essay of 1914. "At present," he wrote, "we are experiencing a new phase of the old struggle—no longer a struggle of contemporary form, filled with life, against the lifeless one, but a struggle of life against form *as such,* against the principle of form. Moralists, reactionaries, and people with strict feeling for style are perfectly correct when they complain about the increasing 'lack of form' in modern life." Bergson's vitalism shattered the mechanistic framework of traditional thinking. In older philosophical systems, Simmel argued, there was "a fixed frame or an indestructible canvas" for thoughts and feelings but in Pragmatism truth became "interwoven with life." That same year, on the eve of World War I, Walter Lippmann surveyed the situation in America: "The sanctity of property, the patriarchal family, hereditary caste, the dogma of sin, obedience to authority,—the rock of ages, in brief, has been blasted for us."[6] Ortega observed that the goals that furnished yesterday's landscape with "so definite an architecture" have lost their hold. Those that are to replace them have not yet taken shape, and so the landscape "seems to break up, vacillate, and quake in all directions." And Yeats, elabo-

rating on the kind of apocalyptic imagery he used in "The Second Coming," recalled: "Nature, steel-bound or stone-built in the nineteenth century, became a flux where man drowned or swam."[7]

The most famous, certainly the boldest, claim about the general breakdown of forms was made by Virginia Woolf. She hazarded the assertion that "on or about December, 1910, human character changed." It was not quite so sudden as the laying of an egg, she conceded, but there was a dramatic change that could be seen everywhere, even in the kitchen. "The Victorian cook lived like a leviathan in the lower depths, formidable, silent, obscure, inscrutable; the Georgian cook is a creature of sunshine and fresh air; in and out of the drawingroom, now to borrow the *Daily Herald,* now to ask advice about a hat." The change affected relations between masters and servants, husbands and wives, parents and children. The prevailing sound of the age was "the sound of breaking and falling, crashing and destruction." Grammar was violated, syntax disintegrated, and Joyce's *Ulysses* was "the conscious and calculated indecency of a desperate man who feels that in order to breathe he must break the windows."[8] Nietzsche had predicted that the great new leaders would philosophize with a hammer, and Woolf insisted that she heard everywhere the sound of axes smashing. She found it "vigorous and stimulating," but others would find it troubling. As in the controversy about the new pace of life, there was only disagreement about the value of it all, for there was no question that the old forms of life and thought were cracking right down to their metaphysical foundations.

∞

If there is no clear distinction between the plenum of matter and the void of space and if matter may be conceived as a configuration of energy alignments, then the traditional understanding of matter as made up of discrete bits with sharply defined surfaces must also be rejected. That conception was undermined by a series of developments in electromagnetic and thermodynamic theory in the latter half of the nineteenth century. In 1896 Bergson surveyed several challenges to the corpuscular theory and argued that the division of matter into independent bodies with absolutely determined outlines is "artificial." He interpreted Faraday's theory of matter as one in

which the atom is a crossing point for lines of energy radiating throughout space: "Thus each atom occupies the whole space to which gravitation extends, and all atoms are interpenetrating." He attributed to Kelvin the view that there is a continuous, homogeneous and incompressible fluid filling space and that what we call an atom is "a vortex ring, ever whirling in this continuity."[9] Challenges to the corpuscular theory also appeared in the popular press. One article on "The Disappearing Line Between Matter and Electricity" argued that matter consists of moving electrical particles. A Michigan physicist dazzled readers of *The Popular Science Monthly* in 1906 with an article entitled "Are the Elements Transmutable, the Atoms Divisible, and Forms of Matter but Modes of Motion?" He answered yes on all counts and argued that the corpuscular theory had become the "phlogiston theory" of his age.[10]

The discovery of radioactive disintegration in 1896 also put in question the stability of matter. Particles of certain elements such as radium disintegrate spontaneously by throwing off energy and in time reveal an appreciable loss of mass. Novelists as well as physicists recognized that the old stable stuff of the universe was no more. The hero of H. G. Wells's *Tono-Bungay* (1909) invents a radioactive potion called "quap," which he bottles and sells as a wonder drug to a public easily manipulated by his advertising gimmicks. The basic elements were once regarded as the most stable things in nature, he noted, but these new elements are "cancerous" substances that live by destroying, and radioactivity is a "disease of matter" that spreads to everything around it. "Quap" (a play on "crap") is an epithet for modern values. "It is in matter exactly what the decay of our old culture is in society, a loss of traditions and distinctions and assured reactions. When I think of these inexplicable dissolvent centres that have come into being in our globe . . . I am haunted by a grotesque fancy of the ultimate eating away and dry-rotting and dispersal of all our old world."[11]

Einstein's relativity theory questioned the stability of all spatially extended forms. He introduced the special theory in an article "On the Electrodynamics of Moving Bodies" (1905) and argued that bodies do change their form when moving with respect to a stationary reference system. A rigid body that has the form of a sphere when viewed at rest will begin to assume an ellipsoid shape when viewed in motion, and all three-dimensional objects will "shrivel up into plane figures" when their relative velocity reaches the speed of light.[12] The general theory of relativity demolished the conventional

sense of stability of the entire material universe. Classical physics taught that all bodies are elastically deformable and alter in volume with changes in temperature. According to Einstein every bit of matter in the universe generates a gravitational force that accelerates all material bodies in its field and modifies their apparent size. There are thus no absolutely rigid bodies. Under these circumstances the grid of a Cartesian coordinate system is useless to plot movement. Einstein suggested using instead a nonrigid reference system which he called a "reference mollusc."[13] As a consequence, that ordered geometrical world, graphically represented in sharp squared Cartesian coordinates, became a complex and unstable world that could only be represented by the matrix of gummy reference molluscs that altered their forms continuously as they were accelerated by the varying masses and proximities of countless moving particles.

X-ray pierced the surface of the human body and other material screens. One commentator noted that it "subverted" all previous conceptions of the action of light by illuminating the interior of opaque objects, and with Edison's discovery of the fluoroscope in 1896 it became possible to view directly the interior of the human body in action.[14] The Futurists saw x-ray as but one more device that broke up the older forms they so detested. Boccioni asked, "Who can still believe in the opacity of bodies?" and he likened the penetrating capacity of x-rays to the "sharpened and multiplied sensitiveness" of Futurist art.[15] Thomas Mann explored its impact in *The Magic Mountain.* The doctor let Hans view his own hand through the fluoroscope screen, and as the x-ray "disintegrated, annihilated, dissolved" the flesh from his living body, for the first time in his life Hans understood that he would die.[16] Mann identified the response that many must have experienced as they peered into the interior of the human body for the first time and observed organs within the shell of flesh that the new technology had suddenly pierced.

The shell of buildings also opened up with the supporting steel frames, walls of glass, and electric lighting that made possible a new interpenetration of indoors and outdoors. The completion of the Eiffel Tower in 1889 announced a new era of construction with steel girders. Traditional distinctions between inside and outside were useless to describe this open structure. Visitors who descended the spiral staircase were outside but at the same time "within," as the continuous spiraling and spinning vistas of sky and surrounding houses produced the inside-outside sensation that Delaunay captured in paintings of the tower knitted into its environs.[17]

Walls had always supported buildings, but around this time architects began to build skyscrapers with weight-bearing infrastructures of steel girders. William Le Baron Jenny's Leiter Building in Chicago (1889–1891) was the first purely skeletal building without any supporting walls. One distinctive feature of the Chicago School of Architecture was the horizontally elongated window that fulfilled the lighting needs of the office buildings. Another was the exterior display of interior skeletal structure in contrast to the tradition of concealing structural forms with façades.

The walls of the Gallery of Machines at the Paris Exhibition of 1889 demonstrated the potential of glass as the Eiffel Tower demonstrated the potential of steel. Liberated from the need to construct solid supporting walls, architects could use sheets of glass to open interiors to the light and scenery of the outside world. The German Expressionist Paul Scheerbart in his *Glasarchitektur* (1914) evaluated its cultural impact. "Brick culture," he wrote, "is depressing." People are stifled in small, closed rooms and poisoned by the "brick bacillus" that he believed inhabited the old masonry. They must open up their lives and their thinking with glass architecture. Then natural light will penetrate interiors to brighten their moods and expand their consciousness. At night buildings will be seen from the outside as forms modeled in light that will illuminate the surrounding terrain, highlight mountains, and make night flying possible.[18]

Scheerbart's vision of a countryside lit up at night was one of the first explicit recognitions of the transformation in architecture brought about by electric lighting. As the American designer Morgan Brooks noted, architects were accustomed to see and conceive structures by daylight and only rarely considered how they would appear at night.[19] Reyner Banham identified some revolutionary implications of electric light: "The sheer abundance of light in conjunction with large areas of transparent or translucent material effectively reversed all established visual habits by which buildings were seen. For the first time it was possible to conceive of buildings whose true nature could only be perceived after dark, when artificial light blazed out through their structure."[20]

The breakdown of closed forms in American domestic architecture was largely the result of the determination of Frank Lloyd Wright, whose "organic style" called for a radical opening of interior living space. A common mid-Victorian home included a breakfast room, dining room, kitchen, pantry, master bedroom, separate children's rooms, maid's room, nursery, den, parlor, bathrooms, living room, sun porch, closets, and nooks set off by doors, halls, small passageways, and walls. Wright's first objective was to reduce the number of such separate parts and make a unified space so that light, air, and vistas permeated the whole. Secondly he sought to integrate the building with its site. The "prairie style" was developed as he flattened out and streamlined the high, bulky Victorian house to conform to the flat sites of the American midwest. Its long horizontal sweep was incorporated into the house with the horizontal lines of continuous windows and the broad eaves of gently sloping roofs. His many windows and exterior doors created the sense of "outdoor living" that became so popular in later years. He identified this motif as "the 'inside' becoming 'outside'."[21] Another objective was to eliminate the box-like room by making all walls enclosing screens and having the ceilings and walls flow into each other.

Although the lines of a prairie house streamed into the environment, Wright strongly believed that the home must be a shelter from the outside, and he concealed the exterior doors behind stone parapets and porch walls. His conviction that interior openness was only possible when the home was safely protected from outside intrusion was part of a broad reconsideration of the requirements for privacy and of the proper relation between public and private spheres.

A number of inventions pierced the shell of privacy. The invention of the microphone around 1877 made it possible for outsiders to listen to private conversations within rooms. Perhaps the first "bug" was made in 1881 by New York prison officials who concealed a microphone in a cell and overheard two inmates discussing a crime.[22] The perfection of dry-plate, fixed-focus photography by Kodak in the 1880s enabled amateurs and journalists to take instantaneous candid snapshots of people anywhere outside of studios without their consent. In 1902 the *New York Times* complained about the invasion of privacy from "Kodakers lying in wait" to photograph public figures.[23] In 1906 G. S. Lee viewed the invasion of the interior of a home by electric doorbells as part of a general internalization of life produced by the new technology. The modern man "abandons the glistening brass knocker—pleasing symbol to the outer sense—for a tiny knob on his porch door and a far-away tinkle in his kitchen."[24] While traveling in the United States, the English writer Arnold Bennett was overwhelmed by the "fearful universality" of the telephone. He found American cities "threaded under pavements and over roofs and between floors and ceilings and between walls by millions upon millions of live filaments that unite all the privacies of the organism—and destroy them in order to make one immense publicity." He also objected to European hotels in which the "dreadful curse of an active telephone" was installed in every room to invade one's privacy.[25] In a short story, "A Rural Telephone," mid-western farmers found the invasion of privacy compensated by the sense of community they got from listening in on a party line.[26] The Telephone Newspaper installed in Budapest in 1893 destroyed the privacy of strangers in hotels by broadcasting their identity and location to every listener each night. In 1912 a popular scientific writer complained that the private communication of the old Morse telegraph was lost with the new wireless messages that could be heard by anyone who wanted to "tap the ether."[27]

The growing intrusiveness of the public sector into private lives can be seen in the passage of legislation to regulate the size and design of buildings, restrict the content and location of advertising, and curb noise. In 1901 a historian of urban design, Charles Mulford Robinson, surveyed a number of rules, "nearly all of recent adoption," regulating the size and height of buildings. In 1887 municipal regulations in Rome established explicit standards that fixed a height limit in proportion to the width of the street. In Belgium, where there were no laws to impose aesthetic standards for buildings, a court decision on June 20, 1890, enabled one town council to demand "correspondence with the site." In Vienna there were new rules to regulate sanitation, the height of buildings, and the location of balconies, and to establish a "general harmony of appearance." In England the Society for Checking Abuses of Public Advertising was founded in 1893 to promote the "dignity and propriety of aspect in towns." Edinburgh, which already had a law prohibiting any advertisements having letters showing against the sky, passed a law in 1899 giving communities the right to say where all advertisements may be placed. In 1895 the Oeuvre Nationale Belge began a crusade to demand that advertisements be a decorative element of the business structure. Several societies for the suppression of street noise were also established. In 1899 the New York City Improvement Society reported that intrusive public noise made up a large majority of the complaints filed with it, and in Boston, Robinson noted, the City Music Commission was charged with seeing that the hurdy-gurdies and hand-organs were in tune.[28]

The invasion of private life is illustrated by the history of privacy legislation in America. E. L. Godkin, editor of the *New York Evening Post,* wrote in "The Rights of the Citizen—to His Own Reputation"

(1890) that Anglo-American law had always recognized the home as a sanctuary where one ought to be able to withdraw to an "inner world of thought and feeling."[29] But modern newspapers invaded it and made gossip into a marketable commodity. In the same year Louis Brandeis and Samuel Warren published a scholarly article on "The Right to Privacy" that elaborated Godkin's polemic into a landmark essay in the history of law. It was triggered by the harassment that Warren suffered in 1883 when a newspaper published lurid details about his married life. By way of introduction Brandeis and Warren traced the progressive recognition of ever more subtle and private injuries sustained by an individual from without, from the earliest laws that protected the individual against harm to body, life, and property to rules against verbal attacks, to copyright laws, and laws to shield reputations against slander and libel. But heretofore legislation focused on an external relation between the individual and the community, whereas Brandeis and Warren insisted that the law must recognize the legal status of the relation of an individual with himself and protect him against loss of self-esteem, against the infringement of privacy itself. "The intensity and complexity of life, attendant upon advancing civilization have rendered necessary some retreat from the world," but recent inventions such as instantaneous photography and newspaper enterprise have invaded the "sacred precincts" of private and domestic life. Man has become more sensitive to publicity precisely in the measure as the techniques for producing unsolicited publicity have become more lucrative. The article concluded with an appeal for legislation to recognize the right to privacy and to make violations of it a criminal offense.[30]

In 1891 the United States Supreme Court held that a person suing a railroad for personal injuries could not be ordered to submit to a surgical examination to substantiate the claim.[31] It is understandable that the first judicial recognition of anything like a right to privacy should have been applied to the more obvious right to the privacy of one's body. In 1902 Abigail Roberson sued the Rochester Folding Box Company for using her photograph without her consent to publicize a product. A lower court ruled in her favor, but the Court of Appeals of New York overturned that decision. This negative verdict inspired an enraged article in the *New York Times* on behalf of the right to privacy, and in the following year the New York State legislature passed a law making it a misdemeanor to use someone's photograph or picture for commercial purposes without first obtaining written consent.[32] In 1905 the Georgia Supreme Court decided on behalf of a man who sued an insurance company for using his photograph for advertising purposes without his consent. This was the first time a court recognized the right to privacy, and its award of damages on the sole ground that this right had been infringed became a legal precedent.[33]

A deeper penetration of the private world by outside forces and an intensification of efforts to protect that world was evident also in psychiatry and literature. Freud concluded that many of his patients suffered from a privatization of older communal celebrations and religious rites. In the modern age public beliefs and public ceremonials were being replaced by the private rituals of neuroses.[34] Individuals were internalizing former public norms and as a consequence were bearing an ever growing load of repression. The progress of civilization itself, he repeatedly stressed, required ever more instinctual renunciation, and that was accomplished as parents, with increasing concern, passed prohibitions on to their children, who internalized them through their developing intrapsychic agencies of censorship and guilt. In therapy Freud encountered growing resistance as his interpretations began to penetrate his patients' defense structures and reveal the unconscious contents of their dreams and symptoms. It was, he said, like a child who clenches his first all the more tightly as a parent tries to open it and see what he is holding. At the broad social level Freud observed people and institutions encroaching upon the private life of individuals, and at the same time he saw people holding tightly to an increasingly intense private world manifested in the growing severity of neuroses.

The intensification of the private sphere also took place in the novel. "The nineteenth-century novel," David Daiches observed, "was anchored in a world of public value agreed on by reader and writer, and its plot pattern was determined by changes in fortune and status on the part of the principal character." People revealed themselves and developed in a public arena of shared principles. They may have clashed with those principles but nevertheless remained defined by them. In *Middlemarch* gain or loss of money and reputation regulated much of the action. In Thomas Hardy's novels marriage, profession, class, and reputation shaped the lives of the characters. According to Daiches, Conrad was the first modern novelist in England, because his finest works were concerned with situations to which public codes were inapplicable. Marlow's journey in *Heart of Darkness* reveals "the inadequacy of all public formulations of human standards and human motives."[35] The assignment of the anarchist hero in *The Secret Agent*—to blow up the Greenwich Observatory—was a repudiation of the public world and the public time that made the coordination of public life possible. In the modern novel outward action does not reveal the essential nature of characters, and public gestures are misunderstood or not understood at all. The private world of Leopold Bloom, instead of collapsing in the face of the assault from without, was galvanized to activity and became the location of the most significant events of the novel. As the public became more intrusive, the individual retreated into a more strongly fortified and isolated private world. That is why we can observe in this period both a greater interpenetration and a greater separation of the two worlds.[36] As we shall see in considering the breakdown of urban-rural, social, and political forms, other seemingly antithetical developments energized each other and occurred in dialectically related pairs.

Throughout history the increase of population has tended to enlarge the size of cities, but a limit was always imposed by the distance workers could travel to work on foot or by horse. With the growth of railways and then of the street railway systems in the late nineteenth century, that limit was extended to include the new "streetcar suburbs."[37] They sprang up around cities all over Europe and the United States and broke down the older spatial forms—whole regions were suddenly abandoned, country villages were inundated by commuters, new neighborhoods sprawled beyond the old city boundaries. A new life style accompanied the new dimensions of the city as suburban housing made it possible for large numbers of people to combine the virtues of urban and rural life. Responses to this phenomenon were divided between tradition-minded observers who bemoaned the loss of rural privacy and those who saw the suburbs as a corrective to both the isolation and provincialism of rural living and to the crowding and corruption of the city.

In 1893 the American science-fiction writer Henry Olerich explored the favorable consequences of the extended city in a utopian novel about a visitor from Mars who comes to earth and tells about his own "cityless and countryless world." In a preface Olerich assessed the benefits that had already resulted from the breakdown of the line between the two realms on earth. Mental activity was "bolder, broader and freer," and misunderstandings and fighting between city and country dwellers were less frequent. The visitor explained to the earth dwellers how the Mars people eliminated urban crowding and crime and the slow, wasteful labor of the isolated farmer by distributing their population over the land to maximize the advantages of both the cooperation that requires close interaction and the independence that requires sizable living space.[38] In 1906 a popular American writer observed how telephones and telegraphs, rural free delivery, and improved roads were mixing city and country life. The expansion of the feeder railroad lines linking small towns was making it possible for workers to commute from the suburbs, and he predicted that soon rural and urban types would "blend into the suburban."[39]

While the suburbs brought city ways to the country, the "garden city" brought the country to the city. In a book of 1898 the English city planner Ebenezer Howard proposed to combine the high wages and employment opportunities of towns with the beauty, low rents, and spacious living of the country by building garden cities.[40] In 1899 Howard founded the Garden City Association and in 1903 the first English garden city was created at Letchworth. His influence quickly extended to the continent, where a number of garden-city movements were formed in the early twentieth century. In 1896, before Howard published his garden-city idea, the German planner Theodor Fritsch proposed building cities "not from the inside out, but rather from the outside in." He conceived of a series of concentric rings zoned only for special kinds of buildings and with a large garden built in the center.[41] The high point of this longing to bring together city and country was the fantastic vision of the German Expressionist architect Bruno Taut. In *Alpine Architektur* (1917) he proposed using electric lighting and glass architecture to transform an entire chain of mountains in northern Italy into a landscape of glittering shrines and crystal-lined valleys. Three years later in *Die Auflösung der Städte* (The Dissolution of Cities) he announced the impending death of those "great spiders"—the old cities.[42]

Some social forms also gave way. Around the turn of the century the division of society into classes became particularly sharp at the lower end of the social hierarchy as the working classes began to come into the political process, but at the same time traditional rigidities loosened at the boundary between the aristocracy and upper middle classes. Once clearly defined by law and preserved by heredity, the aristocracy lost its ability to serve as an exclusive ruling elite and was forced to unite with the upper middle class. The most fervent lamentations about the erosion of class lines were concerned with this particular frontier of the social hierarchy, as many nobles were forced to go into "bourgeois" enterprises or marry for money to preserve their life style.

In 1912 Arthur Meyer, editor of the Parisian high society newspaper *Le Gaulois,* bemoaned the erosion of rigid class lines of the old salons: "Democracy, by breaking down all distinctions, has done away with the barriers which for centuries had guarded the old social hierarchy, and today our salons at their best have little individual character and at their worst are all exactly alike."[43] Snobbery was certainly not unique to this period, but its strongly defensive tone was. Meyer's counterpart in the United States, the racist sociologist Edward Alsworth Ross, blamed the replacement of private cabs by streetcars for a loathsome mixing of upper and middle classes. He also faulted thirty-four American cities for providing in 1911 more than eighteen million free public baths. The flood of democracy ran in public water supplies, contaminating the upper classes with the detritus of the mob. If the trend continues, he warned, "the effect will be a narrowing of the esthetic space between those with position and those without."[44] A British observer, Philip Gibbs, saw the breakdown of classes and neighborhoods as part of a collapse of several traditional forms. The new man has no fixed beliefs, he argued, and everywhere one can see the destruction of old certainties. There is no longer a place for "a class with well-defined boundaries, dividing it from people of poverty on one side and people of wealth on the other." Suburbia has become a "great straggling territory" inhabited by all sorts of people. Modern restlessness has penetrated homes "like microbes through open windows" breeding chaos in the families within.[45] Coming from an upper-middle-class Jewish family and aspiring to be accepted among the highest circles of the French aristocracy, Proust hovered about the frontier between these two classes, precarious as that social no-man's-land had become around the turn of the century because of anti-Semitism and class prejudice brought out by the Dreyfus affair. In the novel, Marcel aspired to be admitted to the exclusive domain of the Princess de Guermantes, but, when he finally arrived, discovered that it had undergone a profound transformation. "A certain complex of aristocratic prejudices, of snobbery, which in the past automatically maintained a barrier between the name of Guermantes and all that did not harmonise with it, had ceased to function. Enfeebled or broken, the springs of the machine could no longer perform their task of keeping out the crowd; a thousand alien elements made their way in and all homogeneity, all consistency of form and color was lost."[46] The imagery of snobbery repeatedly involved the penetration of "pure" classes by foreign elements and the breakup of distinct social strata and rigid class lines.

In Austria class boundaries at the bottom as well as on the upper levels were weakening. Carl Schorske has reconstructed this sweeping transformation. The multinational Habsburg Empire was being pulled apart politically by nationalist movements and socially along class lines as the workers, lower middle class, and peasantry began to challenge a ruling class that was itself an unsteady amalgam of a declining aristocracy and an insecure liberal middle class that wanted to assimilate into the nobility. During the last five years of the nineteenth century the government attempted to hold the empire together by various methods, including promoting artists whose commitment was to the universality of the empire and to the social respectability and stabilizing function of culture. But the structure began to come apart around the turn of the century as the ruling class lost its hold. Schorske traces this spectral social and cultural disordering in Austrian drama, city planning, architecture, psychiatry, and art, culminating with the "explosion in the garden" of rationality with Kokoschka's tempestuous painting and Schoenberg's rejection of tonality that had served as the structural center of music since the Renaissance.[47]

National boundaries themselves had become more porous so that travelers crossed them with exceptional ease. Developments in transportation generated a freer movement of people across national lines. The airplane pierced the wall of frontiers and wiped out the military significance of fixed fortifications. Passports were generally unnecessary. A number of European countries instituted them during the French Revolution, but in the course of the nineteenth century they were gradually eliminated. England, Norway, and Sweden never had them at all. France abolished them in 1843, and except for a brief interval during the war of 1870–71, required no passports from anyone. In 1861 Belgium abolished all passport requirements for travelers entering or leaving her borders. Spain followed in 1863, Germany in 1867, and Italy in 1889. When the German government of Alsace-Lorraine instituted a passport requirement in 1888, there was a violent outcry. One critic protested that passports were inhuman and made it impossible to make surprise journeys or emergency visits. Their institution by the German authorities, he wrote, was an act of "iniquity and torture that put the Alsaciens and Lorrainers beyond the pale of humanity."[48] Widespread protest obliged the German government in 1891 to eliminate passport requirements in Alsace-Lorraine for everybody except French officers and Germans who had not served in the army. The French instituted a similar regulation in 1912 to control travel of foreign military officers, but until the outbreak of war in 1914 most of the countries of western Europe had limited or no passport requirements and allowed almost everyone to travel freely. Stefan Zweig's nostalgia for prewar Europe included memories of the ease of travel in a world without the kinds of traumas experienced at borders during the Second World War. Before 1914 he journeyed to India and to America without a passport and without ever having seen one. "The frontiers," he recalled, "were nothing but symbolic lines which one crossed with as little thought as when one crosses the Meridian of Greenwich."[49]

In addition to the concrete changes that resulted from developments in science, technology, law, and politics, there was also a broad cultural re-examination of older spatial forms in art and theater and of the conceptual forms of philosophy, psychology, and physics. In *Principles of Art History* (1915) the German aesthetician Heinrich Wöfflin identified pairs of contrasting concepts to describe the shifting "apprehensional forms" of art. The pair of open and closed forms, he wrote, is most sharply contrasted in the classical and baroque style. The classical is dominated by horizontals and verticals and contains objects with sharp outlines, firmly grounded in space, harmoniously enclosed by the picture frame. In the baroque style curves and diagonals predominate, and objects and figures are grouped asymmetrically and extend or point beyond the frame. Wöfflin did not treat contemporary painting, but his concepts were well suited to describe its distinguishing characteristics, as forms opened and closed like valves in the engine of modern art.[50]

The Impressionists dissolved forms. They painted blurry "impressions" of objects modified by changing light and atmospheric conditions—drifts of fog, shimmering forest shadows, the glow of gas lamps on rainy streets. Cubists cracked the mirror of art. In their paintings objects open into surrounding space and none has an uninterrupted outline. Parts are broken off, colors bleed into neighboring objects, and translucent facets of space with multiple light sources cut shadows across bounding surfaces. They removed sections of faces and reassembled what remained to create grotesque open forms in defiance of natural appearance. In *Les Demoiselles d'Avignon* the five bodies are from left to right ever less sharply contoured, as though Picasso were giving a step-by-step demonstration of how to dismantle the closed human form. In Braque's *Houses at l'Estaque* (1908) walls fade into vegetation, branches puncture roofs. In *Girl with a Mandolin* (1910) Picasso disjointed the right elbow like a mad surgeon grafting chips of bone onto chunks of space. By 1911, when Picasso and Braque working side by side produced almost identical portraits—*Ma Jolie* and *Le Portugais*—the subjects were alike broken into geometric compositions barely recognizable as people. These portraits proclaimed the triumph of open forms floating in unframed space. As Gertrude Stein explained, in Cubism "the framing of life, the need that a picture exist in its frame, remain in its frame was over."[51]

Of all the changes we have observed, the Cubist assault on the closed form was the most graphic and most significant of this period. It was more than a shift in artistic style such as the one from Realism to Impressionism. It involved a transformation in the very purpose of art from the interpretation of optically perceived reality to the creation of an aesthetically conceived one. The Cubists discovered that they could, and must, deform objects and modify local color in deference to artistic sensibilities alone. If the visual movement of a composition required that an elbow open into the space around it, then they cut it open. For them the breakdown of the closed form was a declaration of independence of art over visual appearance. Their fracturing of objects and splicing them into space can be interpreted as a repudiation of the older conventions that separated subject and background as well as those that insisted that the artist defer to the appearance of objects in reality. The cracked elbow of Picasso's mandolin player broke down the distinction between positive and negative space, proclaimed the autonomy of the artist, and put visual reality in a sling.

Critics came forth in defense of the picture frame and of geometric forms in art, but this served mostly to emphasize the trend to the contrary. In an essay of 1902, "The Picture Frame," Georg Simmel insisted that the essence of a work of art is its self-containment, which requires that it be clearly and completely enclosed in a frame. "The character of a picture frame is to augment and intensify the inner unity of a picture." It leads the eye toward the center and sets off the painting from all extraneous distractions.[52] Three years before the first Cubist painting was exhibited, the French philosopher Paul Souriau published an impassioned defense of "geometric beauty" in an aesthetic treatise on *La Beauté rationnelle.* He argued that rational forms and structures, even rhythms, are proper subjects of art because they are inherent in nature and man and therefore aesthetically satisfying in art. Our *instinct de géomètres* is manifested in the most basic creations of the human mind—buildings, gardens, and decorations—and on a higher level with regular, simple, proportioned forms in art.[53] Simmel and Souriau pleaded for the preservation of traditional, regular forms as if the political and social stability of the world were threatened by the open forms of modern art.

The Futurists blew forms open like blast-furnace doors. In "Technical Manifesto" Boccioni described how "our bodies penetrate the sofas upon which we sit, and the sofas penetrate our bodies. The motor bus rushes into the houses which it passes, and in their turn the houses throw themselves upon the motor bus and are blended with it." Boccioni portrayed this intermingling in *The Noise of the Street Penetrates the House* (1911, Figure 9). The scene is a construction site where buildings *are* open forms, but with Cubist-style fragmentation Boccioni also opened up the houses that surround the site. In line with the title Boccioni graphically depicts the penetration of noise into someone from the house by continuing the outline of a staircase under construction into the left shoulder of a woman standing on a balcony in the foreground, camouflaged against a riot of shapes. Poles of the building under construction stick into her head like Chinese hairpins, and a horse races out of her right hip. The material world penetrates her body as sounds and images penetrate her consciousness. Boccioni explained that the Futurists "shall henceforth put the spectator in the center of the picture."[54] The painting achieves this integration of spectator and spectacle in the figure of the woman and the two observers on balconies flanking her, visually integrated, almost lost, in the world outside. Boccioni elaborated his conception in two painted studies for a sculpture of 1911–12, *Fusion of Head and Window*. One shows a seated woman pierced by sections of a window and by the prismatic effects of light passing through different angles, shapes, and thicknesses of glass. Inside and outside, separated in traditional art by the window, here intermingle in accord with the Futurist design of showing, as Anton Bragaglia expressed it, "the subtle ties which unite our abstract interior with the concrete exterior."[55] In a second study of 1912, ironically titled *Matter*, the woman was dematerialized even further by flickering planes of light reflected in and through sections of glass.

Working with normal three-dimensional matter Futurist sculptors also pierced the shell of conventional closed forms. Again Boccioni announced their purpose: "Let's turn everything upside down and proclaim the absolute and complete abolition of finite lines and the contained statue. Let's split open our figures and place the environment inside them. We declare that the environment must form part

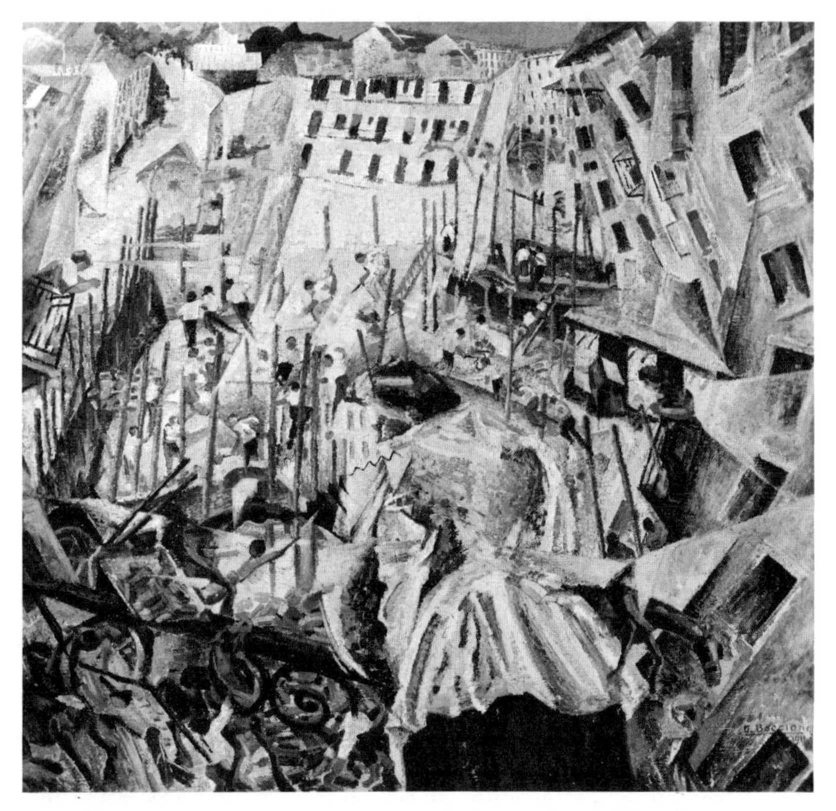

Fig. 9. Umberto Boccioni, *The Noise of the Street Penetrates the House,* 1911.

of the plastic whole, a world of its own, with its own laws: so that the pavement can jump up on to your table, or your head can cross a street, while your lamp twines a web of plaster rays from one house to the next."[56] He attempted to achieve an interpenetration of the traditionally separate elements of base, sculpture, and surrounding space in *Development of a Bottle in Space* (see Figure 7). The bottle emerges from the base as if it were cut out of a block on a potter's wheel, developing in rotation before our eyes. No clear line separates base from bottle, but the whole does not achieve the "complete abolition" of the contained statue. It opens up the form of a conventional bottle but closes around the fractured shapes. Boccioni overstated his objective, since the material substance of any sculpture must have a discrete and continuous bounding surface, however irregular. But the objects depicted in this manner, like the bottle, seem to be open and suggest, even if they do not completely achieve, the plastic value of the environment.

In the theater several technological developments opened new possibilities. The invention of the revolving stage in 1896 by Karl Lauterschläger quite literally revolutionized stage design and, although not widely used at first, suggested new possibilities for quick set changes. The versatile framing of the cinema suggested ways to break up the fixed proscenium of traditional sets, and the replacement of gaslight by electric light in the 1890s made it possible to illuminate different parts of the stage in the course of a single scene and vary the color and intensity of light to create changing chiaroscuro effects. Loie Fuller exploited the new lighting techniques by projecting beams of colored light onto herself as she danced. For a dance called "Radium" light flashed on her costume painted with a luminous substance to give off a phosphorescent glow. One reviewer wrote of a performance in 1900: "Light came from every side. La Loie danced upon glass, from which the vivid splendor of the headlights was reflected, while from the wings, stage and orchestra, wonderful luminous streams seemed to flow toward her."[57] While light played on her body, she also twirled colored banners attached to sword-length poles to extend her body into space. The combination of light and flags created luminescent force-lines, as though a Futurist painting had come to life.

Some daring playwrights and designers tried to eliminate the sharp separation between stage and audience. A precedent for this went as far back as the chorus in classical drama that frequently spoke across the "curtain." In the eighteenth century Rousseau protested the theater's artificial separation between actor and audience, but he had little impact on actual staging. Generally when an actor had something to say to the audience, he had to pretend to be talking to himself, locked in soliloquy. Nineteenth-century dramatists continued to separate these two realms by a wall—an "invisible" wall, but one not to be breached by either actor or spectator.

German and Austrian dramatists took the lead in piercing the fourth wall. The Reformbühne company that opened in Munich in 1889 made a first step by extending the stage into the audience.[58] Then Adolph Appia systematically undertook to transcend the separation between actor and audience. For the staging of some Wagner operas in 1891 and 1892 he raised the house lights and lowered those downstage to soften the transition between the traditionally dark auditorium and the bright stage. He also advocated extending the width of the proscenium to be flush with the side walls in order to eliminate the conventional peephole effect of the stage opening and provide a smoother transition from house to stage.[59] In 1908 a designer for the Shakespeare Theater in Munich explained that he had actors enter on ramps run through the audience to create a greater sense of "organic unity."[60] In 1905 Max Reinhardt brought the performers out from the audience over bridges and ramps, and in 1911 he staged *The Miracle* in a theater that he converted into a cathedral. Actors planted in the audience led crowd episodes and organized religious processions from among the crowd.[61] Open-air theaters, such as the "mountain theater" in the Harz that opened in 1907, were also intended to bring actors and audience closer in a natural setting by reducing the artificial separation between the raised and lighted stage and the fixed rows of seats in a darkened theater.

In a parallel development the emergence of the cabaret in New York after 1910 broke down barriers of class, sex, and race that had dominated the entertainment industry. Previously the upper classes had their theaters, concerts, and operas, while the popular classes enjoyed minstrelsy, variety, and burlesque. By the 1890s movies, vaudeville, and ragtime began to attract people from upper as well as lower classes. The first important cabaret established beyond the vice districts in New York was the Folies Bergère, which opened in 1911 on Forty-ninth Street in the heart of Broadway, and other cabarets quickly invaded New York night life. The layout of the cabaret and its entertainment tended to blur the traditional sharp separation between audience and performer. The stage was as close to the audience as possible, generally surrounded by tables. It was on the same level as the audience, and there were no curtains or footlights to accent the difference between actor and audience. Cabaret performers themselves sometimes dined at the surrounding tables, then rose from among the customers to do their routine, and between numbers patrons were encouraged to dance on the stage-turned-dance-floor. The seating also encouraged fluidity. Dining, drinking, and talking went on while performances were in progress; and the mobility of the chairs allowed customers to create seating arrangements to suit various needs, thereby highlighting what Lewis Erenberg has called "the anarchistic possibilities of the entire room." Many revue numbers encouraged interaction between performers and audience. Chorus girls would tousle men's hair, bump their chairs, or pull them on stage for a dance. In 1912 the editor of a New York society scandal magazine protested that the cabaret "removes the barrier and places the actors in the assemblage, and this familiarity unquestionably has broken down all fortifications of conventionality."[62] His fortress image suggests both the strongly defensive posture of the upper classes and the strength of the conventions that held the older hierarchical society together—and its components apart.

The Futurists also sought to remove the conventional "fourth wall," which they regarded as symptomatic of an antiseptic and moribund society that denatured art. In a manifesto of 1913 Marinetti announced a new theater in which the audience would participate so that the action would develop "simultaneously on the stage, in the boxes, and in the orchestra." Four years later he elaborated on their intentions: "Symphonize the audience's sensibility by exploring it, stirring up its laziest layers with every means possible; eliminate the preconception of the footlights by throwing nets of sensation between stage and audience."[63] During the war they began to stage plays that wove audience and actors into a single theatrical matrix. Part of the action came from performers scattered throughout the audience in Ettore Petrolini's *Radioscopia* in 1917. In Francesco Canguillo's *Lights!*, published in 1919, the audience, inspired by actors planted in their midst, was to perform the entire play. From a darkened theater these actor-spectators were to begin demanding that the lights be turned on, provoking everyone else to join in the protest. When the turmoil peaked the lights were to come on and the curtain to fall.[64]

Making his contribution to the upheaval, Strindberg transformed theatrical forms on stage, spiriting characters about in time and space to replicate the wild gyrations of the dreaming mind. In a preface to *A Dream Play* he explained his method: "In this dream play, as in his former dream play *To Damascus*, the Author has sought to reproduce the disconnected but apparently logical form of a dream. Anything can happen; everything is possible and probable. Time and space do not exist; on a slight groundwork of reality, imagination spins and weaves new patterns made up of memories, experiences, unfettered fancies, absurdities and improvisations. The characters are split, double and multiply; they evaporate, crystallise, scatter and converge."[65] In the first half hour there are nine scene changes. Props and sets are transmuted along with the dimensions of time and space. Backdrops rise or separate to reveal new interior spaces, characters age in a few seconds and exit straight through walls, and a bud atop a castle bursts into a giant chrysanthemum. In successive scenes a tree becomes a hat rack and then a candelabrum; a stage door becomes a filing cabinet and then a door leading into a church sacristy. Strindberg somewhat overstated the accomplishment of the staging, because time and space did not altogether cease to exist. But he did ignore the uniform frameworks of public life and reproduced the malleable frameworks of the private experience of the individual. Without any direct influence, Strindberg's statement of purpose echoed the account of the "primary process" of dream work that Freud presented in *The Interpretation of Dreams,* published just one year earlier. These two pivotal works from widely differing fields showed that the processes of mental life cannot be enclosed in the rigid conceptual framework of traditional psychology nor dramatized convincingly within the rigid unities of traditional theater.

Beside breaking old forms, painters and musicians as well as playwrights found it necessary sometimes to reach out beyond the confines of their respective genres for effective expressive techniques. The term "synesthesia" began to appear in psychiatric literature in the 1890s to describe a sensation, such as a color, produced by a stimulus, such as a sound, generally associated with another part of the sensory system.[66] The idea was not new. The Romantics associated painting and music, and Baudelaire used such "correspondences" in a poem of that title that subsequently became an inspiration for the Symbolists. Wagner sought to achieve sensory correspondences of light and sound in his creation of the *Gesamtkunstwerk,* and Appia tried to stage them for him. Odilon Redon called himself a *peintre symphonique,* and around 1905 the mystical Lithuanian musician M.K. Ciurionis turned to painting to depict colored compositions conceived as symphonic movements.[67]

In one exploration of this cultural exogamy a group of Futurists assigned musical qualities to colors. They constructed a "chromatic piano" that flashed colored lights in sequence triggered from the keyboard, and they made lasting records of these "color symphonies" with cinematography.[68] Carlo Carrà attempted to paint the impression of sound in *What the Streetcar Said to Me* (1911). The forms appear as if chipped out of a transparent solid, revealing facets of noise from the city and the jolting streetcar. Carrà explained Futurist synesthesia in a manifesto of 1913 on "The Painting of Sounds,

Noises and Smells." They will use screaming greens, brassy yellows, and "the rrrrrreddest rrrrrreds that shouuuuuuut"—all the "colors of speed" that suggest carnivals, fireworks, and singing. They will paint the plastic equivalent of the noises and smells of cinemas, brothels, ports, railway stations, hospitals.[69] These Futurists ignored the conventional compartmentalization of each sense and insisted that the mind was capable of creating new aesthetic forms by combining the various receptive capacities of the sensorium.

The most basic requirement of the plastic arts, an identifiable subject, also began to disappear in the early twentieth century as several painters and sculptors experimented with abstraction. In a history of modern art Werner Haftmann concluded that "the entire human structure of the decade—the malaise that sprang from an increasing doubt as to the reality and solidity of the visible world—sought a kind of redemption in abstract painting."[70] The great breakthrough was made by the Russian painter Wassily Kandinsky (living in Munich), who in 1910 produced the first completely nonrepresentational painting. Splotches of color ran through gelatinous globs, like beetles scurrying through the ruins of civilization in the last days of mankind. Although Cézanne turned a bottle into a cylinder, and Gris turned a cylinder into a bottle, both wound up with some recognizable form. Kandinsky began and ended up with mere colored forms, representing on canvas nothing but his inner spiritual life. He saw his work as the triumph of art over the tyranny of the external object, as he explained in his book, *Concerning the Spiritual in Art* (1911). The contemporary age, he wrote, is a "nightmare of materialism." Art imitates nature, the spirit is muted, and modern man is seldom capable of feeling emotions. But painters can retrieve the inner spirit and give it some objective representation by utilizing the natural association between colors and spiritual states. His book is a catalog of synesthetic techniques, a guide to the spiritual significance of colors and forms. Colors suggest sounds, odors, and moods. Some colors are rough and sticky, others are smooth and uniform. They may also be associated with the class structure: "green is the 'bourgeoisie'—self-satisfied, immovable, narrow." In accord with his view that the style of painting is a reflection of the spirit of an age, he linked his painting with the imminent "dissolution of matter." As the structure of matter collapses, artists will be liberated from the obligation to represent things in nature and be free to paint their inner feelings. To achieve this they must break down the barrier between the senses and depict the smells, sounds, impressions, and moods associated with different colors.[71]

Conventional forms of thought also came to be questioned. In one of his bitter shots at academic philosophy, Nietzsche caricatured the reaction to Kant's announcement of his discovery of a moral faculty in man: "The honeymoon of German philosophy arrived. All the young theologians of the Tübingen seminary went into the bushes—all looking for 'faculties'."[72] His remark applied to those philosophers and psychologists who regarded the human mind as a cluster of faculties, operating independently like so many booths at a fair. By mid-century phrenologists had identified thirty-seven of them and assigned a precise anatomical location for each. In addition to the classical faculties of perception, memory, imagination, and will, they found such discrete faculties as secretiveness, conscientiousness, and tune. Philoprogenitiveness was located at the back of the brain wedged between adhesiveness and combativeness. In the 1880s, when Bergson and James began to argue that mental life was a flux with no sharp conceptual or operational boundaries, one of their targets was this kind of ossified faculty psychology. Another was Francis Herbert Bradley's monism.

James spearheaded the attack on Bradley's dismissal of time and change as mere appearances and on the "block universe" of his rigid systematic philosophy. For James only the diversity and movement of experience was real. Modern thinkers should follow not the neat categories and sequential steps of traditional systematic philosophy but the unpredictable and irregular contours of experience. The stream of consciousness cannot be measured in pailfuls, and every single experience has fringes that flutter beyond any single conceptual category. "When we conceptualize we cut out and fix, and exclude anything but what we have fixed, whereas in the real concrete sensible flux of life experiences compenetrate each other." Although in *Principles of Psychology* he used the faculties of traditional psychology as chapter headings, he argued in those chapters that the life of the mind cannot be divided into separate conceptual compartments. Traditional categories provide at best polarities between which run the currents of character, as experience "breaks into both honesty and dishonesty, courage and cowardice, stupidity and insight, at the touch of varying circumstances."[73]

Other old conceptual walls between self and world were pulled down by Husserl and Freud. Husserl rejected the Cartesian notion of a preexisting reality out there waiting to be discovered and argued instead that all consciousness must be "consciousness-of" something. By 1913 he elaborated his revolutionary theory than an act of consciousness and its object are but subjective and objective aspects of the same thing.[74] The immersion of consciousness in the world was Freud's contribution to the new thinking. In an essay "On Narcissism" in 1914,[75] he analyzed the relation between the ego and the external world among "primitives," children, and psychotics (people suffering from what he originally called "narcissistic neuroses"). Among primitives the border between the ego and external world is confused. They believe in the omnipotence of thoughts and words, overestimate the power of wishes and mental acts, and use magic to manipulate the world. A normal infant begins life in a condition of "primary narcissism" in which all psychic processes are bound up with an ego that is undifferentiated and unrelated to anything in the external world. This original narcissism lays a foundation that continues to shape the child's development. A normal person tormented by pain generally gives up interest in others, withdraws libido from love-objects, and reinvests it in his own ego. Those whose lives are a source of continual anguish may withdraw from others completely. The resulting "narcissistic neuroses," Freud believed, detach people from all human relations. This "secondary narcissism," a turning away from object relations, involves at a deeper level a confusion between the self and the world, and in psychotic states of megalomania, paranoia, and hypochondria it obliterates the independent self. People thus emerge out of a normal state of primary narcissism and remain in touch with the external world as long as gratifications draw them toward others and keep them from slipping back into the slough of secondary narcissism.

Freud also believed that the direction of instinctual life may shift between the self and others in normal psychological processes. In "Instincts and Their Vicissitudes" (1915) he explained that the same instinct may be directed at the self or another person to the same psychological effect. Sadism and masochism are, aside from the source and aim, identical. The exhibitionist gratifies his desire to see his own body even as he exhibits it to others. "In these examples," Freud wrote, "the turning round of the subject's self and the transformation from activity to passivity converge or coincide."[76] These versatile mental processes can best be observed in dreams or psychotic symptoms when rational "corrections" are reduced to a minimum and the mind is darting about the universe of experience, mixing self and world in accord with the original narcissistic and undifferentiated state from which it emerged.

While the conceptual forms of thought in philosophy and psychology were being recast and rejected, physics was experiencing similar transformations. In 1907 Einstein theorized that light propagated across a gravitational field is curved. Since the trajectory of light is the shortest distance between two points and the basis for all measurement, his theory altered the very conception of space itself. He concluded that by slowing down the speed of light gravity warps space and time. He also hypothesized that gravity is not a force but an intrinsic curvature of a space-time continuum. Planets move in orbits or "geodesics" around the sun not because the sun attracts them but because in a universe with gravitational force there are no straight lines.[77] The following year, the German physicist Hermann Minkowski speculated on the fate of space and time as separate dimensions of experience. Drawing on Einstein's theory he announced that "henceforth space by itself, and time by itself, are doomed to fade away into mere shadows, and only a kind of union of the two will preserve an independent reality." He suggested calling that union a "world-line," which he described as the "everlasting career" of a point as it exists across what was formerly regarded as the independent dimension of time. All events should be conceived in a four-dimensional continuum represented by coordinates x, y, z, and t, which are to be understood as the same kind of units, not entirely spatial or entirely temporal, not distances or durations but space-time intervals.[78] Eight years later Einstein summarized the shift in world view. "It appears therefore more natural to think of physical reality as a four-dimensional existence, instead of, as hitherto, the *evolution* of a three-dimensional existence."[79]

It was then, and still is, extremely difficult to conceptualize the world as a space-time continuum. The theory broke down the distinction between age-old categories that lay at the foundation of Western thought. The nineteenth century had had trouble enough assimilating Kant's theory that the forms of time and space were determined by the mental apparatus of an observer. But Kant never suggested that these forms were interchangeable, and neither did Newton and the generations of classical physicists who conceived of a universe of masses extended spatially in three dimensions, traveling along the qualitatively different dimension of time. Wyndham Lewis accused Bergson of having "put the hyphen between Space and Time,"[80] but even in the great wash of *durée* they retained their identity as separate dimensions. The two concepts were unified in a single conceptual unit for the first time by Einstein and Minkowski.

∞

No age can sustain continual crisis. The rebuilding of bone cells begins within minutes after a break, and reconstructive processes are at work in the mind of the most crazed psychotic even during moments of panic and acute disorientation. The crises of the generation at the turn of the century were also part of an essentially constructive process, as the most daring innovators put a crowbar to the ironwork of traditional forms to clear the way for all the rebuilding ahead.

While the most sensational accounts emphasized the resulting breakdown, the driving energy was creative. Discoveries about the porous nature of atoms arose out of an effort to understand the composition of matter, and field theory was an attempt to make sense out of the tangle of paradoxes and inconsistencies that resulted from trying to explain light with the old ether theory. For all its interpenetration of inside and outside, the Eiffel Tower was a solid construction, a symbol of the stability of new materials and techniques. Wright's prairie style was intended to make buildings conform to human needs and harmonize with the site. Builders created suburbs not out of any perverse pleasure in forcing people to abandon old neighborhoods but because there was a market for suburban homes among consumers who demanded the advantages of both country and city life. Painters abandoned traditional forms to defend the autonomy of art, as the motto inscribed on the House of the Vienna Secession in 1898 proclaimed: "To the age its art, to art its freedom." Early twentieth-century artists did not reject all forms, only those that stifled their freedom. Picasso did not open up the elbow of the girl with the mandolin to do violence to traditional artistic forms but to arrange the forms in a way that did not do violence to his aesthetic sensibilities.

If the driving impulses for all this were positive and creative, why did general assessments of the period such as those by Hofmannsthal, Musil, Simmel, Ortega, and Woolf emphasize the "sliding," the "bursting," the "lack of form," and the sound of "crashing and destruction"? One reason is that sensationalism attracts critics and historians and makes engaging reading. Picasso's art did scrap the old forms, and his disciples together with his critics were perhaps at first struck more by that than by his construction of new forms. A second, more important, reason was that the old forms were deeply ingrained and required energetic exorcising. Iconoclasm had to be emphatic if it was going to be effective.

The worm of convention was most deeply embedded in the social structure. Deference to rank and hierarchy had been the foundation of European life. The anxiety generated by the growth of democracy and by the erosion of social privilege had its analog in the anxiety generated by the breakdown of old cultural forms. The new technology intensified both forms of anxiety. It cracked open conventional forms of doing work and conducting social relations and broke up conventional habits. Although many dramatic cultural changes came from pressures and problems that arose from within the various disciplines, technology provided a compelling material force for altering old patterns.

The telephone, first used by the rich,[81] soon became a democratic instrument, leveling class lines and binding nations in a single electronic network. Journalists armed with cameras could invade secluded abodes and expose private affairs in their tabloid articles, like the one that angered Warren and Brandeis. The new cinema was a uniquely democratic art form. While the theater was relatively expensive and could not reproduce itself, the technology of the cinema filled hundreds of movie houses with the same big picture for vast working class audiences. Compared to the theater, the cinema was not only far more accessible but, more importantly, enabled all of its viewers to see anywhere that a camera could see. Metaphors reach across hierarchies and link unequals. The powerful metaphorical technic of cinema dramatized the inequalities of old hierarchies and the waste of obsolete conventions. The newsreel, invented around the turn of the century, threaded knowledge across class and national frontiers.

The social and political significance of the cinema was explained in an article entitled "A Democratic Art," that appeared in *The Nation* in 1913. Cheap seats, all at the same price; a wide range of subjects; and its appeal to "all nations, all ages, all classes, both sexes" made the cinema a truly popular art form. In the early nickelodeons of New York that showed silent movies for a nickel, workers from all countries, even those who did not speak English, could mix with the upper classes in the dark with an unprecedented degree of social proximity and intimacy. The cinema makes "a direct and universal appeal to the elementary emotions" and allows everyone to be a critic, as "the crowd discusses the technique of the moving-picture with as much interest as literary salons in Paris or London discuss the minutiae of the higher drama." D. W. Griffith claimed that his stories and his heroes were all democratic. "Are we not making the world safe for democracy, American Democracy, through motion pictures?" he asked. The intrusive camera eye penetrates social barriers, spreads knowledge, shatters superstitions, and extolls "ordinary virtues." In 1918 Herbert Francis Sherwood speculated that cinema is "the language of democracy which reaches all strata of the population and welds them together."[82]

The American architect Louis H. Sullivan argued that modern architecture, if it be true to the spirit of its time, must be democratic. In a series called *Kindergarten Chats* (first published in 1901, revised in 1918), Sullivan characterized modern architecture in political terms, insisting that the "decoys of vestiture" must ultimately give way to a true democratic architecture that will "pierce all feudal screens." With an angry, vivid imagery suggestive of Nietzsche, Sullivan castigated lazy historicists who consult old books for ideas and "chew this architectural cud for a stipend," in "times when the old values must be revalued." In America there is currently a struggle between "aspirant Democracy and the inherited obsession of feudalism." Democracy will dissolve old obstructions and unite the forces of nature with the needs and aesthetic values of the contemporary era. While Sullivan did not detail what specific structures or styles were democratic, he emphasized democracy's function to "liberate, broaden, intensify, and focus every human faculty."[83]

The traditional world was rooted in conventions that dictated how an individual should experience his own self, other people, and objects in the world. It was hierarchically ordered with a proper place for everything. Nobles were superior to bourgeoisie, and the middle to the working class, and everyone was supposed to know his social station and fulfill its duties. Atoms were solid and discrete; space was elastic and inert. The surface of the skin concealed the secrets of the heart, and skeletons became visible only after death. In the home every person had his proper room and corresponding roles, and inside and outside were securely an unambiguously divided by solid walls. No telephones existed to enable one to enter a house in an instant without even being in it. Everything had a separate nature, a correct place, and a proper function, as the entire world was ordered in discrete and mutally exclusive forms: solid/porous, opaque/transparent, inside/outside, public/private, city/country, noble/common, countryman/foreigner, framed/open, actor/audience, ego/object, and space/time. These old scaffoldings had supported the way of life and culture of the Western world for so long that no one could recall exactly how they all started or why they were still there, and it took a generation of restless scientists, artists, and philosophers to dismantle them and begin the great Dionysian "yea-saying" to life and art that Nietzsche had envisioned in his final days of lucidity. The breakdown of old forms, together with the affirmations of perspectivism and of positive negative space, leveled hierarchies. These changes in the experience of space contradicted the notion that convention and habit could dictate privileged points of view, places, or forms. Rather, all had to be tested in the processes of life, selected by the eye of an artist, reconstructed in accord with current values and needs.

8
DISTANCE

On October 2, 1872, some members of the exclusive Reform Club of London were playing a game of whist in front of a fire and debating the chances of a bank robber's making a successful escape. The engineer, Andrew Stuart, maintained that they were good. One of the directors of the bank, Gauthier Ralph, disagreed. A description of the robber had been sent to agents all over the world, and so, "Where could he possibly go?"

Stuart replied, "I don't know about that, but the world is big enough."

"It was once," said another gentleman in a low voice.

Ralph agreed and pointed out that one could now go around the world "ten times more quickly than a hundred years ago."

Another member reported that a new section of the Great Indian Peninsular Railway had opened recently, and according to a calculation in the *Morning Chronicle,* a man could, if he made all the right connections, travel around the world by rail and steamer in eighty days.

They continued to debate the feasibility of actually making such a trip, when, without interrupting his play, the soft-spoken man who first alluded to the shrinking size of the earth brought the discussion to a head around 7:00 P.M. by wagering the others £20,000 that he could tour the world in eighty days or less. He proposed to leave on the Dover train at 8:45 that night, but first insisted on finishing the hand. At 7:25 Phileas Fogg, having won twenty guineas, left the Reform Club and prepared to begin his tour of the world with his French servant Passepartout.

The hero of Jules Verne's classic, *Around the World in Eighty Days* (1873), embodied all the self-assuredness and extravagance of the British empire. The source of his wealth was a mystery, and he used it to indulge tastes acquired during extensive travels. The ice at the Reform Club was "brought at great expense from the American lakes" to keep his drinks fresh. He was polite, taciturn, and punctual—a perfect gentleman whose life style would have pleased Samuel Smiles. He awoke, dressed, breakfasted, and continued his routine throughout the day according to a strict timetable—habits that served him well in the course of his race around the world. His elegant home at Saville Row was outfitted with the most precise new electric clocks and had electric bells and speaking tubes to facilitate instantaneous communication between master and servant. The dash of his successful journey appealed to the masses who quickly made the book an international bestseller. With Fogg and Passepartout they could vicariously race against time and conquer space.

The book was a mixture of fact and fantasy, a compendium of global travel that was actually taking place, and an inspiration for others to follow. Although Verne never admitted to having had a specific model for his hero, two men might well have been. In 1870 the Boston businessman George Francis Train almost toured the world in eighty days. Two years later another American businessman, William Perry Fogg, published a book called *Round the*

World that chronicled a slower 1869 tour. It is most likely that Verne got his eighty-day timetable from one that appeared in the *Magasin pittoresque* in March of 1870, made after the completion of the Suez Canal, the American Transcontinental Railway, and the Trans-Indian Peninsular Railroad.

The fame of Verne's novel made eighty days the time to beat. The first to do it was the American journalist Nellie Bly, who in 1889–90 took seventy-two days. George Train got it down to sixty days in 1892.[1] With improvements in travel, scheduling facilitated by the introduction of World Standard Time, and the invention of the telephone and wireless, the time required to circumnavigate the globe was progressively reduced, and by the turn of the century the hope was to cut Fogg's time in half. When officials from several European countries and China met in 1902 to schedule a railroad journey from Paris to Peking, they also announced that they had "resolved the problem of traveling around the world in forty days."[2] The novel projected a new sense of world unity that became ever sharper in the decades that followed as the railroad, telephone, bicycle, automobile, airplane, and cinema revolutionized the sense of distance.

Railroads were not new, but around the turn of the century their hold on political, military, economic, and private life tightened as the railroad networks thickened.[3] In the metaphorical language of the naturalistic novel, the railroad took on sinister meanings. In Zola's *The Human Beast* (1890) "it was like a huge body, a gigantic being lying across the earth, his head in Paris, his vertebrae all along the line, his limbs stretching out into branch lines, with feet and hands in Le Havre and the other terminals."[4] A stranglehold of iron tentacles on farmers is the central image of Frank Norris's *The Octopus* (1900), and the entire San Joaquin Valley was in its grip: "From Reno on one side to San Francisco on the other, ran the plexus of red, a veritable system of blood circulation, complicated, dividing, and reuniting, branching, splitting, extending, throwing out feelers, offshoots, taproots, feeders—diminutive little bloodsuckers that shot out from the main jugular and went twisting up into some remote county, laying hold upon some forgotten village or town, involving it in one of a myriad branching coils, one of a hundred tentacles, drawing it, as it were, toward that center from which all this system sprang."[5] The railroads ended the sanctuary of remoteness. Wheat farmers were sucked into the mainstream of national and international markets by railroads that united the land masses to sea lanes in a single commercial unit.

In the complex interaction between need and technological invention, it is difficult to identify one or the other exclusively as causal. The railroad responded to economic need and in turn had an enormous impact on economic life. In a similar manner electronic communication related to the creation of worldwide markets. The telegraph, like the railroad, was a product of the early nineteenth century, but the telephone was unique to the last quarter of the century and greatly expanded the range, mobility, and contact points between which messages could be sent, drawing millions of people into an instantaneous communications network. The French were suspicious of the telephone at first and by 1898 had only 31,600 telephones. The more enterprising British had 600,000 in operation by 1912 when the government took control of the system. In Germany there was faster growth. In 1891 there were 71,000 telephones in use, and 1,300,000 before the war. According to one report, over 2.5 billion separate calls were placed in Germany in 1913.[6] The telephone was most enthusiastically received in the United States, where there were 10,000,000 in operation by 1914. Assuming the same ratio of calls per telephone and doubling the figure to allow for the two parties, it can be estimated that in 1914 telephone lines in the United States were used approximately 38 billion times.

Bismarck was the first political leader to grasp the value of long-distance telephone communication. He saw its unifying potential as early as 1877 when he ran a line over 230 miles between his palace in Berlin and his farm at Varzin. By 1885 there were 33 cities linked directly with Berlin. The Russians were quick to perceive the telephone's military use; in 1881 they established a commission to explore the potential and three years later opened a direct line between Moscow and St. Petersburg. The first international line was set between Paris and Brussels in 1887, and overseas telephony began in Europe with the laying of the first underwater telephone cable between England and France in 1891. Long distance service in the United States was inaugurated with the opening of the New York-to-Chicago line in 1892, and by 1915 the first coast-to-coast line opened between New York and San Francisco. The "annihilation of distance" was not a science-fiction fantasy or some theoretical leap of physicists; it was the actual experience of the masses who quickly became accustomed to an instrument that enabled them to raise money, sell wheat, make speeches, signal storms, prevent log jams, report fires, buy groceries, or just communicate across ever increasing distances.[7] An article of 1905 reported that "there are sections where chicken stealing has become a lost art, because the rural telephones make it possible to block every avenue of escape as soon as the crime is discovered."[8] Chicken thieves or Verne's hypothetical bank robber found that the annihilation of distance just increased the space crooks had to go to make their getaway.

Telephony expanded lived space too. "The telephone changes the structure of the brain. Men live in wider distances, and think in larger figures, and become eligible for nobler and wider motives,"[9] wrote one social observer. Conversations whet the appetite for visits. Proust felt both a sense of closeness and separation from telephone conversations. In a letter of 1902 he related how he had spoken to his mother just after she lost her parents, and "suddenly her poor shattered voice came to me through the telephone stricken for all time, a voice quite other than the one I had always known, all cracked and broken; and in the wounded, bleeding fragments that came to me through the receiver I had for the first time the dreadful sensation of all things inside her that were forever shattered."[10] In *Remembrance of Things Past* he translates this experience into a conversation between Marcel and his grandmother and speculates on the essence of telephone communication. It is an "admirable sorcery," which brings before us, "invisible but present, the person to whom we have been wishing to speak, and who, while still sitting at his table, in the town in which he lives . . . in the midst of circumstances and worries of which we know nothing, but of which he is going to inform us, finds himself suddenly transported hundreds of miles." The telephone operators are "priestesses of the Invisible" who bring us the sound of "distance overcome." But the voice of his grandmother also gave Marcel a premonition of an "eternal separation." Her voice was not only distant, but also cut off from the rest of her being—the movements of her body, her facial expressions, the sounds and smells and touches that had completed the *mise-en-scène* of prior conversations. This truncated, abstract individual became a symbol of his own isolation and of the eternal separation that awaits everyone in death.[11]

The distance was also useful to buffer the intensity of face-to-face encounters. Telephone conversation allowed men to take liberties with operators that they would never think of taking in person.[12] Men courted women on the telephone; some even proposed marriage.[13] The telephone also made possible a dispersion of business activity. While in the mid-century company offices were usually located on the premises of factories, around the turn of the century they began to move into urban centers. Central offices began to cluster in New York in the 1880s and in London in the early twentieth century.[14] Telephones also enabled businessmen to get out of cities and run their affairs from the country. An article in *Scientific American* in 1914 titled "Action at a Distance" pointed to the problem of urban congestion and looked forward to a time when the telephone and a picture phone (yet to be invented) would allow business to be conducted at a distance from cities.[15] Telephone connections could have a dual impact. They could disperse business from single-trade neighborhoods or concentrate it. The telephone permitted businessmen to buy and sell from afar without leaving their offices, and at the same time expanded their "territory" and forced them to reach out further. It brought people into close contact but obliged them also to "live in wider distances" and created a palpable emptiness across which voices seemed uniquely disembodied and remote.

The bicycle created another set of distancing effects. In 1893 a new American bicycle cost between $100 and $150. In 1897 it averaged around $80, but by 1902 mass manufacturing had driven the price down to anywhere from $3 to $15, well within the range of the working classes. The "bicycle boom of the 1890s" enabled more and more people to travel freely.[16] In *Voici des ailes!* Leblanc's hero Pascal expressed his exhilaration over the new command of space: "The stifling limits of the horizon are destroyed and nature is conquered."[17] Guardians of morality warned that women were taking advantage of the new mobility afforded by the bicycle and more of them were venturing farther from their homes unchaperoned. The bicycle also bridged social space. One snob writing in the *Fortnightly Review* in 1891 conceded that it was possible for an individual to cycle "without imperiling his or her social status." But the peril was implied with his reassurance that in Dublin, where it was customary for everybody to cycle together, the upper classes "never feel their gentility offended by breathing the same air as the shop-boy from the neighboring grocers."[18] An article in the *Minneapolis Tribune* in 1895 was more positive about the "most democratic of all vehicles" that allowed all ranks to entertain themselves in the same fashion. Another journalist was carried away with praise for the "revolutionary" impact of the bicycle—that "great leveller" and instrument of social equality.[19] Joseph Bishop praised the bicycle for compressing social distances and making it easier for the sexes to meet on a common ground. Bishop also noted the expanded radius of travel and the growing popularity of suburban tours that were "unconsciously but surely bringing city and country into closer unions." His observation that the bicycle had expanded the radius of the social circles from two or three miles to ten or twenty was echoed in similar assessments (for longer distances) about the automobile after the turn of the century.[20]

One early motor-car enthusiast, Alfred Harmsworth, explained in 1902 how the automobile was going to extend social life in the country by making it possible to visit someone 25 miles away without having to arrange a change of horses as would have been necessary ten years earlier.[21] Before the automobile, anyone living more than 8 or 10 miles away was "beyond calling distance." Recalling his childhood, Siegfried Sassoon wrote: "Dumborough Park was 12 miles from where my aunt lived . . . My aunt was fully 2 miles beyond the radius of Lady Dumborough's 'round of calls.' Those 2 miles made the difference, and the aristocratic yellow-wheeled barouche never entered our white unassuming gate."[22] Stefan Zweig observed that "The mountains, the lakes, the ocean were no longer as far away as formerly; the bicycle, the automobile, and the electric trains had shortened distances and had given the world a new spaciousness . . . Whereas formerly only the privileged few had ventured abroad, now bank clerks and small trades-people would visit France and Italy."[23] In *La Morale des sports* (1898) Paul Adam elaborated on the new automobilism. Driving builds skills that require sustained attention and quick responses over large distances. The ease and frequency of traveling engenders an exchange of ideas, stimulates the intellect, breaks up prejudices, and diminishes provincialism.[24] Writing in that same year Henry Adams brooded about the conquest of space as well as time by automobiles that could unite the widely scattered monuments of the past such as the cathedral at Chartres and enable him to pass "from one century to another without a break."[25] Proust made a similar observation. Before the war many people had speculated that speedy railway travel would kill contemplation, but, he observed, "the car has taken over its function and once more deposits tourists outside forgotten churches."[26] However, Proust preferred the railroad. Travel by car was "more genuine" because it allowed one to remain in closer intimacy with the earth and traverse space as a continuum, while trains jumped from station to station and made the difference between departure and arrival as intense as possible. Railway travel was like a metaphor in that it "united two distant individualities of the world, took us from one name to another name."[27] Train travel was one clue to the recapture of time lost. His most famous definition of art centered on the way metaphor united the "distant individualities" of experience: "Truth—and life too—can be attained by us only when, by comparing a quality common to two sensations, we succeed in extracting their common essence and in reuniting them to each other, liberated from the contingencies of time, within a metaphor."[28]

In Verne's story the earth remained the same size throughout the duration of Phileas Fogg's tour; in Proust's, distances contracted and expanded through the mediation of consciousness. Verne used technology to get his hero around the world in eighty days, while Proust used it to illustrate the kinds of spatial and temporal leaps that the mind makes uncontrollably and incessantly. Fogg could plan when he would depart and make connections en route, while Marcel had to wait for involuntary memories to start his journeys and then watch the myriad places in his memory rush past in a flash. Ortega commented on this unique sensitivity to space when he wrote that Proust had invented "a new distance between ourselves and things"—a distance modified by attention and imagination, by love and desire.[29] In the forty years that elapsed between the publication of Verne's travel story and the publication of the first volume of Proust's novel, the sense of distance in literature had been transformed by more than new means of transportation and communication. For Proust, as for Joyce, travel took place in the mind as much as in the world, and distances depended on the effect of memory, the force of emotions, and the passage of time.

The cinema made two unique contributions to the sense of distance in the visual arts with the close-up and the quick cut. The first dramatic use of a close-up was in the last scene of Porter's *The Great Train Robbery* (1903), when the screen is filled with the gun barrel of an outlaw firing point-blank at the audience—a symbol of the explosive impact of the entire medium as well as this technique. Five years later Griffith used the close-up in an adaptation of Tennyson's *Enoch Arden* about a man marooned on an island. In one scene Griffith filled the screen with the face of Annie Lee brooding about her long-lost husband and immediately after cut to her husband alone on the island. The combination of the intimate close-up and quick cut across the ocean shocked the early cinema audiences, but the scene was effective and illustrated how to create new distances and the emotions suggested by them.[30] In 1916 Hugo Münsterberg interpreted the close-up as a device of the cinema to recreate the way the mind selects important objects and fixes on them. If a camera points out that a locket is hung on the neck of a stolen child, it is not necessary to tell the audience that everything will hinge on it when the girl grows up. The close-up increased the size and significance of certain objects, made the audience feel closer to them, and gave film makers one more technique to render the varying lived distances of experience.[31]

While the close-up created intimacy, the quick cut created a dramatic sense of separation and of distance spanned. In an article of 1907 the French critic and novelist Remy de Gourmont announced his passion for the cinema because it enabled him to "tour the world," experience in a flash the radical contrast of desert and ocean, the unpredictable juxtaposition of widely divergent surroundings and their associated emotions. It was a "grand magic lantern" that showed him the most remote and inaccessible spots on the earth. In 1913 another French critic emphasized the great number of places that he had been able to visit so effortlessly on "cinematographic excursions."[32] Münsterberg offered a penetrating interpretation of the historical significance of the unification of distant places through movies. "Events which are far distant from one another so that we could not be physically present at all of them at the same time are fusing in our field of vision, just as they are brought together in our own consciousness . . . Our mind is split and can be here and there apparently in one mental act."[33] In a sequence of shots the cinema was able to approximate the power of the mind to conjure up and juxtapose numerous memories set in the different places where they occurred.

While Proust used technological analogies to illustrate his method of metaphor, the Futurists used technological metaphors to illustrate their method of analogy. In a manifesto of 1912 Marinetti insisted that the modern writer must create a new language and make extensive use of analogy—"the deep love that assembles distant, seemingly diverse things." In the world of the airplane, wireless, and cinema nothing is so distant or diverse as to be beyond analogical connection with something else.[34] With an attack on the old syntax Marinetti appealed for a truly contemporary literary style that he characterized as "Imagination Without Wires—Words-in-Freedom." The modern writer must "cast immense nets of analogy across the world. In this way he will reveal the analogical foundation of life, telegraphically, with the same economical speed that the telegraph imposes on reporters and war correspondents in their swift reporting." Marinetti mixed arguments as recklessly as he mixed metaphors—the innovation of a concise, telegraphic style was certainly not compatible with the extravagant and enigmatic analogies he called for in this manifesto. But he believed that the speed of communication between distant places also called for poets to "weave together distant things with no connecting strings by means of essential free words."[35] While the lack of connecting strings made wireless communication more versatile, the lack of logical connections made Futurist writing more muddled, although it seemed to them to be the only kind of writing that reflected the analogical foundation of life.[36] Other critics were not as insistent about the essential function of analogy but saw connections nevertheless between the new command of distance and modern sensibilities. In 1914 Alexander Mercereau, a promoter who brought several Cubist exhibitions to eastern Europe, wrote: "Our own age, too—an age of dynamism and intense drama, of electricity, motor cars and great factories, in which new inventions rapidly proliferate—multiplies our faculties of seeing and alternately contracts and extends the scope of our sensibilities."[37] The cultural record is replete with such observations of the connection between the new technology, new ways of seeing, and a new sense of distance.

In addition to the unique spatialities created by technology, two major socioeconomic developments also affected the sense of distance between people and between nations, and two social sciences sharpened their analytical tools for evaluating their significance. Sociology investigated the proximities of crowded urban life, while geopolitics studied the vast expanses and separations of imperialism.

Despite the loss of 30 million people to emigration, the population of Europe from 1890 to 1914 increased from 370 to 480 million. The expansion of suburbs, facilitated by the new technology, was necessitated by the growing population and the growing concentration of people in cities. The most rapid urbanization was in Germany, where, between 1880 and 1913, the number of cities with populations of over 100,000 increased from fourteen to forty-eight. Even though French population remained almost unchanged—38.3 million in 1891 and 39.6 million in 1911—by 1914 Paris and its suburbs reached over 4 million. That year the population of London was a crushing 7.5 million. In Vienna, "many people were forced not only to let all their spare rooms, but also to rent bed space to *Bettgeher,* who enjoyed no privileges whatsoever in the apartment, not even the use of any closet space."[38]

Sociology is the geometry of the social sciences, the one most concerned with space. Its central analytical impulse is to understand the spatial distribution of social forms. After thirty years of historical orientation under the aegis of evolutionary theory, sociologists in the 1890s became increasingly interested in the structure of societies; and the most pressing problem of the period, the one on which modern sociology cut its teeth, was the problem of "the crowd."

In 1890 the French sociologist Gabriel Tarde published the results of ten years' research on *The Laws of Imitation.* He assumed that ideas are transmitted from person to person and from generation to generation by imitative diffusion. He observed that there has been a "lessening of distances through more rapid means of locomotion" and an "increasing density of the population," which have facilitated the process of imitation.[39] In 1893 Durkheim introduced an analogous notion of contracting social distances to explain the division of labor. He wrote: "If we agree to call this relation and the active commerce resulting from it dynamic or moral density, we can say that the progress of the division of labor is in direct ratio to the moral or dynamic density of society." The diminution of the "real distances" between individuals in society had been produced by a rise in population, the formation of cities, and an increase in the number and rapidity of the means of communication and transportation.[40]

In 1893 the Italian sociologist Scipio Sighele identified collective psychological processes that occur in crowds. Like Tarde and Durkheim, Sighele noted the growing density of population and consequently of crowds, but he rejected any spatial reductionism and insisted that special psychological characteristics are operative among individuals in large groups that cannot be explained by the objective social circumstances of crowding alone. The "collective soul" of a mass of people is very different from a simple sum of individuals. It involves a leveling of individuality and the emergence of instinctive over "more cultivated" character, bringing forth aggression and crime, wild bursts of irrationality, and slavish devotion to a leader.[41]

Two years later the French sociologist Gustave Le Bon announced in his classic, *The Crowd,* that it was "the era of crowds." He was particularly alarmed by the craze of popularity for General Boulanger, who in 1886 began to incite mass demonstrations against the government of the Third Republic, and he worried that this was a new trend leading eventually to the rule of nations by the "voice of the masses."[42] Like Sighele, Le Bon believed that the growing density of people was a necessary precondition, but not the crucial determining factor, for the creation of a crowd. There must also be present a unique psychological reciprocity among the people in it. The mind of the crowd is a "collective mind" in which the individual is swamped by mass emotions and in which ideas spread by a kind of "social contagion."[43]

Tarde, Sighele, and Le Bon tried to interpret the processes of transmission over distances and the kinds of messages that were being sent. Their approaches differed but their focus was the same, so much so that an argument over priority broke out when Sighele claimed that *The Crowd* was, "in large part, a clever reconstruction of my book."[44] The intensity of the priority squabble and the immense and immediate popularity of Le Bon's book suggest that the three were responding to issues that deeply concerned many people.

Whether they saw crowding as the cause of a leveling of individuality, a breakdown of the family, an epidemic of crime, a disruptive acceleration of the pace of life, or a multiplication of sensory stimuli that was too much for the nervous system to handle, their sociological judgments about crowding were generally negative.[45] Positive assessments came from those who saw rather a multiplication of opportunities, a shattering of provincialism, or a unification of humanity in the great urban crowds. In 1913 Gerald Stanley Lee published a popular evocation of the crowd. At times given to metaphorical excess, his exuberance is revealing, as is his title—*Crowds: A Moving-Picture of Democracy.* Lee's excitement was intensified by the new technology that brought the crowd together, leveling former hierarchical distinctions. Proximity to the center of the city was an indication of vitality: "The power of a man can be measured today by the mile, the number of miles between him and the city; that is, between him and what the city stands for—the centre of mass." The great masses that inspired horror in Tarde and Le Bon were to Lee a vision of democracy. The modern hotel was a twentieth-century Parthenon, "not of the great and of the few and of gods, but of the great many, where, through mighty corridors, day and night, democracy wanders and sleeps and chatters." Electricity he saw as the current of the republican ideal because "it takes all power that belongs to individual places and puts it on a wire and carries it to all places." He interpreted the elevator, paradoxically, as a great leveller, "giving first floors to everybody and putting all men on a level at the same price."[46] Lee set aside a century of Malthusian forebodings about the evils of overpopulation and crowding, convinced as he was that the reduction of living space between people would also reduce their misunderstandings and conflicts.

The Belgian poet Émile Verhaeren envisioned new dimensions of city life in his epic of 1895, *Les Villes tentaculaires.* The central image allowed him to suggest the suffocating aspect of the city as well as its all-embracing, unifying function. In an interpretation of Verhaeren's conception of the new sense of space in the modern age, Stefan Zweig concluded: "We measure differently with these different velocities of life. Time is more and more the victor of space. The eye, too, has learned other distances . . . and the human voice seems to have grown a thousand times stronger since it has learned to carry on a friendly conversation a hundred miles away." To comprehend the "other" distances people must develop new feelings, a new morality, and a new art. Zweig believed that Verhaeren was the first to put the modern dimensions into verse. His poetry was "a new outlook which not only sweeps the distance, but has also to reckon with height, with the piled tiers of houses, with new velocities and new conditions of space." Verhaeren's "tentacular city" was the heart of modernity, a melting pot where different people and customs "dissolve into one."[47]

This image of unification, another consequence of the diminution of lived distances, was the inspiration for Jules Romains's philosophy of *unanimisme,* which he embodied in a collection of poems, *La Vie unanime* (1908). With the exuberance of a Futurist, he envisioned a world of shared sensations in which the territorial integrity of people and objects was dissolved in *unanimisme.* "The child upstairs suffers in me; / His body sends obscure messages to mine, / And, like a haunting procession of automobiles, / Its nightmares pass under my moist brow." The entire world is as unified as a modern theater audience, inhabiting the same proximate space, breathing the same air, responding to the same noises and gestures, attending to the same words. Churches, those older sanctuaries of unified beings, have been replaced by modern factories that unify thousands and send a signal of smoke more powerful than the ringing of the old bells.[48]

The other branch of social science that emerged in this period, the study of geopolitics, provided a new language of discourse about the sense of distance. It was not a study of distances alone, although one investigation of world trade routes in 1888 was subtitled "Science of Geographical Distances."[49] It developed along with the great expansion of empires in the late nineteenth century and was especially concerned with the way the size of states, their location, and the distances between them shaped their politics and history.

A good deal of nineteenth-century geography was a mixture of map making, travel accounts, gazetteers, and regional descriptions.

Modern systematic geopolitics was pioneered in Germany by Friedrich Ratzel, who became professor of geography at Leipzig in 1886. Trained in geology, zoology, and history, Ratzel sought to create a comprehensive method for studying the connection between all features of the earth's surface and human history.

Ratzel called attention to a dual focus in the title of his first major work, *Anthropogeographie* (1882). The numerous fragmentary studies of recent years, he claimed, failed to identify the "spatial unity of life," which is the land. It is the material link among people and forms an organic unit with them that develops according to evolutionary laws. The word "culture" also denotes tillage of the soil, and in the course of history a people becomes ever more deeply tied to the land. Although this early work is more restrained than the geopolitical propaganda that he wrote after Germany entered the imperalist scramble in the mid-1880s, in it Ratzel allowed himself to moralize on behalf of bigness and expansion in general. The larger the expanse of a state, the more freely may its population develop. His chapter on "The Progress from Small to Large Spaces" articulated what became the ethical imperative of imperialism all over the world—that big is good. "All people," he wrote, "who remain at lower stages of cultural development are also spatially small [*kleinräumig*]. Their field of living is small, as is their field of action and their circle of vision . . . All the qualities of a people that facilitate political expansion must always be of special value because of the abiding tendency toward the formation of bigger spaces." Geography must become a "science of distances" to investigate the way the earth's surface influences culture and politics.[50]

In 1896 Ratzel codified his science in an article on "The Laws of the Spatial Growth of States." These laws were: the space of a state increases with the growth of culture; the growth of states follows other signs of development such as production and commerce; the growth of a state proceeds by the amalgamation of smaller units; the frontier is the peripheral organ of the state; a growing state strives to include valuable sections such as coast lines, river beds, natural resources; the first impetus for territorial growth comes to the primitive state from without, from a "higher civilization"; the tendency toward territorial growth intensifies as it passes from state to state.[51] The realization of these laws was the result of an instinct for expansion that Ratzel believed to be inherent in every living cell as well as in individual organisms and nation states. He fleshed out this theory in a massive tome of 1897, *Politische Geographie*. It is impossible to exaggerate the pervasive influence of evolutionary theory in this period and the appeal of the biological metaphor. Ratzel applied the organic analogy to the state, which he interpreted as a "rooted organism" that must grow or die. Qualitative development may occur within its borders, but the most palpable growth is territorial. The development of a world politics (*Weltpolitik*) is a "common stage" in the development of a nation. The British became mighty by expanding around the world, and all major powers are driven by similar spatial instincts and aspire to bring the entire globe into a single political system. "Modern people are becoming ever more conscious of spatial relations," and modern politics has become a "school of space" which teaches the geopolitical law of natural selection—that among nations the struggle for existence is a struggle for space. In the great new empires that sprawl over ever greater distances, spatial extension is a source of spiritual rejuvenation and national hope. It reduces friction among its inhabitants and minimizes the intrusiveness of its government. Ratzel also saw some disadvantages. Large spaces cut people off from others and intensify racial conflicts; they reduce the frequency of creative interchange and engender a tedious repetition of traditional customs. But his bias was on the side of largeness, as is evident from a discussion of "the political effects of smaller spaces." Smaller states develop prematurely and do not reach the higher cultural levels attained in larger states. *Kleinstaaterei* kills off everything manly and important and generates a "miserable local vanity," a uniformity of values and goals that is mirrored in the sameness of the landscape and the uniformity of vegetation and climate.[52]

Two turning points marked the emergence of German imperialism. In 1884 Bismarck abandoned his opposition to overseas territorial expansion, and Germany quickly staked out all four of its African colonies within a year. Then in 1897 the state began to build up a high-seas battle fleet to protect that empire and secure Germany's place as a world power. In 1890 the American naval captain Alfred Thayer Mahan published *The Influence of Sea Power upon History*, an enormously popular book which argued that sea power was the key to national prosperity. When Ratzel published *The Sea as a Source of the Greatness of a People* in 1900, his country was committed to the goals of a *Weltpolitik*, and he indulged in some explicit propagandizing on behalf of German expansionism, drawing from Mahan to support his urgings. The development of all states "stands under the law of progress from small to big spaces." The sea is the new realm for such development in overseas colonies, and therefore "Germany must be strong on the seas to fulfill her mission in the world."[53]

Ratzel was not only concerned with space and distance. He considered the importance of environmental influences such as terrain, resources, vegetation, and climate; and his more exclusively spatial factors included location, borders, and surrounding territory. But there is a distinctively abstract emphasis throughout, as if the scientific validity of his discipline depended on the reduction of geopolitical phenomena to purely spatial terms. Much like Durkheim, who tried to factor out subjective considerations in his analysis of the division of labor by emphasizing the social impact of a reduction of the "real distance" between people, Ratzel sought to get beyond examining how different cultures used the land and focus on the more objective factor of space itself as a political force. His rhapsodic praise of large space was typical of a great number of geopolitical thinkers in Germany and elsewhere up to the outbreak of World War I. There were opponents to expansion and imperialism, to be sure, but among most geopolitical thinkers there was a tendency to equate bigness not only with national political strength and economic prosperity but also, following Ratzel, with such individual character traits as boldness, openmindedness, and farsightedness.

In America Ellen Churchill Semple restated Ratzel in 1911. Convinced by Turner that the frontier had been a major factor shaping the American character and political system, Semple enthusiastically translated Ratzel's notion of space, *an sich,* as the essential factor in the formation of a nation. "Man is a product of the earth's surface," she began, and "big spacial ideas, born of . . . ceaseless regular wandering, outgrow the land that bred them and bear their legitimate fruit in wide imperial conquests." For a country that pubesced territorially in the fulfillment of a "manifest destiny" to stretch across the North American continent, the Ratzelian geoethical concept that big is good made sense, and she elaborated Ratzel's dictum as if it were a universal law: "The larger the area occupied by a race or a people, other geographic conditions being equal, the surer the guarantee of their permanence, and the less the chance of their repression or annihilation."[54]

While Semple reflected the values that guided the American imperial venture, a more critical commentary on Ratzel by Camille Vallaux in the same year reflected the French experience. Vallaux charged that Ratzel's theory of pure space (*l'espace en soi*) was nothing but a scientifically worded justification for German imperialism.

Pure space is "passive and neutral" and does not determine the fate of nations. The ability of a state to react quickly and coherently determines its evolution more than its sheer size. The Russian empire was even weakened by its vast space, which made it incapable of prompt, concerted action, as demonstrated by its disastrous defeat by the Japanese in 1905. Some small spaces can have great political significance, while large ones can have little. The English Channel is only 40 kilometers wide but has had an enormously beneficial effect on the development of England, while the Sahara desert, which extends over two million square kilometers and which the French annexed after much travail, has had a comparatively trivial effect. It makes impressive political rhetoric and is imposing when viewed on a map but in reality has done nothing to promote the power, security, or prosperity of France. There is a "collective consciousness of occupied space" that unites the people in it and that is partly determined by size, but it is mediated by other factors that determine its historical and political impact. Although Vallaux rejected pure space as an independent geopolitical variable, he accepted "distance"—a geopolitical concept that included the time it took to traverse a space. He produced an isochronic map of 1906, showing the world marked in zones determined by the number of days' travel between them and London or Paris and concluded that "it is the notion of time and of distance, not space, that Ratzel ought to have emphasized."[55] There is some equivocation in his criticism of Ratzel's concept of pure space, because, as he concedes, Ratzel was aware of how national character, location, and technology affected the significance of any geographical space. Theirs is a difference largely of emphasis. Ratzel tried to establish geopolitics as an objective social science and affirm environmental over the "racial" determinants that were frequently invoked by historians of his time. Vallaux found this spatial reduction too abstract and called for an approach that took into account the mediation of "human" factors. Their writings also reveal their respective national orientations. Ratzel polemicized on behalf of German expansion, while Vallaux criticized French imperialism in the Sahara, which he viewed as of no economic, and of little political, value. In 1911 the great space in Equatorial Africa was *an sich* quite worthless.

While theoretical and national biases generated varieties of emphasis about the significance of geopolitical space, some views were shared. Most people agreed that geography played an important role (much neglected until then) in the politics of nations and that the space commanded by a country was an important factor in the determination of its political and economic life. While some pointed out that size was not an unequivocal determinant of power and prosperity, most generally accepted the notion that all things being equal, bigness tended to promote freedom, prosperity, and security. No one argued that smallness was a source of national greatness. In an age of the most rapid expansion of territorial empires in the history of mankind, it is no wonder that the social sciences developed to interpret this phenomenon should bow before its most visible and palpable achievement.

Rivers of geopolitics coursed all over the European cultural terrain. They started in the high reaches of theoretical tomes such as Ratzel's two major works and cascaded through volumes of the new periodicals that were founded—*National Geographic Magazine* (1889), *Annales de géographie* (1891), *The Geographical Journal* (1893), *Geographische Zeitschrift* (1895). They ran down through the flatlands of popular consciousness in countless magazine and newspaper articles and welled up everywhere, from formal diplomatic pronouncements to barroom banter. The most controversial geopolitical theory of the age was proposed by the Reader in Geography at Oxford and Director of the London School of Economics, Halford J. Mackinder, in an essay of 1904, "The Geographical Pivot of History." With the closing of the frontiers, the development of worldwide electronic communication, and the extension of railroad networks, he argued, it had become necessary to view the world as a single organism that will respond as a whole to power shifts anywhere on the globe. "For the first time we can perceive something of the real proportion of features and events on the stage of the whole world, and may seek a formula which shall express certain aspects . . . of geographical causation in universal history." Also for the first time in history there was a potential for an "empire of the world" dominated by whoever controlled the pivot area or "heartland" of Euro-Asia.[56] Mackinder's premise that it was possible, indeed necessary, to think in worldwide terms is characteristic of a number of observations made at that time. Some saw the global perspective as a force for peace, others as a cause of war, but all shared the idea that the world was becoming smaller and more unified.

Marshall McLuhan's more recent formulation that "the medium is the message" focused attention on the dimensions and structures of experience mediated by technology. He argued that the impact of the printed word was felt—more than in the content of any printed page—in the extension of the visual faculty, the intensification of perspective and individualism, and the cultivation of the linearity, precision, and uniformity of thought. The message of the railroad lay not in anything it transported but in the acceleration of movement and the dimensions and structures of the cities it created: "the 'message' of any medium or technology is the change of scale or pace or pattern that it introduces into human affairs." In broad, bold historical terms McLuhan interpreted the message of electronic media. "After three thousand years of explosion, by means of fragmentary and mechanical technologies, the Western world is imploding. During the mechanical ages we had extended our bodies in space. Today, after more than a century of electronic technology, we have extended our central nervous system itself in a global embrace, abolishing both space and time as far as our planet is concerned."[57] His famous characterization of the impact of electricity—its creation of a "global village"—suggests only the positive communal side, but it is a view that was anticipated by many who saw a growing sense of community and unity in the world brought about by the reduction of effective distances between people and nations and by the extension of experience beyond traditional horizons.

Already in 1891 the editor of *Revue scientifique* wrote that "to say that there are no longer distances is to utter a very banal truth."[58] He was as impressed by the new proximities as was H. G. Wells, who in 1901 observed that "the world grows smaller and smaller, the telegraph and telephone go everywhere, wireless telegraphy opens wider and wider possibilities to the imagination." Technology demolishes "obsolescent particularisms" such as national boundaries and will someday lead to the creation of a "world-state at peace with itself."[59] In the same year a Russian internationalist, J. Novicow, surveyed the divisive forces in a world mesmerized by an "idolatry of square kilometers," where nations had become susceptible to a new disease that he called "kilometritis" (*kilométrite*). But he emphasized the unifying forces, citing a remark by William Gladstone that "each train that passes a frontier weaves the web of the human federation."[60] Electronic communication had made possible a vital unity of nations by accelerating communication among them. The scaling down of effective distances and the expansion of mental horizons would, he hoped, lead to the creation of a Federation of Europe. In 1903 a prominent American businessman, George S. Morison, predicted peace and unity among nations resulting from the increasing availability of energy and its magnified application by technology. "The manufacture of power has given us the means of traversing the entire globe with a regularity and speed which brings all races together, and which must in time remove all differences in capacity . . . It is gradually breaking down national divisions [and] will finally make the human race a single great whole."[61]

Hopes for international cooperation ran high when forty-five nations sent representatives to the second International Conference at The Hague in 1907.[62] Although their four months of meetings failed to make any substantial progress in the three major areas of discussion—disarmament, arbitration, and laws of war—it was, like the outbreak of war in 1914, a collective failure. As in a tightly knit neurotic family, antagonisms can be as binding as harmonies, even more so. In the years before the war Europe was indeed, as many historians have written, an "armed camp," but a camp nevertheless. A year earlier G. S. Lee had announced that mankind was ready for someone who would address himself to "a world which for the first time in its history has at least the conveniences for listening all over." Lee sensed "a kind of soul-suction—a great pulling from the world" that was readying itself for a Great Author.[63] He did not anticipate that the world might instead turn to a Great Dictator, but he was correct in anticipating that the technology of electronic communication had made a worldwide audience possible.

Around this time global travel was becoming more accessible to the ordinary tourist.[64] I have not been able to determine the origin of the term "globetrotter," but in a popular German travel guide of 1907 it is cited as new jargon to describe a type of traveler that was appearing in increasing numbers.[65] The French poet Valéry Larbaud, in a collection of poems and sketches of 1908, created the fictional traveler A. O. Barnabooth to express the new sensibilities of worldwide travel:

> When we wish, we enter virgin forests,
> Desert, prairie, colossal Andes,
> White Nile, Tehran, Timor, southern seas,
> And the entire surface of the planet is ours
> when we want!
> For me,
> Europe is like one great city . . .[66]

The popular German cultural historian Karl Lamprecht also felt a new sense of ubiquity in the world. Railroads, steamships, and telephones were "homeless" technologies that created vastly extended horizons for everyone and made it possible, in a sense, to be everywhere at once.[67] Higher literacy standards and a growing popular press expanded individual horizons in another way, and with the extension of manhood suffrage the masses began to affect national politics and international relations to an unprecedented degree.

In England between 1890 and 1914 "the publicity given to politics and foreign affairs was increased a thousand times" as more and more people became better informed about conditions all over the world.[68] Herbert Feis reconstructed the history of economic internationalism in *Europe the World's Banker 1870–1914,* where he observed that "in the sphere of financial interest and calculation, distance lost its meaning; along all lines of latitude and longitude British capital worked its way."[69] In 1916 Lenin assessed the extraordinary international scope of production and distribution, which he saw as the most distinctive characteristic of capitalism in the late nineteenth and early twentieth century, when finance capital "spreads its net over all the countries of the world."

> When a big enterprise assumes gigantic proportions, and, on the basis of exact computation of mass data, organizes according to plan the supply of primary raw materials to the extent of two-thirds or three-fourths of all that is necessary for tens of millions of people; when the raw materials are transported to the most suitable place of production, sometimes hundreds or thousands of miles away, in a systematic and organized manner; when a single centre directs all the successive stages of work right up to the manufacture of numerous varieties of finished articles; when these products are distributed according to a single plan among tens and hundreds of millions of customers . . . then it becomes evident that we have socialization of production.[70]

Such worldwide planning required international organizations. A model was the St. Petersburg Convention of 1875 which worked out procedures for international telegraphic communication. There was also international cooperation in pure science; at a Parisian congress in 1881 the ohm, volt, coulomb, farad, and ampere were settled as standard units. A second congress in 1889 adopted the watt and joule, and a third, also in Paris in 1900, added the maxwell and gauss.[71] The International Geodetic Association was formed in 1867 to coordinate geological surveys in Europe and was extended in 1886 to include the entire world. The International Association of Seismology was formed in 1903 to study earthquake activity, and in 1904 an International Committee for the Map of the World was founded. A historian of internationalism, F. S. L. Lyons, discovered an "astonishing" number of such new organizations after the turn of the century—119 from 1900 to 1909 and an additional 112 in the five-year period prior to the war.[72] The belief that the world was moving toward greater cooperation was expressed repeatedly, as in a 1913 publication of a plan to create a World Centre of Communication in Paris. "It is being realized more and more that nations can never again be entirely separated. Impassable walls to enclose and protect them are things of the past."[73] The eminent Belgian internationalists Henri La Fontaine and Paul Otlet echoed that sentiment in an article in *La Vie internationale,* a journal founded in 1912 to publicize international meetings, projects, and organizations. "From ever greater contact among nations and the sharing of their experiences and works internationalism will find its greatness and its force; thus will arise the universal civilization of all the countries reconciled and united."[74]

But another response to the shrinking distances and the growing proximity was conflict. Its most visible manifestation was imperialism, and its most explicit written expression, the imperialist tracts of its defenders. The English and French who first staked out empires and the Germans and Americans who came later shared the same assumptions, values, and fears—that the new technology had greatly facilitated the exploration and seizure of vast empires, that the command of territorial space was essential to national greatness, that such command was morally defensible, and that the imperialist scramble would eventually lead to war, especially after all the "virgin land" had been taken. Internationalism and imperialism coiled around the staff of the new technology and around one another like the snakes of a caduceus.

Until 1880 France, England, and Portugal were content to just nibble at the African cookie, about 1,000,000 square miles of coastland. By 1890 Italy, Germany, Belgium, and Spain had taken a total of about 6,000,000 square miles; and by 1914 all but Abyssinia and Liberia was under European domination, which extended over 11,500,000 square miles: France had 4,238,000; Britain 3,495,000; Germany and Italy about 1,000,000 each; Belgium 800,000; Portugal 780,000; and Spain 75,000. Almost all of this had been acquired by 1900, although much of it was not ratified by treaty or officially recognized by the other powers until the early twentieth century. The major European powers had greedily transformed the heart of darkness into their place in the sun. Acquisitions in Asia in this period did not involve quite such enormous areas of land but were in response to the same impulse. Novicow's notion of the idolatry of square kilometers was, like Marx's notion of the fetishism of commodities, a collective mystification about value that at once magnified material worth and concealed human misery. It glossed over the suffering of native populations and glorified the prize of empire. The rhetoric varied slightly from country to country, but the idea that large national territory betokened greatness was, as even that optimistic pacifist Norman Angell conceded, a "universal assumption."[75]

J. R. Seeley's *The Expansion of England* (1883) was not so much a plea for expansion as an explanation and defense of it in terms of unifying forces at work. Seeley noted that "in the modern world distance has very much lost its effect." Steam and electricity had made possible the organization of enormous territorial states such as Russia and the United States, and the growth of trade and emigration had set in motion "vast uniting forces."[76] He did not consider the potential rivalry in empire building because there was little of it when he wrote; but his complacency conceals the aggressive nature of the British empire, not merely against native populations but against every other country that aspired to national greatness, especially Germany, which began to expand in the following year and discovered everywhere the threatening reality of "Greater Britain."[77] James Froude's *Oceania* (1885) was concerned with the preservation of the British Empire, which was then already vast. He challenged the separationist argument with a classic formulation of arboreal analogy:

> The oak tree in the park or forest whose branches are left to it will stand for a thousand years; let the branches be lopped away or torn from it by the wind, it rots at the heart and becomes a pollard interesting only from the comparison of what it once was with what fate or violence has made it. So it is with nations. The life of a nation, like the life of a tree, is in its extremities. The leaves are the lungs through which the tree breathes, and the feeders which gather its nutriment out of the atmosphere. A mere manufacturing England, standing stripped and bare in the world's marketplace, and caring only to make wares for the world to buy, is already in the pollard state; the glory of it is gone for ever.

"The world has grown small," he explained, the British islands are "full to over-flowing," and the great challenge for the English people in 1885 is therefore to provide living space for the millions to come by maintaining the unity of Oceania.[78]

Even before Germany began directly to challenge their sea power, the British realized that preservation of the empire was going to be difficult, especially as improvements in the construction of warships eroded their former naval advantage. During maneuvers in 1888 and 1889 it became apparent that in an age of the steamship successful blockades would become increasingly difficult and that a great power must maintain safe ports and coaling stations around the world. Sir Charles W. Dilke elaborated on these problems in another imperialist tract of 1890 that anticipated a showdown. Dilke's vision illustrates the arrogance of empire and the fixation on size. "The world's future," he wrote, "belongs to the Anglo-Saxon, to the Russian and the Chinese races, [and] before the next century is ended, the French and the Germans seem likely to be pigmies when standing by the side of the British, the Americans, or the Russians of the future."[79] Such declamations on behalf of bigness were typical and abundant. In the words of one history of British imperialism: "Expansion in all its modes seemed not only natural and necessary but inevitable; it was pre-ordained and irreproachably right. It was the spontaneous expression of an inherently dynamic society."[80]

All of the imperialist apologists shared the same imagery and rhetoric—a mixture of organic analogy and natural selection, and later, some ideas derived from Nietzsche and Bergson. The French, like the English, acquired an empire early and relatively easily, so it seemed to them at first to be a natural political process. The influential political economist Paul Leroy-Beaulieu wrote in 1874 of imperialism as if it were a continuation of human embryological development: "Colonization is the expansive force of a nation, its power of reproduction, its dilation and multiplication across space; it is the submission of the universe or of a vast part of it to the language, the manners, the ideas and laws, of the mother country." It is unjust and unnatural for civilized countries to suffocate in restricted spaces and leave over half the world to "ignorant" and "decrepit" peoples. Colonial empire is vital: "either France will become a great African power, or in a century or two she will be no more than a secondary power; she will count for about as much in the world as Greece or Roumania."[81] In *L'Expansion de la France* (1891), Louis Vignon, a professor at the Ecole Coloniale, wrote that every nation in the world maintained the territory that it had conquered by its military might and hard work, by its wealth and genius, "and the energy that it unceasingly develops in the universal struggle which is the great law of nature."[82] Another propagandist argued in 1897 that colonial expansion was a vital necessity and functioned like a "dynamometer" for measuring the respective vitality of nations. Twenty years later Jules Harmand thought that expansion was a universal law of nature, an energy charged with *élan vital,* the spirit behind an instinctual "will to power."[83]

The moral commitment to expansion in England and France makes it necessary to reevaluate German war guilt. Because Germany was the leader of the defeated Central Powers, its archives were the first to be opened for countless studies made by its own and foreign investigators. In Germany the accusations and apologies were intense and, in the German scholarly tradition, abundantly documented. In 1961 Fritz Fischer reopened the discussion with *Griff nach der Weltmacht* (Grab for World Power), which revived the case against Germany, and he continued in 1969 with *Krieg der Illusionen* (War of Illusions), documenting even more thoroughly the expansionist groups and imperialist thinkers who set the Second Reich on a course for war. His study is exhaustive and convincing. However, a survey of the same kinds of documents of the other belligerents would reveal a similar number of expansionist groups and leaders; similar weakness and ineffectiveness among the socialists, liberals, and internationalists who opposed expansion; and a similar willingness to risk war in the interest of national greatness. The political situation in each country necessitated different actions and different rationalizations of them, but the underlying values and long-range goals were the same. All nations assumed that they must, like any organism, continue to grow or die. By 1900 it was clear that territorial expansion would involve a higher risk of European war than it would have done twenty years earlier, and so many imperialists shifted the goal of expansion to the commercial sphere. All nations assumed that growth was good, and none questioned the wisdom of trying to achieve it. They all realized the dangers of challenging another country that had already staked out a territorial or commercial empire, but all were convinced that it was a risk that had to be taken in the life-and-death struggle that was the reality of international relations. When viewed in the context of the governing political values of the time, the existence of the British empire was just as much a threat to the peace as Germany's program to match it. No *moral* distinction can be made between the building of the British or French empires in Africa and Germany's challenge to them, unless a distinction is made between taking land from African natives and taking land from European conquerors. The ethic is the same—that big is good—as are the implied political imperatives—to expand for greatness or fall into mediocrity. For Germany it was the spectre of winding up like Sweden; for France, as we have seen, winding up "second rate" like Greece or Roumania. It is possible to attribute responsibility based on the way each country carried out its respective national policy, but it is essential to recognize that those policies were framed around shared values concerning the control of space. Germany's failure was essentially a failure to accomplish at the same time as other countries what everyone in positions of power and a vast majority of the population believed to be a political necessity and a moral good. No leader at that time could achieve or maintain power in his own country if he was not committed to growth and national might. There were opponents to imperialism, to be sure, but they failed to affect the behavior of nations, and the empires grew.

In a famous memorandum in 1907 the Assistant Undersecretary of the British Foreign Office, Sir Eyre Crowe, evaluated German *Weltpolitik* in the moral-political context of the time: "A healthy and powerful State like Germany, with its 60 million inhabitants, must expand, it cannot stand still, it must have territories to which its overflowing population can emigrate without giving up its nationality . . . it would be neither just nor politic to ignore the claims to a healthy expansion which a vigorous and growing country like Germany has a natural right to assert in the field of legitimate endeavor."[84] Since there were limited resources and limited space into which a "healthy" nation could expand, the fulfillment of German "national right" would have to clash with British interests; thus Crowe's hope that German expansion would remain "in the field of legitimate endeavor," by which he meant not threaten England, was wishful thinking, if not self-contradictory. The document is otherwise one of the most lucid analyses of the imperialistic impulse of all nations at that time and of their commitment to a body of ideas that led them into continual conflict.

Reverence for the very concept of expansion was as old as the former empires of Greece and Rome, Portugal and Spain, England and France. The expansion of Christendom to the Near East, New World, Asia, and, finally, Africa was for the European mind a model of the success of an idea, a triumph of goodness and holiness. Europeans did not need to be convinced that the command of space was desirable: it was embedded in their historical consciousness. Its distinctive manifestation in the late nineteenth century was materially and intellectually dependent on the new technology that knitted distant places together with unprecedented speed and efficiency. At all times in European history smallness and contraction were associated with disease and death, and Germany's "grab for world power" was but a modern national variant of an ancient and universally accepted principle. If the literature seemed more strident and urgent in Germany, it was due in part to its late start and fear of being left out. The values that shaped German policies were the same as those that had inspired the English and French in earlier decades. Fischer has surveyed the writings of such influential historians and political thinkers as Gustav Schmoller, Hans Delbrück, Max Sering, Otto Hintze, Erich Marcks, Paul Rohrbach, and Rudolf Kjellén, who elaborated on Ratzel and staked out the arguments and justifications for German *Lebensraum, Schicksalsraum,* and *Weltpolitik.*[85] It is not necessary to duplicate this work, but a brief consideration shows the rhetoric of German imperialism to be similar to that of the other major powers, who shared the "universal assumption" that bigness was essential to national might, prosperity, and hope for future generations.

Two books by Kurt Riezler, private secretary to German Chancellor Bethmann Hollweg, recapitulated many of the arguments of German imperialism and reveal one line of direct influence on the decision makers of the crisis of July 1914. Bismarck had described diplomacy as the "art of the possible," but for Riezler it was the achievement of the impossible, as the title of his first book spells out: *Die Erforderlichkeit des Unmöglichen* (The Requisiteness of the Impossible; 1913). Bismarck's restrained colonial policy until 1884 had violated the law of nature that every living thing must grow. Riezler believed that Germany must expand or die, and he ransacked the catalog of natural symbols to illustrate why. The germination of seeds, the processes of nutrition, the spreading of branches, the tumescence of organs, the reproduction of species—for two hundred pages he worked the argument from analogy to show how every living thing reached beyond itself in an endless and irresistible process of entelechy. Bergson's notion of a vital energy driving organisms to achieve ever more complex forms was incorporated in Riezler's "physiology of states," which viewed Germany as an organic unity of individual souls bound together by a common striving for the impossible. There were four elements of national unity: space, race, culture, and nation. The development of spatial unity necessarily led to conflict with outsiders. "The organism is a process that must unfold in action and reaction against a hostile surrounding world." Riezler supplied the biometaphysics of German *Einkreisung* (encirclement).[86] In *Grundzüge der Weltpolitik in der Gegenwart* (1914) he analyzed two opposing tendencies in the world that regulated the distance between people and structured international relations. The cosmopolitan tendency drew people together and fostered international rights; the national tendency distanced people and generated hostility, and both seemed to be on the rise. "The age of international trade and the exchange of men, goods, and ideas is also the age of increasing national tendency and the estrangement of peoples." The national tendency is to grow and become autonomous like individuals; nations must expand in space as the branches of trees need to spread out to absorb sunlight through their leaves. The will to dominate surroundings is unstoppable and insatiable, a principle of life that necessarily leads to clashes. Although his biometaphysics pointed to conflict, Riezler also observed that the rise of commercial traffic had drawn nations together, that the international political matrix was so interconnected that every movement unsettled the entire structure, and that with the powerful new weaponry it was doubtful whether even the victor would benefit from war.[87] From Ratzel to Riezler German geopolitical thinking was preoccupied with the frustration of confining frontiers, the ambition to expand, the fear that the country might not develop properly, and the growing conviction that its expansionist foreign policy might lead to a dangerous war.

Long before the branches of American expansionism spread over the Pacific and the Caribbean, imperialism was deeply ingrained in popular consciousness. In 1845 the editor of the *New York Evening News,* John L. O'Sullivan, introduced the term "manifest destiny" to describe the taking of land inhabited by Mexicans and Indians in the westward expansion of the United States: "Away, away with all these cobweb tissues of rights of discovery, exploration, settlement, contiguity, etc . . . [It] is by the right of our manifest destiny to overspread and to possess the whole of the continent which Providence has given us for the development of the great experiment of liberty and federative self government entrusted to us. It is a right such as that of the tree to the space of air and earth suitable for the full expansion of its principle and destiny of growth."[88] His rhetoric is indistinguishable from later European imperialist tracts.

Americans expanded into "virgin land" with the confidence that they were carrying out a higher destiny, if not God's will. German geopoliticians no doubt had that impressive transcontinental expansion in mind when they envisaged Germany swelling into a vast continental empire or crossing the seas in the search for *Lebensraum.* The frontier, that crucible of American democracy, was also a setting for the destruction of a people who, until recently, were depicted in popular culture as yelping scalp hunters who squandered the wide open spaces on bizarre tribal rituals and the worship of strange gods; and the "pioneer spirit" remains synonymous with courage and resourcefulness. In no other country in the world was expansion conceived to be as natural and as inevitable a part of the fulfillment of a nation's destiny as it was in America.

Eventually American expansion reached across the seas: Hawaii, Puerto Rico, Guam, and the Philippines were annexed in 1898; Cuba became a protectorate in 1901, the Canal Zone was in effect annexed in 1903; and the Virgin Islands were bought from Denmark in 1917. The initial phase coincided with the emergence of the corporate system in America that necessitated an aggressive foreign policy to secure outlets for investment. One of the architects of that policy, the leading banking authority Charles A. Conant, analyzed American expansion in an important article of 1898. Once again the arboreal imagery is brought out to make the point. "The irresistible tendency to expansion, which leads the growing tree to burst every barrier, which drove the Goths, the Vandals, and finally our Saxon ancestors in successive and irresistible waves over the decadent provinces of Rome, seems again in operation, demanding new outlets for American capital and new opportunities for American enterprise." The historical examples suggested that the tendency to expand was universal, as indeed Conant believed it was, but its manifestation in his time was specific to a unique crisis of capitalism. The formal political relation with the other countries was relatively unimportant, but the securing of some arrangement that would allow investment of surplus savings was essential, "if the entire fabric of the present economic order is not to be shaken by social revolution."[89] The universal commitment to at least commercial expansion was attested to by the former Secretary of State John W. Foster, who wrote in 1900: "Whatever difference of opinion may exist among American citizens respecting the policy of territorial expansion, all seem to be agreed upon the desirability of commercial expansion."[90] William Appleman Williams has surveyed the "contours" of American history shaped by an "expand-or-stagnate" approach to foreign policy that was supported by presidents McKinley and Roosevelt, implemented by political leaders John Hay and Henry Cabot Lodge, and interpreted by the historians Frederick Jackson Turner and Brooks Adams. As Turner wrote in a popular version of his frontier thesis published in the *Atlantic Monthly* in 1896, "For really three centuries the dominant fact of American life has been expansion."[91] Adams observed a more historically specific manifestation in an essay of 1900 on *America's Economic Supremacy*: "All the energetic races have been plunged into a contest for the possession of the only markets left open capable of absorbing surplus manufactures, since all are forced to encourage exports to maintain themselves."[92]

The modern period had a new sense of distance, created by technology and mediated by urbanism and imperialism. Lines of communication and transportation were extended over unprecedented distances, spreading out and at the same time bringing people into closer proximity than ever before, and evaluations of the effects were as diverse as the spatial dynamics involved. In 1903 the German historian Erich Marcks observed: "The world has become harder and more bellicose, even more exclusive; more than ever it has become a greater unity in which everything touches, everything interlocks, but also in which everything pushes and bangs together; in this unity there are few traces of the unprejudiced harmony of basic ideas and the free complement of rivalries that we used to dream of."[93] Technology tightened the skein of nationalism and facilitated international cooperation, but it also divided nations as they all grabbed for empire and clashed in a series of crises. It is one of the great ironies of the period that a world war became possible only after the world had become so highly united.

9

DIRECTION

The drive to expand and control space was universal, but the direction of that drive varied from country to country as each nation jockeyed for position in the scramble for empire, trade, and allies. Travel moved more freely along the east-west axis on the Trans-Siberian and Berlin-Bagdad railways, and the dynamics of European politics lined up the same way from the alliance systems, which created the two great fighting fronts that eventually bracketed the Central Powers in 1914. Before considering national spatial orientations, however, we must first examine the one universal directional shift that came about with the airplane.

Its cultural impact was ultimately defined by deeply rooted values associated with the up-down axis. Low suggests immorality, vulgarity, poverty, and deceit. High is the direction of growth and hope, the source of light, the heavenly abode of angels and gods. From Ovid to Shelley the soaring bird was a symbol of freedom. People were divided in their response to flying; some hailed it as another great technological liberation and some foresaw its destructive potential. H. G. Wells was among the latter, although at first he thought that flying machines would have little impact. In 1901 he made his worst prediction: "I do not think it at all probable that aeronautics will ever come into play as a serious modification of transport and communication."[1] He made up for that mistake seven years later with a science-fiction novel, *The War in the Air,* which detailed the potential dangers of aerial warfare and accurately anticipated many of the events and emotions of 1914–1918, when airplanes did play a significant military role. Wells envisioned greater destruction from aerial bombings than actually occurred in the First World War, but the novel identified growing fears among his contemporaries and foreshadowed the horrors of Guernica and Dresden. Its hero, Bert Smallways, grew up in a small English town and became intrigued with the "ideal of speed" of the early twentieth century. During a balloon ride he drifted over Germany and observed the assembly of an air fleet that was preparing to launch a war to give Germany world supremacy. In his novel Wells commented on the unique new dimensions of the age: natural bonding instincts among all men had once held the world together; "but with the wild rush of change in the pace, scope, materials, scale, and possibilities of human life that then occurred, the old boundaries, the old seclusions and separations were violently broken down." The airplane was but one variant of the technology that sundered protective frontiers and created new spatial dynamics. "The development of science altered the scale of human affairs. By means of rapid mechanical traction it brought men nearer together, so much nearer socially, economically, physically, that the old separations into nations and kingdoms were no longer possible." People began to elbow one another in the crowded cities, and nations clashed around the world, "slopping population and produce into each other," wasting their energies on military armaments. There were six great powers—America, Germany, England, France, Russia, and an alliance of China and Japan—and war began when Germany rejected the Monroe Doctrine and decided to challenge American hegemony in the western hemisphere.[2]

With an uncanny prevision of the rush toward war in July 1914, Wells described in his novel of 1908 a scurrying of diplomats and military men who began the fictional war in the air. There was an aerial version of the Schlieffen Plan requiring that the German air fleet strike the Americans swiftly before they could launch a counterattack. Kaiser Wilhelm's famous comment, made two days after the assassination at Sarajevo, that it was "now or never" (referring to the opportunity to crush the Pan-Serbian movement and preserve the Austro-Hungarian Empire),[3] was prefigured by Wells, who wrote, " 'Now or never,' said the Germans—'now or never we may seize the air—as once the British seized the seas!' " The chain reaction of alliance systems and military necessities that rapidly drew Europe into war was anticipated on a truly global scale in the novel. "There was no time for diplomacy. Warnings and ultimatums were telegraphed to and fro, and in a few hours all the panic-fierce world was openly at war." Cities were leveled, the political and social structure of nations disintegrated, and "the fantastic fabric of credit and finance that had held the world together for a hundred years strained and snapped."[4] There were no battle fronts. The men in airships who were fighting for the first time from the third dimension could peer down and hurl fire into the heart of cities without having to penetrate their massive walls or well guarded perimeters. Wells concluded with a vision of panic and pestilence, a collapse of civilization brought about by the war. After witnessing the massacre of New York, Bert Smallways realized that the same could happen to London, "that the little island in the silver seas was at the end of its immunity," that nowhere was safe.

In 1909 C. F. G. Masterman elaborated a vision of the future when bombs would fall from airships "like the fire and brimstone that rained down upon the cities of the plain." Then England would suddenly be "helpless and vulnerable before armies dropping from the skies."[5] In that same year Paul Scheerbart, who designed open buildings with glass architecture and open cities running over mountains radiating light for night flight, speculated on the disastrous effect of aerial warfare on fixed fortifications. Airplanes would be able to drop "dynamic torpedoes" at will on any stationary target. "The mere contemplation of such arts of war," he wrote, "can cause a nervous breakdown."[6]

But other nerves were positively stimulated. In July of 1909 the

English Channel was crossed in an airplane in thirty-six and a half minutes. Popular response was wildly enthusiastic, and the French aviator Louis Blériot became an international hero. Stefan Zweig interpreted the reaction in Vienna. "We shouted with joy when Blériot flew over the Channel as if he had been our own hero; because of our pride in the successive triumphs of our technics, our science, a European community spirit, a European national consciousness was coming into being. How useless, we said to ourselves, are frontiers when any plane can fly over them with ease, how provincial and artificial are customs-duties, guards and border patrols, how incongruous in the spirit of these times which visibly seeks unity and world brotherhood!"[7] In August there was an aviation contest at Rheims that drew participants and spectators from many nations. The flimsy planes of canvas and wire raced around hastily erected pylons, and half a million spectators for the week's events crowded into grandstands to watch the display of technology and skill. In November the Grand Prix race of the Automobile Club de France was canceled because there were too few entries.[8] The Rheims competition helped shift attention from the roads to the sky, and for the moment the frontier of adventure and the new direction of human attention was up.

The airplane's uplifting of human consciousness, its capacity to unify people and nations was noted the next year by the American journalist Victor Lougheed, who claimed that airplanes wiped out "the artificial barriers of nations' boundaries" and predicted that "the eye-like gleams of their searchlights will mingle to the uttermost ends of the earth beacons of science and romance and progress and brotherhood."[9] The historical record abounds with such rhapsodic praise for the transcendent qualities of flight. In 1913 G. S. Lee ventured the thought that it created a new direction of human consciousness. It revolutionized thinking "as if a lid had been lifted off the world." Wilbur Wright's flight around New York drew hundreds of thousands of people onto the rooftops and "left the entire city with its heads up."[10] Even Proust, who hardly left his house, became tragically involved with the passion for flying in France when Alfred Agostinelli, his one-time lover and model for the fictional Albertine, died in a crash. In March 1913 Agostinelli ventured over the sea during a training flight and on a low turn caught the water with his wing. Horrified observers on shore a few hundred yards from where he crashed watched as Agostinelli, who could not swim, stood on his seat in the sinking wreck and drowned as a rescue boat drew near.

But in the novel Proust emphasized the upward dynamic. "The airplanes which a few hours earlier I had seen, like insects, as brown dots upon the surface of the blue evening, now passed like luminous fire-ships through the darkness of the night . . . And perhaps the greatest impression of beauty that these human shooting stars made us feel came simply from their forcing us to look at the sky, towards which normally we so seldom raise our eyes."[11]

Another spiritual high came from looking down. For Marinetti it evoked the broad vision of Futurism that helped free the poet from "wingless" tradition and the "limited syntax that is stuck in the ground." In flight he saw bold analogies. "As I looked at objects from a new point of view, no longer head on or from behind, but straight down, foreshortened, that is, I was able to break apart the old shackles of logic and plumb lines of the ancient way of thinking."[12] Gertrude Stein speculated that the Cubists' breakup of the old ways of seeing was suggested by aerial vision, even though none of them had been up in a plane. She commented that landscapes seen from an automobile are essentially the same as those seen walking or from a carriage or train, "but the earth seen from an airplane is something else."

> So the twentieth century is not the same as the nineteenth century and it is very interesting knowing that Picasso has never seen the earth from an airplane, that being of the twentieth century he inevitably knew that the earth is not the same as in the nineteenth century, he knew it inevitably he made it different and what he made is a thing that now all the world can see. When I was in America I for the first time travelled pretty much all the time in an airplane and when I looked at the earth I saw all the lines of cubism made at a time when not any painter had ever gone up in an airplane. I saw there on the earth the mingling lines of Picasso, coming and going, developing and destroying themselves . . .[13]

The Cubist reduction of depth, elimination of unessential detail, composition with simplified forms, and unification of the entire picture surface are pictorial representations of the view of the earth's surface from an airplane in flight. Although there is no evidence that either Picasso or Braque flew or was influenced by aerial photographs, Stein's comment is at least an appropriate metaphor if not an account of direct influence.

The American military analyst Henry Suplee thought that the very possibility of aerial bombings would be a deterrent to war. He argued that wars must cease because no ruler would dare to undertake them, since all were equally vulnerable. He captioned an aerial photograph of ships in New York harbor: "Defenseless New York—From an Aeroplane Speeding a Mile above Brooklyn Bridge." Neither ships nor fixed fortifications could protect a city. "Anyone who has had the opportunity of gazing from above upon an unarmed and unprotected city realizes almost instantly the hopelessness of any attempts at armed protection."[14] But not everybody was as confident that the fear of destruction would stop wars. At the Hague Peace Conferences of 1899 and 1907 it was debated whether the discharge of projectiles from balloons during wars ought to be permissible. The first conference instituted a five-year ban, and it was renewed in 1907. An article of 1912 urgently called for a general protest against aerial bombings but suggested that the question would probably not be resolved until the meeting of the third Hague Peace Conference, scheduled for 1915, which, of course, was not held.[15]

When war came in 1914 it did not bring the rain of fire and destruction that many predicted. There was some aerial bombing, but airplanes were used largely for reconnaissance and to direct artillery fire. In a book on the airplane in war published in 1916, the British military historian William A. Robson could still invoke the ecstatic rhetoric of flight common in prewar years—"the wonderful panorama spread beneath and all round, the exhilarating rush of air in the face, the consciousness of speed . . . the extraordinary sensation of buoyancy." There is no friction in the air, no jarring, no dust, none of the obstacles from contact with earth. An aviator is "filled with the indescribable impression of spaciousness, grandeur, dignity, . . . uplifted in mind and body." In that age of unprecedented destruction on the ground Robson insisted that the airplane would speed up international travel and communication and reduce the ignorance and misunderstanding that cause war. Even as German Fokkers and English Havillands were dogfighting in the European skies he wrote: "there is such an infinite amount of room aloft that there is certainly enough space not only to satisfy the need of all, but to enable every nation to move in sufficient breathing space."[16] Robson concluded his hopeful prediction for peace and greater understanding in the age of flight with the hope that aircraft would in time "enable earth and sky to stand close enough together to fulfill the conditions necessary for the breaking down of the barrier between East and West."[17]

In the years before World War I the airplane did not have the range to link East and West, but it did begin a revolution in geopolitical relations among the nations of Europe, as, for example, Blériot's flight diminished England's strategic advantage as an island and required that its statesmen be more intimately involved with continental affairs. Aside from the new sense of upwardness created by the airplane, however, the sense of direction for individuals did not change significantly in this period. But on the larger geopolitical level the directional bearing of nations shifted decisively as new technology and political maneuverings repeatedly shocked international relations.

There is a striking contrast between the spatial orientations of the countries of Europe in the two opposing alliance systems. Of the five major belligerents in the war of 1914, England had two unique geographical features, being the smallest nation and the only island. The compactness of population and proximity of cities enabled it to act in a more concentrated manner than Russia, with its vast size and sparse population, or Austria-Hungary, which had numerous disruptive internal divisions. As an island-nation with secure borders England was able to secure a far-reaching empire and had a distinctive outward orientation. British commerce radiated from a compact core and reached around the world, criss-crossing the globe with territory and corridors of influence in every direction. Interests in the American colonies, Canada, Australia, and India spanned the latitudes, and in Africa the British moved longitudinally and almost secured enough territory to run a railroad line south to north from the Cape to Cairo. While pursuing a traditional policy of isolation from the Continent, the English were always attentive to developments across the Channel, especially when they began to threaten British commerce and the British navy, as they did in the 1890s when German production challenged British industrial supremacy (German steel production surpassed British in 1893) and the German government began to build a battle fleet. In a few years splendid isolation turned into a troubling obsession, as Ernest E. Williams elaborated in his alarmist bestseller of 1896, *Made in Germany.* In an inventory of English life Williams pointed out German goods lurking everywhere. "Made in Germany" could be found stamped on clothing, toys, dolls, embroidery, needlework, cottons, leather goods, books, newsprint, china, pencils, prints, engravings, photographs, pianos, mugs, drain pipes, ornaments, wall hangings, iron goods, electrical appliances. German opera, instruments, and sheet music dominated the English musical world. He completed this litany of commercial doom with the most terrifying fact of all—that in 1893 the port of Hamburg surpassed Liverpool for the first time in total tonnage of shipping.[18] And that statistic pointed beyond the area of commercial concern toward the battle fleet that challenged British naval supremacy and threatened the security of the homeland.

England's growing antipathy against Germany was played out on a backdrop of increasing friendship with France. In 1904 the two countries formalized their accord by agreeing to divide up North Africa and acknowledge each other's respective spheres of influence there with the Entente Cordiale. Stretching even farther in space, the British entered into an Entente with Russia in 1907 and soon resolved their long-standing conflicts in Persia, Afghanistan, Tibet, and China. British continental ties intensified in the years after the turn of the century, polarized by the positive and negative currents that coursed across the Channel with growing frequency in a series of diplomatic crises that tightened the bonding within and the conflicts between the two great alliance systems.

The powers composing the Triple Alliance lacked the internal coherence enjoyed by nations of the Triple Entente, especially England and France. Austria was an unwieldy patchwork of different nationalities, cultures, and languages; Italy was plagued with strong regionalism and almost unintelligible dialect differences; Germany was insecure about its wholeness. Although it could be argued that Russia's vast size and large number of nationalities made it as vulnerable to internal chaos as Austria-Hungary, the Russian Empire was centuries old (as were the English and French monarchies) and projected a strong national sense. English unity was imposed by spatial compactness and the insulation of the sea, while the French were galvanized from a long process of the centralization of authority in Paris.[19] French power radiated to the provinces from a capital that was centrally located and was able to act in a far more unified manner than was possible in Austria-Hungary, Italy, and Germany.

Mountains and sea protected French frontiers in three directions, and so its open face was destined to look east. National consciousness had turned outward to empire after defeats on the nearby land mass, but Frenchmen were still traumatized by Germany's victory in 1870 and the collapse of their army at Sedan. Abutting the new powerful Reich, the French were always conscious of the eastern frontier and were fired by a spirit of revenge for their loss of Alsace and Lorraine. That territorial amputation left a wound that would not heal for forty-four years. The attention riveted on the east in the early twentieth century was graphically represented in the textbooks of military schools that prepared future soldiers for another east-west showdown. Up to 1906 many exercises included English landings in the north, but these hostile expectations soon disappeared, as did those designed to counter the newer, Italian adversary, expected to invade from over the Alps. The Pyrenees clearly defended the French against any serious threat from the west, and so military minds fixed on the Rhine.[20]

The defeat that cut deep into French confidence puffed German pride, but the Germans had other borders on their mind. In *The Geographical Causes of the World War* (1920) the German geographer Georg Wegener assessed the unique role that location and topography played in European history: "In scarcely any other country of the world has geography destined such a pitiful tragedy as it has in Germany." As an emigration center for Indo-Germanic people, a crossroads of trade, and a strategic divider of nations lacking secure frontiers, Germany had always been a buffer or a battlefield.[21] In 1912 General Friedrich von Bernhardi went beyond neutral characterizations and charged that the Reich was a "mutilated torso"; Admiral Alfred Tirpitz viewed it, without a navy, as a "mollusc without a shell."[22] Its strategic location appeared to make it inevitable that surrounding peoples and armed forces would repeatedly penetrate its frontiers or blockade it. Moreover, beginning in 1879 Germany embarked on a diplomatic course that assumed its isolation, thereby turning into a self-fulfilling prophecy and creating the myth and the reality of *Einkreisung* (encirclement).

By 1882 Germany, Austria-Hungary, and Italy consolidated the center of Europe in an alliance, setting up the geopolitical conditions for the counterformation of the opposing Triple Entente. They bonded together even more tightly after the Franco-Russian Alliance of 1894 raised the specter of a two-front war. On England's behalf Kipling may have been pleased by the meeting of East and West, as he applauded "two strong men" standing "face to face," but the Germans saw themselves caught in a deadly trap. They commissioned Count Alfred Schlieffen to create a battle plan to keep the two surrounding armies from meeting on German soil. In 1894 the count reckoned that Germany could survive only with a speedy and decisive victory, which he planned to achieve by launching a holding action against Russia in the east while defeating France in a lightning-quick attack. A massive German army would wheel through Belgium, Luxembourg, and southern Holland to Paris, while one-eighth of the main force in the west would act as a decoy for, and contain any attack from, the bulk of the French army stationed across their common frontier. For ten years Schlieffen worked out details that required unprecedented precision and coordination to realize the day-by-day mobilization and battle objectives. In a memorandum of 1905 he drew up the final form of this plan, which dominated strategic thinking for the next nine years as Germany prepared for war.[23] During the last twenty years of the Second Reich the decisive battle on the western front became the *idée fixe* of the German general staff.

The German fear of encirclement became more complete and more severe as England left the orbit of friendly powers and eventually became a partner in the hostile camp of the Triple Entente. Kaiser Wilhelm bungled Anglo-German diplomatic relations in 1895 with a congratulatory wire to President Kruger of the Transvaal, who was fighting the English. In 1905 Chancellor Bernhard von Bülow engineered another diplomatic crisis by having the Kaiser stop in Tangier to protest French designs in Morocco and test the new Anglo-French Entente. But an international congress held the following year supported French rights in Morocco and increased Anglo-French understanding and, in German eyes, tightened the ring around Germany. By far the most decisive cause for hostility in Anglo-German relations was the buildup of the German navy—both in terms of the propaganda of the Navy League formed in 1897 to push for it and the growing number of ships that challenged British national security. German national consciousness oscillated between grandiose visions of *Weltpolitik* and paranoid fears of *Einkreisung* in a political paradox that brought forth a variety of explanations and interpretations. General Bernhardi shifted between pathos and panic, claiming that Germany had been robbed of its national frontiers, was surrounded by enemies, and was open to attack by "Slavonic waves" dashing furiously against the coast of Germanism.[24] Riezler explained the intolerability of encirclement on the eve of the Great War. Unlike all other states, which have essentially one border to defend, Germany was in the center of Europe and had to prepare for battle on all sides. He added a temporal twist to *Einkreisung* when he observed that his country was surrounded by countries with older and more highly developed cultures.[25] Germany was thus cramped in time as well as in space, vulnerable in the infancy of its nationhood and in being at the geographic center of Europe. The Swedish geopolitician Rudolf Kjellén summarized the concerns of Bernhardi and Riezler: "Germany is the 'Kingdom in the Middle' in Europe. This is the central fact of its unique position. It has no free side like the other great powers, no field of expansion on its own door. On all sides it is fenced in by older and larger cultural lands; its essence is to be surrounded by other great powers."[26] This kind of mental state consistently shook German consciousness and traces of it are abundant. During the diplomatic crisis of July 1914, the Kaiser responded to dispatches with marginal notes that were then circulated among government officials. On one urgent telegram from his ambassador in St. Petersburg that he received on July 30, notifying him that Russia was preparing to mobilize, he jotted down a final elaboration of *Einkreisung.* It reads as if he was almost relieved that twenty years of a self-fulfilling prophecy had at last turned a paranoid fear into reality.

> For I have no doubt about it: England, Russia and France have *agreed* among themselves—after laying the foundation of the *casus foederis* for us through Austria—to take the Austro-Serbian conflict for an *excuse* for waging a *war of extermination* against us . . . So the famous *"encirclement"* of Germany has finally become a complete fact, despite every effort of our politicians and diplomats to prevent it. The net has been suddenly thrown over our head, and England sneeringly reaps the most brilliant success of her persistently prosecuted purely *anti-German world-policy,* against which we have proved ourselves helpless, while she twists the noose of our political and economic destruction out of our fidelity to Austria, as we squirm *isolated* in the net.[27]

After the war a debate raged between those who saw *Einkreisung* as a reality and an inevitable cause for war, and those who saw it as a myth and an unnecessary provocation of the peace.[28] While British radial consciousness moved freely outward from a coherent and naturally defended core, German radial consciousness was split between a bombastic outgoing *Weltpolitik* and a paranoid reflex that weakened a heterogeneous and geographically indefensible Fatherland.

Drang nach Osten—there was an urgency about that expression of the German drive to the east. It focused at the turn of the century on two long-term projects, one of which was the building of a Berlin-to-Bagdad railway. In 1899 the Turkish government granted an exclusive concession to a German syndicate, the Anatolian Railway Company, to build a railway to connect Constantinople with Bagdad across Asia Minor. Since the time of Alexander, control of that route had been the key to political and commercial expansion in the east. England saw the completion of the railway as threatening its trade with India and reacted by officially protesting German presence in the Persian Gulf and sending troops to occupy Kuwait. Russia intensified its interest in the Balkans and, together with France, worked to delay financing and construction of the line. In Germany of course there was strong support for this avenue of commercial expansion. In 1900 General Colmar von der Goltz wrote about the prospect of fulfilling Germany's eastern policy and bringing German *Kultur* and goods to the Persian Gulf. "Accordingly," one historian recounted, "many a Reich-German finger followed the thin black lines from Berlin southeastward and many a patriotic heart beat faster at the mention of *'unsere Bagdadbahn.'* "[29] Others envisioned colonies blooming in the desert, making room for the ever expanding German population. In 1902 Paul Rohrbach popularized the railroad project by drawing attention to the political and military as well as the economic advantages of the railroad.[30] As the railroad stretched across the strategic lands toward Bagdad, it emerged as a major cause of international conflict.[31]

In another expression of its drive to the east, Germany envisioned a central European empire, cutting deep into eastern Europe and the Near East. In a famous statement of 1890, the influential economist Gustav Schmoller predicted that in the twentieth century the course of history would be determined by the Russian, English, American, and perhaps Chinese empires and that only a central European customs federation would save the European states from second-rate status.[32] German interest in controlling an economic *Mitteleuropa* was cultivated by a number of economic societies patterned after the one founded in Berlin in 1904 by the German economist Julius Wolf. The most outspoken proponent in the prewar years was the theologian and social critic Friedrich Naumann, who envisioned an enlightened German empire expanding into Eastern Europe, combining the virtues of capitalism, nationalism, and social reform.[33]

Not everyone in Germany looked east with unmitigated enthusiasm. Although the Germanic populations scattered throughout the Austro-Hungarian Empire—the country's only reliable ally—seemed to beckon to be included in a territorially expanded Germany, such an expansion would involve enormous political conflicts. Commercial opportunities opened up by the favorable relations with

Turkey and made possible by the construction of the Bagdad railway seemed attractive, yet it was an alien country. To the southeast lay the troubled Balkans that threatened the peace and territorial integrity of the Austro-Hungarian Empire. Beyond that was the greatest cause for concern, Russia, an empire of yet other alien peoples. Drawing lessons from defeat in the 1905 war with Japan, the Russians were rebuilding their army, pointing it toward the west. And as that army grew in size, it became apparent that the Austro-Hungarian Empire could do little more than slow down the Russian steamroller should it ever be set in motion toward Germany.

Whereas England, France, and Germany all looked outward to empire, the Austro-Hungarian Monarchy focused inward on the numerous internal divisions that threatened its existence. Irredentist movements pulled at it like taffy. Germany attracted its "people" in Austria, Russians looked protectively on the northern Slavs, Italy wanted to include nearby ethnic settlements, Rumania coveted Transylvania, and the Serbians within the empire sought to be incorporated into the independent kingdom that bordered it to the south. Czechs, Poles, and Hungarians clamored for autonomy from within; and the Armenians, Bulgars, Greeks, Albanians, Gypsies, and Jews, honeycombed about the empire, added to its inner tensions. Since the thirteenth century the Habsburg dynasty had spread through a series of marriages and wars. The full imperial title gives a sense of its heterogeneity:

> We, . . . by God's grace, Emperor of Austria; King of Hungary, of Bohemia, of Dalmatia, Croatia, Slavonia, Galicia, Lodomeria, and Illyria; King of Jerusalem, etc.; Archduke of Austria; Grand Duke of Tuscany and Cracow; Duke of Lothringia, of Salzburg, Styria, Carinthia, Carniola and Bukovina; Grand Duke of Transylvania, Margrave of Moravia; Duke of Upper and Lower Silesia, of Modena, Parma, Piacenza and Guastella, of Ausschwitz and Sator, of Teschen, Friaul, Ragusa and Zara; Princely Count of Habsburg and Tyrol, of Kyburg, Görz and Gradiska; Duke of Trient and Brixen; Margrave of Upper and Lower Lausitz and in Istria; Count of Hohenembs, Feldkirch, Bregenz, Sonnenberg, etc.; Lord of Trieste, of Cattaro and above the Windisch Mark; Great Voyvod of the Voyvodina, Servia, etc., etc.[34]

The Hungarian historian Oscar Jászi argued that the monarchy was doomed to death from the centrifugal forces of national particularism; another observer predicted that after the death of Emperor

Francis Joseph it would fall apart "like an old barrel robbed of its hoops."[35]

The internal weaknesses of the empire monopolized the attention of its leaders and created a power vacuum that regulated foreign policy. These conditions "intensified the greediness of neighbors and made their claims bolder and more arrogant."[36] The greatest trouble spot was the Balkans. The Austrian annexation of Bosnia in 1908 unsettled the entire region and charged the Russians with added determination to prevent further expansion by Austria-Hungary. The Pan-Serbian movement became particularly threatening as Serbs within the empire and in neighboring Serbia sought to unite. Breaking off a chunk of Austria-Hungary would invite further nationalist uprisings and, it was feared, begin the complete dissolution of the Habsburg Monarchy. The Austrians were reasonably certain that in case of war Germany would honor its treaty obligations and protect the monarchy's western border, but their anxiety mounted as they peered east and contemplated what would happen if the chronic Balkan crisis should some day draw them into war with Russia. And the rest of Europe wondered the same thing.

As early as 1883, from the security of the British isles, J. R. Seeley speculated: "Russia already presses somewhat heavily on Central Europe; what will she do when with her vast territory and population she equals Germany in intelligence and organization, when all her railways are made, her people educated, and her government on a solid basis?"[37] Following a trip through Russia in 1901, a pro-imperialist senator from Indiana, Albert J. Beveridge, warned about the "gray-clad, militant figure" standing on the frozen shores of the Pacific that in the last five years "has claimed the acute attention of every cabinet in Europe and of every thoughtful American citizen." The Trans-Siberian railway, "by far the greatest single work of construction recently accomplished anywhere in the world," had overnight created a powerful and menacing Russian presence in the Far East.[38] Henry Adams imagined the huge bulk of Russia combining with that of China to constitute a "single mass which no amount of new force could henceforward deflect." The Trans-Siberian, like the cathedral at Chartres and the dynamo, was the material representation of an irresistible expansive force—the *vis inertiae* of the Russian Empire.[39]

Germans viewed Russia with envy and fear; the political analyst Wolf von Schierbrand surveyed the popular expression of these sentiments. The "huge and increasing bulk" of the Russian Empire, its "dangerous" policy of expansion, its seemingly "inexhaustible" resources made it a threat to German continental hegemony and world political aspirations.[40] In *Beyond Good and Evil* Nietzsche commented on the vast space that makes Russia the country of the future, "an empire that has time and is not of yesterday."[41] Russia was, in a literal sense, big with future. In *The Magic Mountain* Mann observed the same thing. After several years at the Berghof, Hans Castorp began to give himself so fully to the timelessness that even Settembrini, who at first chided him about his punctuality, remarked that "there is something frightful in the way you fling the months about." He warns Hans not to succumb to the prodigious wastefulness of the Asiatic world that comes from having an abundance of space.

> This barbaric lavishness with time is in the Asiatic style; it may be a reason why the children of the East feel so much at home up here. Have you never remarked that when a Russian says four hours, he means what we do when we say one? It is easy to see that the recklessness of these people where time is concerned may have to do with the space conception proper to people of such endless territory. Great space, much time—they say, in fact, that they are the nation that has time and can wait. We Europeans, we cannot. We have as little time as our great and finely articulated continent has space, we must be as economical of the one as of the other, we must husband them, Engineer![42]

In the language of contemporary geopolitics Mann interprets the broad reaches of the Russian continental empire as the basis for an expansive temporal thrust toward the future.

Russians were acutely aware of the way the vast space shaped their character and historical destiny. One telling example of how fully some of them came to rely on that geopolitical factor is a remark made in August 1915 by the War Minister, Aleksei Polivanov, about the disastrous situation at the front: "I place my trust in the impenetrable spaces, impassable mud, and the mercy of Saint Nicholas Mirlikisky, Protector of Holy Russia."[43] The main target of the Russians' spatial reach was Constantinople and the Straits of the Dardanelles, where they hoped to secure free access to the lucrative and strategically crucial sea lanes of the Mediterranean. Russian foreign policy fanned over the region and into the Balkans where, in league with Serbia, Russia drew the line against the Central Powers and helped create the first fighting front of the Great War.

While foreigners viewed the "inertial force" of Russia as promoting an outward-moving continental surge, the Russians became ever more conscious of the propinquity and ubiquity of hostile nations. Germans were perhaps more outspoken about their sense of encirclement, but Russians saw an even larger number of potential enemies in every direction. In the west, Norway and Sweden harbored old hostilities, while Germany dominated the Baltic Sea and buttressed the most powerful coalition of continental armies in Western Europe. Austria-Hungary and Rumania completed the frontier of unfriendly neighbors in the west, while the opening of the Black Sea in Russia's southwest corner was locked shut by the united arms of Turkey and Germany. Lining their own side of the western frontier was a chain of potentially rebellious suppressed minorities of Finns, Estonians, Latvians, Lithuanians, Poles, Jews, and Ukranians. To the south were old sites of feuding with Persia and Afghanistan. From the southeast Mongol hordes could still give nightmares to the Russian soul, and off the east coast was Japan and the memory of a fresh military defeat. Nature itself bore down menacingly from the north with rough Arctic terrain and ice.

The diversity of Russian topography, climate, and people was manifested in a diversity of national images. Ladis Kristof has identified four, centered in different regions and pointing in different directions. The first focused on medieval Kiev as the heart of a Russia that derived from "the original, European channel." It turned its back on Mongol-Tartar Russia and looked to the west for cultural historical origins. The second located the heartland around Moscow and saw Russia as distinctly separate from Western Europe. It was an inner-directed national image, centered on the vast territory of Russia itself as the source of identity and autonomy. The third image was associated with the city of St. Petersburg, linking Russia culturally and historically with Western Europe. This tie began to tighten with the Franco-Russian Alliance of 1894. The fourth image was of a Russia spread over the vast stretches from the Carpathians to the Pacific, a cultural melting pot and geographical meeting point of the east and west occupying the region of the Eurasiatic steppe.[44] As Western European countries built large colonial empires in the late nineteenth century, Russia began to focus on the resources and opportunities of its eastern parts. The opening of China to world trade and commercial investment in the late 1890s and the simultaneous emergence of Japan as a power to be reckoned with pivoted the attention of the world to the east and attracted the attention of Russians to that flank of their continental empire. On the eve of war in 1914 Russia, like Germany, was torn between two spheres of interest, mindful of the possibility of war at either extremity of the east-west axis.

Throughout the period the attention of the world powers was constantly drawn to the new dynamics that emerged along that axis. The new east-west railroad lines, the rise in global travel, the division of the world into geographically precise and temporally ordered time zones, the alignment of the alliance systems, and the battle plans of generals underscored the ancient and universal significance of the east-west axis as the direction of the earth's rotation and the location of dawn and sunset so deeply imbedded in the poetry and imagery of human consciousness. European scholars had long cultivated fantasies of the east as a place of "romance, exotic beings, haunting memories and landscapes, remarkable experiences."[45] It now also emerged as a place of speculation, exotic raw materials, risky and lucrative investments, fabulous profits. The political and economic ties between east and west were also the subject of bold hypothesizing.

The most novel geopolitical thesis of the early twentieth century focused on the Eurasian heartland. In opposition to traditional speculation about the westward march of empire, Sir Halford Mackinder looked east and considered Europe as subordinate to Asia, because "European civilization is, in a very real sense, the outcome of the secular struggle against Asiatic invasion." The age during which sea power had determined the sources of national might and the distribution of wealth had come to an end. A generation earlier steamships and the Suez Canal increased the mobility and significance of sea power, but the transcontinental railways were transforming the significance of land powers, especially in the heartland of Euro-Asia, and henceforth land power would dominate. Russia occupied the central strategical position, equal to that held by Germany in western Europe and, with a fully developed railroad system, would be able to dominate the entire continent. The resurgence of land power was dividing the world into distinct geopolitical regions. An "outer crescent" included Britain, South Africa, Australia, the United States, Canada, and Japan; inside that was the "inner crescent" of Germany, Austria, Turkey, India, and China. In the center was the "pivot area"—a large wedge-shaped section of Russia broadest in the frozen north and narrowing toward the Persian Gulf. Surprisingly the heartland excluded almost all of the European north, about half of Central Russia up to the outskirts of Moscow, all of the Ukraine, the Baltic provinces, and the entire Pacific coast. In his desire to emphasize the new strategic potential of the land-locked core, Mackinder discarded the perimeters, even though they included the most highly populated and industrialized regions and those that were strategically located in contact with neighboring states or the sea. Mackinder was convinced that the advantages of internal lines of communication and transportation would enable Russia to pick off enemies one by one, control the heartland, and, if allied with Germany, secure a transcontinental power base from which it might control the "empire of the world." The westward movement of empire, he concluded, was but "a short rotation of marginal power round the south-western and western edge of the pivotal area."[46]

The spectre of a world empire dominated from the heartland did not become a reality in that generation. Mackinder greatly overestimated the role that the railroads could play, as was evident in the following year when Russia's first military foray over the Trans-Siberian railway resulted in a catastrophic defeat by Japan. He tended to see the sprawling Eurasiatic land mass in Western European terms and hence anticipated a far greater coherence of action than Russia could possibly achieve at that time. The Russians themselves ignored his thesis, because it left some of the most important parts of their country out of the crucial pivot area and looked to desolate backwater areas to be the core of a world empire.[47] But the theory did attract widespread attention in Western Europe and the United States, where global thinking conformed to political reality. Mackinder argued that for the first time the world could, and indeed must, be considered as a whole in which important national developments had worldwide ramifications. His thesis had an unsettling revisionistic thrust, displacing the seat and future of world power away from the west to the east.

The shifting directional orientation of nations was but one aspect of the complex history of diplomacy that led to war. In the thirty-seven days between the assassination of the Archduke Francis Ferdinand on June 28 and the last declarations of war on August 4, the period of the "July Crisis," these orientations repeatedly affected negotiations as leaders tried to prevent, and at the same time prepare for, the outbreak of hostilities. And the sense of direction among nations was but one element of the general sense of space that shaped the identity and bearing of nations. Changes in the sense of time were manifested in the pace of diplomacy during that period and the pace of fighting afterwards.

10

TEMPORALITY OF THE JULY CRISIS

Out of the millennia over which he traced the rise and fall of civilizations, Spengler identified the unique temporality of the period that precipitated the decline of the West: "In the Classical world years played no role. In the Indian world decades scarcely mattered; but here the hour, the minute, even the second is of importance. Neither a Greek nor an Indian could have had any idea of the tragic tension of a historic crisis like that of August, 1914, when even moments seemed of overwhelming significance."[1] The crisis would also have been unfathomable to anyone who had lived before the age of electronic communication. In the summer of 1914 the men in power lost their bearings in the hectic rush paced by flurries of telegrams, telephone conversations, memos, and press releases; hard-boiled politicians broke down and seasoned negotiators cracked under the pressure of tense confrontations and sleepless nights, agonizing over the probable disastrous consequences of their snap judgments and hasty actions. During the climactic period between July 23 and August 4 there were five ultimatums with short time limits, all implying or explicitly threatening war if the demands were not met. In the final days the pressing requirements of mobilization timetables frayed the last shreds of patience. And even before mobilizations were formally announced, armies prepared for war, making a sham of the efforts of diplomats, who continued to go through the motions of negotiating as time and the peace slipped away.

The assassination of Archduke Francis Ferdinand, heir to the Austro-Hungarian throne, triggered an international diplomatic crisis that was a mere prelude to the coming war. Political leaders received the horrendous news via quick telegraphic communications, and the shocked public read about it in the daily press. Newspapers immediately and directly affected reactions, creating an enormous audience. Popular anger encouraged the indignation of the Dual Monarchy and reduced the flexibility of all diplomats who had to worry about their own governments as well as their opposition.[2] Although the cinema did not play as immediate a role as the press in whipping up war hysteria, over the years it had stimulated popular interest in political issues. American and English film makers used the medium in its infancy to propagandize about the Spanish-American and Boer wars.[3] By 1913 a French movie critic observed the enormous impact of the cinema on politics, and especially on the career of President Poincaré, whose popularity in France was greatly enhanced by his frequent appearances in newsreels. He believed the cinematograph was "a kind of popular annex to the Elysée Palace."[4] Stefan Zweig noticed the influence of the cinema on politics and was struck by the hatred of the people of Tours who jeered when Wilhelm II appeared in a French newsreel.[5] The technology of mass communication had become a factor in political and diplomatic affairs and directly accelerated popular response to the already frenetic diplomatic activity.

European capitals responded to events in 1914 as if they were so many outlets along a party line, jumping at the jingle in every foreign office. But there was a sequence of pivotal events focused in single locations. Immediately after the assassination the Dual Monarchy held the center stage as it poised to devise the best means to crush the Pan-Serbian movement. Most of the Austrian leaders realized that it was essential to react swiftly. The Chief of the General Staff, Conrad von Hötzendorf, called the situation acutely threatening, and said that "a decision could no longer be postponed."[6] One of the few officials in Vienna who spoke out for restraint at the outset was German Ambassador Tschirschky, who advised his Chancellor, Bethmann Hollweg, on June 30 "against too hasty steps." When the German Kaiser read that report on July 2, he was enraged and jotted in the margin, for Bethmann and other top officials: "Now or never. Who authorized him to act that way? That is very stupid!" And he added, "Tschirschky be good enough to drop this nonsense! The Serbs must be disposed of, *and* that right *soon!*"[7] Tschirschky got the message in Vienna and on July 4 told the *Frankfurter Zeitung* that "Germany would support the Monarchy through thick and thin in whatever it might decide regarding Serbia, [and that] the sooner Austria-Hungary went into action the better. Yesterday would have been better than today, and today would be better than tomorrow."[8] This suggestion was formalized in a private meeting at Potsdam on July 5, when the Kaiser issued the "blank check," personally assuring Austrian Ambassador Szögyény of Germany's full support for whatever measures his country contemplated in response to the assassination and urging him to tell Vienna that "action must not be delayed." The Kaiser shared a notion, widely held among military leaders everywhere, that Russia was not yet prepared for war, and therefore the present moment would be the most advantageous to settle the Serbian problem. On July 6 Szögyény telegraphed Berchtold, the Austrian Foreign Minister, that Bethmann also considered "immediate action on our part as the best solution of our difficulties in the Balkans."[8] During that first week in July even the English were in favor of a speedy reaction by Austria to the assassination in order to contain the problem.[9]

At a meeting of ministers on July 7 Berchtold proposed that Austria move quickly to come to terms with Serbia, but Hungarian Prime Minister Tisza was at first reluctant to present an ultimatum to Serbia. By July 14 he was converted to the tactic of an ultimatum with a time limit, but other developments persuaded the Austrians not to deliver it until July 23. During the lull, while the ultimatum was being debated and then drafted and then held for delivery, there were repeated urgings for haste. On a communication from Tschirschky of July 10, reporting the Austrian decision to put an end to the trouble in Serbia, the Kaiser penned an impatient, "it is taking a long time," and suggested demanding that Serbians "clear out of the Sanjac! Then the row will be on at once!" On July 11 Tschirschky reminded Berchtold "that quick action was called for."[10] That same day Riezler noted in his diary that the Austrians would take "frightfully long" to mobilize—sixteen days. "That is very dangerous. A fast *fait accompli,* and then friendly with the Entente, then they can stand the shock."[11] Thus Bethmann, the Kaiser, Riezler, Szögyény, Berchtold, Tschirschky, and even Grey all thought along similar lines—a speedy blow against Serbia would work to contain the scope of any hostilities that might result. Even Conrad was aware of the need to be quick. On July 12 he wrote to Berchtold:

> . . . in the diplomatic field everything must be avoided in the nature of protracted or piecemeal diplomatic action which would afford our adversary time for military measures and place us at a military disadvantage. . . . Hence it would be wise to avoid everything that might prematurely alarm our adversary and lead him to take counter-measures; in all respects a peaceable appearance should be displayed. But once the decision to act has been taken, military considerations demand that it must be carried out in a single move with a short-term ultimatum which, if rejected, should be followed immediately by the mobilization order.[12]

By July 18 the Austrians had drafted an ultimatum and had decided on a forty-eight hour limit for the Serbians to consider it.

A great many factors led to the breakdown of peace, but the sheer rush of events was itself an independent cause that catapulted Europe into war. The kind of temporal precision that Spengler saw in retrospect as unique to this period was particularly evident in German Secretary of State Jagow's adjustment of the exact hour for delivery of the ultimatum. On July 21 he learned that President Poincaré of France was scheduled to depart from St. Petersburg by ship at 11:00 P.M. after a state visit, which was later than the 5:00 P.M. Central European Time when the ultimatum was scheduled to be presented in Belgrade. He immediately sent a telegram to Tschirschky in Vienna, who passed along this information just in time to arrange to have the Austrian officials delay the *befristete Démarche* by an hour until 6:00 P.M., when Poincaré would definitely be at sea, unable to confer face to face with the Russians about a possible course of action.[13]

When Serbian Minister Paču received the Austrian Ambassador Giesl at the Foreign Office in Belgrade on July 23 and received the text of the Austrian note demanding a response by July 25, he replied that some of the ministers were absent from the capital and a quick response was impossible. Giesl, the modern diplomat, countered that "the return of the ministers in the age of railways, telegraph and telephone in a land of that size could only be a matter of a few hours and that he had already that morning suggested [Prime Minister] Pašić's being informed."[14] His cool response that Serbia should have no trouble rounding up its ministers within 48 hours was almost as arrogant as the ultimatum itself. He could not have been more wrong about the suitability of the time limit, but he was right about one thing—it was an age of the telegraph and telephone. It was ironic that one diplomat should remind another of that fact at that moment, as if the availability of time-saving technology justified such precipitous diplomacy.

As soon as the terms of the ultimatum became known, once more disseminated by speedy modern communication, leaders in every European capital realized that it meant war. Grey described it as "the most formidable document he had ever seen addressed by one state to another that is independent"; the Russian Foreign Minister Sazonov exclaimed "C'est la guerre européenne"; and within the Austro-Hungarian Empire itself preparations were immediately taken to evacuate the entire legation in Belgrade. The content of the ultimatum was most certainly one of the reasons that it was at once interpreted as a provocation for war, for items 5 and 6 demanded that Austrian officials participate in the suppression of subversive movements in Serbia and take part in the investigation relating to the trial of the assassins; but the exceptionally short time limit was equally important. The English, Russian, and French diplomats immediately reached this conclusion and tried to extend the ultimatum's terminal date. On July 24 the German Ambassador in London, Lichnowsky, telegrammed Jagow in Berlin: "What Sir Edward Grey most deplored, beside the tone of the note, was the brief time-limit, which made war almost unavoidable. He told me that he would be willing to join with us in pleading for a prolongation of the time-limit in Vienna, as in that way *perhaps a way out* might be found." The same day Sazonov sent a telegram to missions in Vienna, Berlin, Paris, London, Rome, and Bucharest that opened with a plea for more time:

"As Austria-Hungary has only addressed herself to the Powers twelve hours after the delivery of her ultimatum to Belgrade it is impossible for the Powers in the short time remaining to undertake anything useful towards the settlement of the complications that have arisen. Therefore, in order to avoid the innumerable and universally undesirable consequences to which Austria's course of action would lead, we should think it necessary that she should first of all extend the time limit which she has set Serbia for a reply." The following day the French Ambassador in Berlin told Jagow, somewhat more obliquely, that "the shortness of the time limit given to Serbia for submission would make an unpleasant impression in Europe."[15] And from aboard the *France*, voyaging in the company of Poincaré, French Minister of Foreign Affairs Viviani, after learning the substance of the ultimatum by a radiogram from Russia, sent a wireless message to St. Petersburg, London, and Paris urging that Serbia "request an extension of the twenty-four hour [sic] time-limit" and try to work out an acceptable response compatible with its honor and independence.[16]

An accurate reconstruction of events up to July 25 requires a temporal precision accurate to the day; after that the hour, sometimes even the minute, becomes crucial.

On July 25, while the Entente powers were still straining to get the time limit extended, Austria and Germany seemed to be of one mind on the need to move quickly after the anticipated rejection of the ultimatum in the hope of containing the scope of the war. Around four hours before the expiration of the time limit, Szögyény telegrammed to Berchtold the thinking in Berlin: "here the general belief is, that if Serbia gives an unsatisfactory answer our declaration of war and war operations will follow immediately. Here every delay in the beginning of war operations is regarded as signifying the danger that foreign powers might interfere. We are urgently advised to proceed without delay, and to place the world before a *fait accompli.*"[17] And even Tisza, who at first had been the voice of caution, now took the lead in urging Emperor Francis Joseph to order immediate mobilization in the event of an unsatisfactory reply from Serbia. "The slightest delay or hesitation," he wrote, "would gravely injure the reputation of the Monarchy for boldness and initiative and would influence the attitude not only of our friends and foes but of the undecided elements and result in the most fatal consequences."[18]

On the same day, before the Serbian reply was known, Russia ordered the commencement of a "period preparatory to war"—a partial measure that involved the mobilization of four military districts, activation of the Baltic and Black Sea fleets, and acceleration of army supply. The justification was that since Russia required more time to mobilize than any of the Western powers, it was necessary to set in motion some preliminary operations in order to put the preparedness of Russia on equal footing with the others should war break out. This partial or precautionary mobilization became an ingredient of diplomacy in the race against time as the other countries grew increasingly alarmed at Russia's jump-start and began to worry about the pressing timetables of their own mobilization plans. The first country to order full mobilization was Serbia—at 3:00 P.M. on July 25, three hours before it was obliged to hand its reply to the Austrian ambassador.

At a time when communication technology imparted a breakneck speed to the usually slow pace of traditional diplomacy and seemed to obviate personal diplomacy, one piece of machinery failed to work. As the Serbian reply was being dictated, the only remaining typewriter in the Serbian Foreign Office broke down and the final text was copied by hand—an eerie omen of the impending hostilities, fought with human hands and paid for with human lives no matter how impressive the mechanical weaponry.

The exit of the Austrian legation from Belgrade was a prelude to the last frenzied days. Giesl knew before the deadline arrived that the Serbian reply would not be unconditional acceptance, and following strict instructions from Berchtold he burned the code books, prepared to close the legation, and packed his bags. At 5:55 P.M. Pašić handed him the Serbian reply. Giesl quickly determined that it was indeed unsatisfactory, sent Pašić a note severing diplomatic relations, drove to the Belgrade station in time to catch the 6:30 P.M. train, and ten minutes later crossed the Austrian frontier, where he immediately sent telegrams to Berchtold and Tisza. He then telephoned Tisza, who was waiting in Budapest and had kept the lines open to receive the call and relay the report to Vienna.[19]

There is a good deal of evidence that Austria did not regard the Serbian reply as requiring an immediate declaration of war. When Francis Joseph learned of the rupture of diplomatic relations, he said that that "is not necessarily a *casus belli*!" And in a conversation with Giesl after his arrival in Vienna, Berchtold emphasized that "the breaking off of relations is not by any means war."[20] But on July 26 Germany pressed for an immediate declaration of war to forestall drawn-out mediation by other powers and to keep the war localized between Austria and Serbia. On the morning of the 26th Berchtold received Szögyény's telegram of the preceding day, reporting that following an unsatisfactory Serbian reply Germany wanted no delay in the declaration of war. Berchtold then met with Tschirschky and Conrad. The German ambassador reaffirmed the German position, while Conrad, who had earlier been eager to fight, urged caution, pointing out that a full Austrian mobilization would take sixteen days. After Tschirschky left, Berchtold told Conrad: "We should like to deliver the declaration of war on Serbia as soon as possible so as to put an end to diverse influences. When do you want the declaration of war?" Conrad replied: "Only when we have progressed far enough for operations to begin immediately—on approximately August 12." And Berchtold responded: "The diplomatic situation will not hold as long as that."[21] This was one instance when a mobilization timetable might have had a dilatory effect; but the diplomatic situation was getting worse every day and Austria finally bowed to German pressure. On July 27 Conrad agreed to an immediate declaration of war. At 4:37 P.M. Berlin received the news in a triumphant telegram from Tschirschky: "It has been decided here to send official declaration of war tomorrow, at the latest the day after, in order to cut away the ground from any attempt at intervention."[22]

The Austrian declaration of war was sent just before noon on July 28. For Tschirschky it was in the nick of time, or it might have been thwarted by a most unexpected source. Only an hour earlier Kaiser Wilhelm, having just returned from a trip to Scandinavia, learned of the Serbian reply early on the morning of the 28th and wrote in the margin of his copy: "A brilliant performance for a time-limit of only 48 hours. This is more than one could have expected! A great moral victory for Vienna; but with it every reason for war drops away, and Giesl might have remained quietly in Belgrade. On the strength of this *I* should never have ordered mobilization!" He immediately sent a handwritten letter to Jagow, repeating that "every cause for war falls to the ground" and urging that Austria merely occupy Belgrade as a "hostage" and guarantee for the carrying out of Serbian promises.[23] But his "Halt in Belgrade" proposal (as it came to be known) was too late, as were all the others made in the next few days. Since the assassination Germany had been rattling Austria's sabre ever more furiously, and the Kaiser's eleventh-hour change of heart was not sufficient to stop the outbreak of war.

Throughout the 28th Bethmann neglected to reinforce the Kaiser's plan and try to block the declaration of war and instead occupied himself with numerous other matters that came into his office from the foreign offices and governments of all countries concerned. Only at 10:15 that night, after Austria had already declared war on Serbia, he telegrammed to Tschirschky in Vienna a version of the Kaiser's plan for a Halt in Belgrade, but omitted to mention the Kaiser's more important instruction to tell Vienna that "no more cause for war exists." His skimpy translation of the Kaiser's mediation proposal caused it to be disregarded, like all the others that were made in the next few days.

The Kaiser, staying out at Potsdam, did not use the telephone to communicate with his ministers, and so he was not aware of what was, or was not, happening in Berlin on July 28 to implement his peace plan. But he did use the telegraph in a last-minute exchange with his cousin Tsar Nicholas II of Russia to try and avoid a European-wide war. At 1:45 A.M. on July 29 he sent a telegram to his cousin, reminding him of the "dastardly murder" that sparked the crisis and appealing for help to "smooth over the difficulties that might still arise." That telegram crossed one from Nicholas to him sent at 1:00 A.M. The Tsar made a similar appeal, urging the Kaiser to stop Austria from "going too far" and to help him resist the pressures that were building up in Russia that were about to force him "to take extreme measures which will *lead to war.*" At 8:20 P.M. Nicholas sent another cable to Wilhelm suggesting that the Austro-Serbian problem be resolved by the Hague Conference. That evening Wilhelm replied to Nicholas's first telegram with a suggestion that must have seemed equally preposterous to the Tsar—that Russia "remain a spectator" in the Austro-Serbian conflict. The Tsar responded early on July 30, justifying the Russian military measures under way as necessitated by Austria's own preparations. At 3:30 P.M. of the 30th the Kaiser telegrammed back, warning that Russian mobilization threatened the peace and concluding with a final jab: "The whole weight of the decision lies solely on your shoulders now, who [will] have to bear the responsibility for peace or war." The Tsar's reply at 2:55 P.M. on the 31st reveals the extent to which military necessities had encroached on diplomacy: "It is *technically* impossible to stop our military preparations which were obligatory owing to Austrian mobilization." On August 1 Nicholas telegrammed his cousin for the last time. He conceded that Germany was obliged to mobilize in turn but requested that Wilhelm give him the same guarantee that he had offered—"that these measures do not mean war." But the Kaiser could not give such a guarantee, because mobilization in Germany was linked inextricably with war on two fronts. At 10:30 P.M. on the evening of August 1, the Kaiser closed this unique exchange with a short telegram informing the Tsar that he had ordered mobilization of the German army.[24] It contained no guarantees.

The telegrams exchanged between the Tsar and the Kaiser constituted a small fraction of the hundreds sent during the negotiations. Because they took place between the two monarchs of rival powers, they highlighted the strength and the weakness of telegraphic communication. The telegraph was unquestionably speedy. As war seemed imminent, both men, within forty-five minutes of one another, independently decided to reach across the vast distances to make a direct and personal appeal to the family ties, the traditions, and the shared values of the two monarchies in order to try and save the peace. And the telegraph made it possible. But the mechanical impersonality of this exchange excluded the expression of human sentiments that could have emerged in a face-to-face meeting. Proust had a vision of death when he first spoke to his grandmother over the telephone; perhaps the Kaiser and the Tsar heard a death rattle of diplomacy in the clicking of the telegraph key. This telegraphic exchange at the highest level dramatized the spectacular failure of diplomacy, to which telegraphy contributed with crossed messages, delays, sudden surprises, and the unpredictable timing. Throughout the crisis there was not just one new faster speed for everyone to adjust to, but a series of new and variable paces that supercharged the masses, confused the diplomats, and unnerved the generals.

The arrogance, the lack of safety precautions, the reliance on technology, the simultaneity of events, the worldwide attention, the loss of life all evoke the sinking of the *Titanic* as a simile for the outbreak of the war. The lookouts on the *Titanic* were blinded by fog, as the political leaders and diplomats and military men were blinded by historical shortsightedness, convinced that even if war came it would not last long. On the eve of disaster they shared a confidence that the basic structure of European states was sound, able to weather any storm. Europe, they were certain, was unsinkable. The concentration of wireless messages from the sinking ship, the rescue ships, and the coastal stations suggests the flurry of telegraph messages and telephone conversations exchanged during the July Crisis. Even the icebergs floating in the path of the liner had an analog in the eight assassins who lay in wait for Francis Ferdinand at various points on his parade route the day he was murdered. He had ignored warnings about the danger from terrorists in the streets of Sarajevo, as the captain of the *Titanic* had ignored wireless messages about the dangerous waters ahead. The captain raced against time for the fastest Atlantic crossing. Another race took place between the armies of the great powers, which rushed to mobilize toward the end of July as the diplomacy began to founder.

The importance of speed in the mobilization and concentration of men on the battlefield were lessons learned from German victories against France in 1870. The French had been overwhelmed by numbers, unable to field more than 240,000 soldiers to oppose the force of 370,000 that General Moltke was able to get across the frontier efficiently with the highly developed German railroad system. "Build no more fortresses, build railways," Moltke ordered, and he shaped German strategy accordingly.[25] In a book on modern warfare, published in 1885, the French General Victor Derrécagaix agreed that the first concern of a nation must be "to cover its territory with a network of railways which will ensure the most rapid possible concentration." The mobilization of mass armies requires the simultaneous use of countless national resources: "the installation of new authorities; the formation of new agencies; the organization of depots, interior garrisons, commands, special governments, and station services; the creation of new staffs; the organization of trains, parks, convoys, and accessory field services; the assembling of horses, provisions, munitions, means of transport, etc."[26] The lesson of 1870 was not lost on other European powers, which soon commenced the construction of railway lines in coordination with military needs. The Russians, whose vast distances made it impossible to compete with the Europeans in mobilizing and moving their forces quickly, tried to use railroads to thwart an invasion of their homeland. They constructed lines on a broader gauge (5 feet instead of the European gauge of 4 feet 8.5 inches) to make invasion difficult for an enemy with railroad superiority.

By 1914 armies were vastly larger than those fielded in the Franco-Prussian war; the problems of staffing, supply, and concentration of men much greater; and the need for temporal precision even more acute. One observer of the French mobilization in 1914 explained the kind of timing that was necessary: "From the moment mobilization is ordered, every man must know where he has to join, and must get there in a given time. Each unit, once completed and fully equipped, must be ready to proceed on a given day at the appointed hour to a pre-arranged destination in a train awaiting it, which in its turn must move according to a carefully prepared railway scheme. Each unit has also to drop into its place in the higher formations, and these again must find themselves grouped in position according to the fundamental plan. No change, no alteration is possible during mobilization. Improvisation when dealing with nearly three million men and the movements of 4,278 trains, as the French had to do, is out of the question." The German mobilization timetable was even more precise, for it led immediately to war on two fronts and required the element of surprise to succeed. The French and the Russian army made a real distinction between mobilization and war, but not the Germans. As A. J. P. Taylor noted, Schlieffen "had no plan for mobilization."[27] Once the Germans began to mobilize war inevitably followed, since the execution of the plan in the west called for the immediate invasion of Belgium. After the outbreak of hostilities, the timetable of the battle offensive was equally exacting. The Germans had to reach Paris quickly and roll the French army toward the German frontier to prevent the French forces from retreating behind Paris, which would prolong the fight in the west and delay the transfer of troops to the Russian frontier, where they would be needed to deal with the enormous Russian army. The arrogance of Schlieffen's plan is evident in the general's memorandum of 1905 that cooly assumed German forces would proceed "undisturbed." "It can be assumed," he wrote, "that the German deployment takes place undisturbed . . . It is important that north of the Meuse the defile between Brussels and Namur is passed *before* a clash with the enemy, so that beyond it the deployment of 9 army corps can develop without interruption."[28] The only way to maintain the timetable was to act with utmost speed and overwhelming force to assure success in all phases, never let the enemy recoup, and stay on the offensive from beginning to end. Almost every trained soldier was to take the field in the first weeks with no reserves left for a prolonged fight. Schlieffen staked everything on a quick victory and planned his whirling right hook with meticulous precision. The other belligerents could try to bluff by means of mobilization orders in the hope that a show of force might achieve diplomatic success, but the Germans could not. For them mobilization was war.[29]

One reason for the diplomats' failure to avert the war was that they lacked clear understanding of the various kinds of mobilization. Sazonov in particular appears not to have grasped the full implication of Russia's "partial mobilization" against Austria—that it was in fact a mobilization of all Russian troops slated to march against Austria under a complete mobilization, that Austria would therefore have to counter with full mobilization, that Russia would then be forced to mobilize fully, which in terms of the alliance between Austria and Germany would call for a total German mobilization and hence war. Bethmann had some sense of this at least by July 26 when he instructed his ambassador to Russia, Friedrich von Pourtalès, to tell Sazonov that "preparatory military measures on the part of Russia directed in any way against ourselves would force us to take counter measures which would have to consist in mobilizing the army. Mobilization, however, means war, and would moreover have to be directed simultaneously against Russia and France, since France's engagements with Russia are well known."[30] But neither Bethmann nor Pourtalès knew that the German plan called for moving troops across the frontier into neutral Belgium only a few hours after the order for mobilization. As the diplomats parleyed over mobilizations and partial mobilizations, time was quickly running out: for Germany there was no time for diplomacy in their mobilization plan.

The drama preceding the order for full Russian mobilization illustrates the close timing of events in which the telephone played a special role. On the 29th the circuitry was buzzing all over Europe, but especially in St. Petersburg. The bombardment of Belgrade signaled to the Russians an extreme danger, and Sazonov and the Tsar succumbed to pressure from the Chief of the General Staff, Ianushkevich, who was afraid of being caught off guard because of the slowness of Russian mobilization. That morning the Tsar gave him the order for full mobilization. To make it official required signatures by the ministers of War, Marine, and Interior. Ianushkevich delegated that task to the Chief of the Mobilization Section, General Dobrorolski, who secured the signatures and proceeded to the central telegraph office in St. Petersburg to dispatch the order throughout the empire. At 9:30 P.M., as the telegraph operator was typing up the order for transmission, Ianushkevich telephoned Dobrorolski at the telegraph office and instructed him to hold the telegram until the arrival of a messenger. A few minutes later the messenger arrived and explained that the Tsar had had second thoughts and was ordering him to declare a partial mobilization instead. The Tsar had changed his mind after he received Wilhelm's telegram suggesting that Russia might "remain a spectator" in the Austro-Serbian conflict. Partial mobilization would get in the way of any subsequent full mobilization, and Dobrorolski, thinking only of the military danger, became wild with frustration. The morning of the 30th Ianushkevich attempted to intervene. Over the telephone he urged Sazonov to remind the Tsar of the technical problems of a partial mobilization and its political consequences—that France might regard Russia as failing to honor her alliance obligations and that Germany might persuade the French to remain neutral and then fall on the Russian army, which would be in disarray changing from partial to full mobilization. He implored Sazonov to call him at once when the Tsar changed his mind. "After this," he added, "I will retire from sight, smash my telephone, and generally take all measures so that I cannot be found to give any contrary orders for a new postponement of general mobilization." In an interview with the Tsar that afternoon Sazonov pressed the issue, reminding the Tsar of the "thousands and thousands of men who will be sent to their death" if Russia is attacked unprepared and urging him to think of the safety of the empire. The Tsar finally agreed to an immediate general mobilization. Sazonov rushed to a telephone on the ground floor of the royal palace and relayed the news to Ianushkevich, adding: "Now you can smash your telephone. Give your orders, General, and then—disappear for the rest of the day."[31] Once again Dobrorolski got the three signatures and hurried to the telegraph office with the signed ukase. Dobrorolski recalled: "Every operator was sitting by his instrument waiting for the copy of the telegram in order to send to all the ends of the Russian Empire the momentous news of the calling up of the Russian people. A few minutes after six, while absolute stillness reigned in the room, all the instruments began at once to click. That was the beginning moment of the great epoch."[32] For the next several hours the clicking went on as the orders were sent from major military districts to local centers and as confirming telegrams came back to the main office in St. Petersburg. Thus began the greatest simultaneous movement of forces so far in history—744 battalions and 621 cavalry squadrons—about 2,000,000 men.[33] Dobrorolski reflected on its temporal exactitude. "When the moment has been chosen, one only has to press the button and the whole state begins to function automatically with the precision of a clock's mechanism . . . The choice of the moment is influenced by a complex of varied political causes. But once the moment has been fixed, everything is settled; there is no going back; it determines mechanically the beginning of war."[34]

Immediate engagement of all major powers was postponed briefly by last-minute negotiations centering on two ultimatums with time limits that Germany presented to France and Russia in an effort to avoid the moral responsibility for starting the war. At 7:00 P.M. on July 31, the German Ambassador in Paris, von Schoen, delivered one to the French government, requiring that they remain neutral in a war between Germany and Russia. France was given eighteen hours to reply. (In the unlikely event that France agreed, Bethmann instructed Schoen in secret to require in addition as a pledge of neutrality that France turn over the fortresses of Toul and Verdun until after the completion of the war. France would be given three additional hours to decide that.) At midnight the Germans issued an ultimatum to Russia to suspend all mobilization against Austria-Hungary and themselves. Russia was given twelve hours to respond. That same evening General Joffre urged Poincaré to mobilize French troops on their eastern front: "Every delay of twenty-four hours in calling up reservists and sending the telegram for *couverture* means a retardation of the concentration of forces, that is, the initial abandonment of fifteen to twenty kilometers of territory for every day of delay."[35] When urging the Tsar to mobilize, Sazonov had translated delay into lives lost; when urging Poincaré, Joffre translated it into kilometers. In July 1914, time was equated with life and space. The morning of August 1, Joffre threatened to resign his command unless France mobilize immediately. At noon the time limit for Russia expired, and one hour later the eighteen hours given to France were up. Poincaré finally assented to Joffre's urgings and gave the order, which was carried out at around 3:45 P.M. Germany ordered full mobilization fifteen minutes later and an hour after that declared war on Russia.

There were two more ultimatums, and as time was running short, so were their time limits. On July 29 Jagow sent to the German Minister at Brussels, Klaus von Below, a copy of the ultimatum to be delivered to the Belgian government. In this original version of the ultimatum Belgium had twenty-four hours to deliberate opening its territory for the free passage of German troops, but at 7:00 P.M. on August 2, when Below handed the note to the Belgian foreign minister, the time limit had been cut in half. Twelve hours later Belgium rejected the German demand. The German invasion of Belgium the following morning was the occasion for the fifth ultimatum. At 7:00 P.M. the British ambassador in Berlin delivered it to the German government, demanding that they withdraw their ultimatum to the Belgian government and withdraw their troops from Belgian territory. Germany was given five hours.[36] When the clock struck midnight on August 4 and there was no satisfactory response, Germany and England were at war.

Diplomacy is an art of timing. A common assumption throughout the early nineteenth century was that "time alone is the conciliator."[37] It was an age of slow communication when many ambassadors, fearful of exceeding their instructions, merely passed on information and compiled massive reports about circumstances that may have changed completely by the time the reports arrived at the seat of government. Others took advantage of the distance to take responsibility and act forcefully. But with the introduction of the telegraph in the late 1840s, the power of plenipotentiaries and the speed of their operation changed. In response to an inquiry about its impact in 1861, the British diplomat Sir Arthur Buchanan replied that "it reduces, to a great degree, the responsibility of the minister, for he can now ask for instructions instead of doing a thing on his own responsibility."[38] A similar observation appears in a popular handbook for French diplomats of 1899.[39] In 1902 the British ambassador in Vienna deplored "the telegraphic demoralisation of those who formerly had to act for themselves and are now content to be at the end of the wire."[40] Another effect was the acceleration of the pace of diplomacy. In 1875 the French historian Charles Mazade urged that no serious issue involving a question of war or peace ought to be deliberated over the telegraph and concluded that the Franco-Prussian War of 1870 could have been avoided if diplomats had taken the time to pursue a normal pace of negotiations and not placed themselves at the mercy of the telegraph.[41] More recently the French historian Pierre Granet argued that the introduction of the telegraph brought distant events to the foreground with "burning actuality" and necessitated rapid and often ill-considered responses. The outbreak of the Franco-Prussian War was again a case in point, as the publication of a telegram, suitably edited by Bismarck, inflamed the French against the Germans and precipitated the declaration of war. "Functioning ceaselessly between the European capitals, permitting news to travel at an accelerated rate, the telegraph significantly heightened the intensity of public opinion, which ultimately led the governments into what was an avoidable conflict." The telegraph did not allow time for tempers to cool: "The constant transmission of dispatches between governments and their agents, the rapid dissemination of controversial information among an already agitated public, hastened, if it did not actually provoke, the outbreak of hostilities." He too regretted "the disappearance of the time factor in the nineteenth century, which until then had often soothed international relations."[42]

Observers during and after the First World War agreed that the telegraph and telephone had shaped the pace and structure of diplomacy during the July Crisis. Ironically, although both inventions were used more to bring on war than to keep the peace, all the leading diplomats at that time failed fully to appreciate their effect on the conduct of diplomacy. In 1917 the author of *A Guide to Diplomatic Practice,* Sir Ernest Satow, suggested that there may have been a fateful time lag between technological innovation and the response of the diplomatic corps. "The moral qualities—prudence, foresight, intelligence, penetration, wisdom—of statesmen and nations have not kept pace with the development of the means of action at their disposal: armies, ships, guns, explosives, land transport, but, more than all that, of rapidity of communication by telegraph and telephone. These latter leave no time for reflection or consultation, and demand an immediate and often a hasty decision on matters of vital importance."[43] A historian of the telephone contrasted the peaceful potential of the device and its provocative use after the assassination of Archduke Ferdinand. "All the world's telecommunications facilities . . . which should have been turned to peaceful uses, were set to the frantic uses of war."[44] And while Marshall McLuhan was not referring specifically to the telephone or telegraph, his observation about war in the modern age, although overstated, suggests another way they might have structured the July Crisis:

> War is never anything less than accelerated technological change. It begins when some notable disequilibrium among existing structures has been brought about by inequality of rates of growth. The very late industrialization and unification of Germany had left her out of the race for staples and colonies for many years. As the Napoleonic wars were technologically a sort of catching-up of France with England, the First World War was itself a major phase of the final industrialization of Germany and America.[45]

There is abundant evidence that one cause of World War I was a failure of diplomacy, and one of the causes of that failure was that diplomats could not cope with the volume and speed of electronic communication. Most of the aristocrats and gentlemen who made up the diplomatic corps in 1914 were of the old school in many respects, as wary of new techology as some generals were wary of newfangled weapons and strategies. And as the generals failed to appreciate the significance of long-range artillery and machine guns and continued to think in terms of the glory of the cavalry charge and the "terror of cold steel,"[46] the diplomats failed to understand the full impact of instantaneous communications without the ameliorating effect of delay. They still counted on the ultimate effectiveness of "spoken words of a decent man"[47] in face-to-face encounters but were forced to negotiate many important issues over copper wire. The piles of futile telegrams (like the later rows of dead soldiers) were the tangible remains of their failure. The circumstances of the delivery of the Austrian ultimatum reveal the demands that the new speedy technology placed on the old-style diplomacy. The ultimatum was drafted in a spirit of extraordinary haste and given a time limit that was unthinkable before the age of the telegraph and telephone, but it called for a careful response that required consultation and weighing of choices that was impossible in the time allowed. At a moment of one of the most bitter confrontations of the entire crisis, when Austria was ready to rupture diplomatic relations and make war, Giesl felt obliged to justify *the short time limit* by reminding Paču that it was an age of the railroad, telegraph, and telephone; and afterwards he found his explanation significant enough to include it in a report to Berchtold in Vienna. Perhaps Giesl was also thinking of the train ride across the frontier and the telephone call that he was expecting to make in two days, after the time limit ran out. When the Austrians finally declared war on Serbia on July 28, they did it, as no country had ever done before, by telegram.[48] Riezler's diary entry of July 25 illustrates the feverish tempo after the delivery of the Austrian ultimatum:

> The last days the Chancellor [Bethmann Hollweg] almost constantly on the telephone. Apparently preparations for all eventualities, conversations with the military about the unspeakable. The merchant fleet is warned. [Rudolf] Havenstein [President of the Reichbankdirektoriums] financial mobilization. Until now nothing had to be done about what happened from without. Great movement. The Austrian note not smart, much too long. The first telegrams about the reaction of the great powers to the Austrian ultimatum arrived, in an hour they go to Berlin. What will our

> destiny be? But destiny is for the most part completely stupid and uncertain and all entangled in pure chance. Whoever grabs it, gets it. This damn crazy world has gotten too confused to comprehend or predict. Too many factors all at once.[49]

Bethmann was but one of many tragic heroes of the drama of simultaneity of the July Crisis. Wires hummed all over the world as European leaders were transmitting desperate appeals for time, last-minute offers, and, finally, mobilization orders and declarations of war.

∞

The self-images of nations in the full spectrum of time—past, present, and future—changed after 1880. A nation's sense of the future, which determines long-range foreign policy as well as present diplomatic maneuvers, is based partly on its sense of the past. Examination of the sense of the past of the major belligerents in World War I reveals a striking contrast between the temporalities of the nations of each alliance system and underlying causes of resentment and misunderstanding.

England's past was shaped by a popular myth that there had always been an England, even before 1066. There was a real sense of national unity already in the Elizabethan age, buttressed by long unbroken traditions such as the common law and the parliament. During the Great War English soldiers sang derisively, "England was England when Germany was a Pup."[50] France also had a long history of unity; the territorial extent of France in 1914 had been almost completed by Louis XIV in the seventeenth century.[51] The French even traced their national tradition back to St. Louis in the twelfth century; and full constitutional unification came during the revolution of 1789, when everyone became a citizen with equal rights under the law. In 1913 Russia celebrated the 300th anniversary of the founding of the Romanov Monarchy, which had remained in power without interruption in spite of the furious internal conflict among various national minorities.

While England, France, and Russia could look to long and relatively unbroken pasts, Germany, Austria, and Italy could only glance back two generations to the wars that brought them into being as modern nation states. Hence it was impossible for their citizens to believe, as could the English, that their country had always been and always would be. The wars that brought these nations into being were still vivid in popular consciousness. Austria was born after the shameful defeat by Germany in 1866; Italy became a nation after a long and difficult travail assisted by French and German armies. The creation of Germany after three wars between 1864 and 1870 left it fixated on military victories, linking national unification with war against surrounding enemies. The last-born of nations aspiring to a world empire, it eyed the senior world power and naval rival across the Channel with envy and resentment. Perhaps some of the anti-Semitism in Germany and Austria can be attributed to resentment of a people that had maintained a strong sense of identity for several thousand years without a territorial base. In contrast, Germans had been occupying the same land for centuries but had not been able to pull it together into a nation until recently. For Germany and Austria the past was a confusion of invasions, conquests, and internal struggles.

In the traditional assessment that old is good, England, France, and Russia would appear morally superior to the newer creations of the decade of the 1860s. No doubt one of the reasons German leaders wanted to place responsibility for the outbreak of the war onto either Russia or France came from a deep sense that they had to establish their moral credentials. As heirs to the Holy Roman Empire, Germany and Austria had some sense of their ancient heritage, but that lineage had been repeatedly interrupted by dynastic and territorial changes. The historicism of Hegel, Darwin, and Marx had taught that all things emerge and ripen in time as a composite of what has gone before. England, France, and Russia could look back on relatively continuous growth of national institutions and draw strength from their long and rich patrimony. The patina of time lay thick on the older nations of Europe and gave them an aspect of genuineness; the Central Powers were newcomers held together by the newly fabricated skin of German armor.

While distinctive past orientations of the five major belligerents shaped their expectations about the future, the exigencies of the present during the July Crisis tended to work on all of them more or less alike. The present was a parenthesis of time between a glorious and prosperous European past and an uncertain future, when diplomats spoke and acted for nations under circumstances of extraordinary temporal compression. This was the climax of the age of simultaneity. Telegrams sent to half a dozen capitals triggered a variety of responses simultaneously. Telephones carried instantaneous two- and three-way conversations. The compacting of events in time was best suited for the one new art form of the period—the cinema—that was able to suggest the multiplicity of occurrences in many distant places in a single moment. The July Crisis was a montage of the simultaneous activity of scores of diplomats and later, of millions of soldiers. Each nation had a unique perspective and a different stake in the outcome, but the sense of the present was generally similar—thickened for all by the sheer density of events and expanded spatially by the new technology. Far different was the sense of the future, which determined why and how the various nations finally went to war.

Of all the belligerents England was the most conservative and had the least expectation of a radically new future. Its interest was in preserving the status quo and the greatness of its empire. The English were hopeful, if not absolutely certain, that the future would continue to resemble the past. As A. J. P. Taylor observed, "In England landowners granted leases for 99 or even 999 years, equally confident that society and money would remain exactly the same throughout that time."[52] The English were concerned about the growth of the German economy and especially about the buildup of the German navy, but they were nevertheless still leaders in the naval race and expected to continue controlling a rich colonial empire. They had little immediate concern for the future strength of the Russian army, because in 1914 Russia was an ally whose main target for expansion was the Balkans, and whose main military thrust was directed against the Central Powers. There is no question that Great Britain was the most reluctant of the major powers to become involved in war in 1914. British diplomacy throughout the July Crisis sought to avoid major changes, slow things down, mediate between rival powers, delay the outbreak of hostilities, and limit the scope of the war once it started.

In France a livelier interest in the future was divided sharply between active and passive modes. All of the military preparations spoke of the great *élan* of the French soldier, which would carry armies to victories, even against long-range artillery and machine-gun fire. Some of that bombast, however, expressed the underlying fear that in the coming war against Germany, France would again be on the defensive. The repeated urgings always to remain on the offensive reveal the suspicion that Germany might attack first and hold the offensive.[53] The "10 kilometer withdrawal" that the French executed on July 30 was designed to force Germany to attack first and assume moral responsibility for starting the war.[54] In 1914 the French were holding a vast colonial empire of their own, but one far less profitable than the English administered. Their ability to control the future and preserve that empire was dependent on the help of the British and Russian armies to stop Germany and on the good graces of the British navy to allow the French free passage in and out of the Mediterranean through the Strait of Gibraltar. Thus in 1914 the French were more apprehensive about the future than were the English. Poincaré feared the continual growth of the German army and judged that the time was right to capitalize on the alliance with Russia and England to crush Germany before its army got too strong and before French socialists, who were protesting the enormous burden of military expenditure, succeeded in repealing the three-year military service that had been enacted in 1913.

While the sense of the future in England and France was moderately hopeful, the third nation of the Triple Entente had a strong feeling that the future would bring it a commanding position in European affairs when its army came to full strength. In 1906 Rainer Maria Rilke had sensed that Russia's future was as vast and as potentially fruitful as its great open spaces: "there one notices so little of the times, of the temporal, because there it is always future already and every passing hour is closer to eternity."[55] Thomas Mann had expressed a similar notion through Settembrini, who speculated that the Russians' squandering of time was made possible by their great space and that consequently "they are the nation that has time and can wait." Vast human resources were to be tapped for the rearmament program inaugurated following the defeat of Russia in the Russo-Japanese War and scheduled to be completed in 1917, when Russia expected to be able to field the largest army in history. That target date must have been on the mind of the Austrian ambassador in Berlin on July 12, 1914, when he wrote to Berchtold in Vienna summarizing German views on the suitability of war at that time: "Germany has recently found its conviction confirmed that Russia is preparing for a war with its western neighbors, and does not regard war as a possibility of the future, but positively includes it in the political calculations of the future. This is important: it intends waging war, it is preparing for it with all its might, but does not propose it for the present, or we should rather say, is not prepared for it at the present time."[56] Six days later Jagow wrote to the German ambassador in London: "According to all competent observation, Russia will be prepared to fight in a few years. Then she will crush us by the number of her soldiers; then she will have built her Baltic Sea fleet and her strategic railroads. Our group, in the meantime, will have become weaker right along."

In that same letter Jagow assessed the dismal future of the Austro-Hungarian Empire if it should fail to crush the Pan-Serbian movement. "We neither could nor should attempt to stay her hand. If we should do that, Austria would have the right to reproach us (and we ourselves) with having deprived her of her last chance of political rehabilitation. And then the process of her wasting away and of her internal decay would be still further accelerated. Her standing in the Balkans would be gone forever."[57] In his retrospective assessment of the July Crisis and Austria's role in it, Conrad observed that Austria had for years given an impression of "impotence" and made its enemies more aggressive by "continual yielding and long suffering." A failure to act after the assassination "would unleash those tendencies within the empire which in the form of Southern Slav, Czech, Moscowphil, Roumanian propaganda and Italian irredentism are already shaking at the foundations of the historic structure . . . The Sarajevo assassination has toppled over the house of cards built up with diplomatic documents, in which Austro-Hungarian policy thought it dwelt secure . . ." The monarchy has been "seized by the throat" and is forced to take a stand to prevent its ultimate destruction.[58] Everywhere in the Austrian capital there was a sense of imminent doom. Shortly after the assassination, on July 2, the Emperor confided to Tschirschky: "I see a very dark future . . . what is particularly disquieting to me is the Russian practice mobilization which is planned for the fall, just at the time when we are shifting our recruit contingents."[59] On July 18 Schoen reported a conversation with German Under-Secretary of State Zimmermann about the Austrian situation. Zimmermann stated "that Austria-Hungary, thanks to her indecision and her desultoriness, had really become the Sick Man of Europe as Turkey had once been, upon the partition of which, the Russians, Italians, Roumanians, Serbians and Montenegrins were now waiting." A decisive and forceful move would make it possible for Austrians and Hungarians "to feel themselves once more a national power." But, he warned, "in a few years, with the continuance of the operation of the Slavic propaganda, this would no longer be the case."[60]

The pervasive sense that the Austro-Hungarian Empire was sick and impotent, wasting away from internal decay and surrounded by enemies, is metaphorically suggestive of a case history that Minkowski discussed in 1923. The link between the two is Minkowski's diagnosis of a connection between a blocked future and a sense of helplessness in the face of imminent destruction. The patient was a foreigner who reproached himself for not having taken out French citizenship. He was convinced that an atrocious retribution was in store for him: someone would cut off his limbs. The entire world will witness the punishment, which will also involve all of the filth and refuse of the universe being poured into his stomach. When Minkowski first saw him, the man announced that his execution would take place that night, and he fell into panic. He kept expecting it to happen on several succeeding nights. Minkowski concluded that the core of the disorder was a blockage of the flow of time, which in normal persons goes from past to future in a continuum. A normal person can draw from past experience to assuage momentary panic or a sudden sense of helplessness, but his patient confronted every experience anew, as though each day he were beginning life all over again. "Life, our personal *élan,* lifts us and carries us . . . toward a future which opens its doors wide to us." But for his patient that movement was lacking. "No action, no desire emerged which, emanating from the present, could go toward the future across this succession of dull and similar days. Because of this, each day had an unusual independence. [It] did not vanish into the sensation of the continuity of life. Each one emerged as a separate island in the dark sea of becoming. [And] the future was blocked by the conviction of a destructive and terrifying event." The focal delusion of persecution, the idea of execution, was actually an attempt by the rational part of the mind to establish a logical connection "between the various sections of a crumbling edifice." The personal *élan* not only determines our orientation toward the future but also our relation with the environment. A normal person knows where self leaves off and the surrounding world begins, but in pathological states that frontier becomes confused. Hostile inner impulses are experienced as if they have been acted upon in the world, triggering in turn the retribution of the world and manifesting themselves as delusions of persecution. The sense of guilt is further intensified by the blocked future, because there is no opportunity to rectify errors or expiate evil deeds, since all change, all activity, is impossible. Mental life dims as the future closes, and the individual becomes dominated by a static sense of fixity and evil.[61]

The collective psychology of the Austro-Hungarian Empire in

1914, as evinced in statements of its leaders, suffered some of the same morbid thoughts. The pervasive sense that the empire had no future brought out deep feelings of impotence and helplessness. The leaders were also preoccupied by the fear of internal disintegration ("gnawing away," "Sick Man of Europe"), analogous to the "crumbling edifice" of the personality of Minkowski's patient. Spokesmen for the Habsburg Empire were obsessed with the thought that it would be dismembered by irredentist movements. The assumption of a hostile environment ("seized by the throat") can also be a consequence of the pathological state of a blocked future, as Minkowski explained. No doubt the short time limit that Austria set to Serbia's ultimatum was a projection of its own sense that time was short. This act of revenge for its own lack of a future robbed it of the worldwide moral support that it enjoyed at first as the target of an assassin's bullet, and created instead a hostile reaction among all of the potential belligerents. The one course of action that Austrian leaders could see involved the ultimate acting out of hostile impulses—war. As A. J. P. Taylor quipped, "the Habsburg Monarchy brought on its mortal crisis to prove that it was still alive."[62] The causes of pathology in Minkowski's patient and the problems facing Austria were certainly different, but the case history helps to illustrate the sense that the future was closed and time running out.

Germany's view of its future was not hopeless like Austria's, but there was widespread concern that the coming years might bring an eclipse of the military superiority on the European continent that it had enjoyed since 1870. In 1911 Moltke urged a preventive war because Germany was "in a condition of hopeless isolation which was growing ever more hopeless."[63] In a conference with the Kaiser and military officials in December 1912, he expressed the conviction that war was inevitable and "the sooner the better." Admiral Tirpitz recommended postponement until naval preparations were further along, but Moltke replied angrily that "the Navy would not be ready then either and the Army's position would become less and less favorable."[64] On June 1, 1914, Moltke is reported to have said, "We are ready, and the sooner it comes, the better for us."[65] In a conversation on July 1 with the Chief of the Cabinet in the Austro-Hungarian Foreign Ministry, Count Alexander von Hoyos, the German publicist Victor Naumann related that in German military circles and the foreign ministry "the idea of a preventive war against Russia was regarded with less disfavor than a year ago." In a report the following day, the Saxon minister in Berlin wrote of "renewed pressure from the military for allowing things to drift towards war while Russia is still unprepared."[66] The observations of Jagow and Zimmermann noted above on the urgency of Austrian action in the face of growing Russian might reflected widely held sentiments in Germany about the Russian menace. Even Bethmann, who emphatically rejected the notion of a preventive war when it was suggested to him in June 1914, came to embrace the idea, however reluctantly, a month later as he contemplated Germany's ever diminishing military advantage vis-à-vis Russia. On July 7 Riezler recorded the cause of Bethmann's brooding: "The future belongs to Russia which is growing and growing and is bearing down on us like an ever increasing nightmare." On July 20 Riezler jotted an addition about Russia's "growing demands and enormous explosive force." "In a few years there will be no defense against it."[67]

Although Germany's future was not felt to be nearly as bleak as Austria's, a number of influential leaders during the July Crisis anticipated growing problems in domestic and foreign affairs that might best be settled by having a "preventive war" at that time. Conservative forces attached to the army and the monarchy feared the rising strength of the Social Democrats, who in the election of 1912 had become the biggest single party in the Reichstag. They argued that war would reduce worker support for the Social Democratic party as the armaments industry created jobs, defuse its program of pacifism and internationalism, and stop the erosion of aristocratic privilege as the nobility carried out its ancient function of waging war and defending the fatherland. The chance of victory in a war between the two alliance systems would be better in 1914 than later, they reasoned, before Austria and the Triple Alliance became too weak and before the Russian army got too strong. The Kaiser's "now or never" referred specifically to the Austrian situation, but it could well have applied to Germany, and it set the tone for a number of similar remarks made by high German officials during the July Crisis. Together with the Kaiser, the chancellor, the secretary of state, the under-secretary of state, the chief of staff, a number of key ambassadors, and several lesser officials believed that war was inevitable and that the sooner it came the better. They did not fear, as did the Austrians, that this would be a life-and-death struggle for the preservation of their empire, but they shared the same sense of urgency in their conviction that the opportunity for building and maintaining a world empire was diminishing with the passage of time.

Their different attitudes were manifested in the battle plans by which Austria and Germany commenced hostilities. Austria's sense of helplessness and its self-doubt, one might even say guilt, over the suppression of nationalist aspirations of subject minorities were acted out during the first days of war, which Austria began by shelling Belgrade across their common frontier without marching any troops into Serbian territory. It was a hesitant start, suited to a country that was uncertain of the future and had been passively awaiting its imminent dissolution. In contrast the Germans conceived and attempted to execute an active and overwhelming battle plan, the most ambitious project ever undertaken for controlling the immediate future of so many people. The essential feature of the Schlieffen Plan was the manipulation of events by the concentration of overpowering force and the relentless and speedy exploitation of victory at every juncture. The plan was worked out day by day, railroad car by railroad car, army by army, as the German forces in the west had to remain on the offensive continuously in order to insure success. It was a precisely timed, daring operation for the one and a half million men who were to carry it out on the western front, where the schedule required that the roads through Liège be opened by M-12 (the twelfth day after mobilization), Brussels subdued by M-19, the French frontier crossed by M-22, and Paris defeated by M-39.

Austria's war began tentatively as if it were not entitled to invade Serbia, even though Serbia had been threatening the territorial sovereignty of Austria for some time. Perhaps the Germans believed in their own future more than in Austria's and therefore felt more entitled to seize it. This speculation might explain Germany's having had the audacity to recommend the Halt-in-Belgrade tactic for Austria (for the presumed reason of avoiding the moral disapproval of other countries), when their own battle plan called for smashing their way through neutral Belgium from the opening shot. The Schlieffen and the Halt-in-Belgrade plans were spatial manifestations of a different sense of the future. Austria waited for inevitable dissolution to come to it, while Germany actively brought on a cataclysmic war on two fronts, making certain to drag neutral Belgium and equivocal England in as enemies from the outset.

The temporal orientations of each of the major belligerents in past, present, and future thus shaped the way they made diplomacy and prepared for war. They all shared a similar experience of the present. The strong sense of the past among the nations of the Triple Entente was a stabilizing force that made them less eager for radical change. Relatively longer and more stable national traditions in

England and France brought feelings of confidence in the future despite growing military and economic rivalry from America and Russia, as well as Germany. The two great colonial empires believed that the future would continue to resemble the past and that they would continue to be major powers. The future ascendency of Russia seemed to be undisputed—viewed with pride at home, with a sense of security among Russia's allies, and with fear and apprehension among its enemies.

The pace of events was fast. Soldiers of the belligerents trained for war in theory with the precision of a military parade, the steady beat of marchtime; but as the possibility of war began to turn into reality, diplomacy and mobilization requirements clashed, creating unexpected surges and retardations. Time limits and timetables alternately accelerated and impeded the driving pulse of diplomatic activity throughout the July Crisis. The armies finally marched, but their progress was interrupted with unpredictable halts and withdrawals. Europe went off to war ragtime.

11
THE CUBIST WAR

In 1906 Picasso completed a portrait of Gertrude Stein. Her eyes stare into the world slightly askew in anticipation of the more radically reconstructed eyes with frontal and profile views that he would combine in single faces in his fully Cubist portraits. In fact she viewed the world as if through the eyes of that portrait—always rearranging things to reflect her constantly shifting perspectives. When she first looked down from an airplane flying over America she saw a Cubist landscape with "the mingling lines of Picasso coming and going, developing and destroying themselves." And when she looked back on World War I she saw a Cubist war:

> Really the composition of this war, 1914–1918, was not the composition of all previous wars, the composition was not a composition in which there was one man in the center surrounded by a lot of other men but a composition that had neither a beginning nor an end, a composition of which one corner was as important as another corner, in fact the composition of cubism.[1]

Stein believed that the spirit of an age shaped all things from "the way roads are frequented" to painting and war. The changing dimensions of life and thought that we have observed emerging in the generation prior to the outbreak of war were also manifested in the "composition" of the fighting itself. The war embodied most of the transformations in time and space of the prewar period and allowed her the bold metaphor.

The war imposed homogeneous time. In 1890 Moltke had campaigned for the introduction of World Standard Time, and in 1914 he used it to put into effect a war plan that required all the men to be in the right place at precisely the right time. In the prewar years wrist watches were thought to be unmanly; during the war they became standard military equipment. Before the battles wrist watches were synchronized so that everyone went over the top at the correct time. Edmund Blunden recalled how before an offensive a runner distributed watches which had been synchronized at field headquarters.[2] The battle of the Somme began on the morning of July 1, 1916, as hundreds of platoon leaders blew their whistles when their synchronized watches showed that it was 7:30 A.M., sending the soldiers of the Third and Fourth British armies up the scaling ladders and over the parapet into no-man's-land.[3] The delicate sensitivity to private time of Bergson and Proust had no place in the war. It was obliterated by the overwhelming force of mass movements that regimented the lives of millions of men by the public time of clocks and wrist watches, synchronized to maximize the effectiveness of bombardments and offensives. This imposing coordination of all activity according to a single public time reversed the dominant cultural thrust of the prewar years that explored the multiplicity of private times.

The war experience intensified both sides of the controversy about time's texture as atomistic or a flux. At the front there was an expecially sharp division between night and day—one, a time of movement and activity, the other, of inactivity and waiting. Cecil Lewis described that contrast just before the battle of the Somme. "By day the roads were deserted; but as soon as dusk fell they were thick with transport, guns, ammunition trains, and troops, all moving up through Albert to take their positions in or behind the lines. . . . Yet when dawn came, all signs of it were gone. There was the deserted road, the tumble-down farmhouses, the serene and silent summer mornings. Never do I remember a time when night so contradicted day."[4] The transition between night and day was marked with a regular ritual. Just before dawn all troops stood in appointed places with rifles in hand and bayonets fixed in readiness for any enemy action, and they remained in that position for about an hour until the dangerous half-light had passed. The same procedure occurred at dusk. As David Jones recalled, "that hour occurring twice in the twenty-four, of 'stand-to,' was one of peculiar significance and there was attaching to it a degree of solemnity, in that one was conscious that from the sea dunes to the mountains, everywhere, on the whole front the two opposing lines stood alertly, waiting any eventuality."[5] Preparations for fighting divided time into discrete units as neat as the lines of position and advance on the commanding officers' maps, from which the planners attempted to construct the sequence of events in battle. One British admiral distributed to the men under him a poem whose central metaphor linked military preparedness with exact intervals of time and which underlined the virtues of a highly regimented and punctual fighting force.

The Seconds that tick as the clock moves along
Are Privates who march with a spirit so strong.
The Minutes are Captains. The Hours of the day
Are Officers brave, who lead on to the fray.
So, remember, when tempted to loiter and dream
You've an army at hand; your command is supreme;
And question yourself, as it goes on review—
Has it helped in the fight with the best it could do?[6]

But the armies did not always click with clockwork precision. Minutes and seconds could lose their hold as ordering units in the wild chaos of events, and in the long stretches of time between battles life became a monotonous flow. Minkowski described that kind of life as a flux, but one without a rich past or future and thus not a healthy streaming of *élan vital.*

> The monotonous life in the trenches sometimes made us forget the date or the day of the week . . . We substituted another "calendar" for them, more appropriate to the situation: we simply counted the days that had passed since we came to the front and those which separated us from our return to the rest camp . . . We succumbed to the tedium and the monotony of the succession of days, and we fought off the boredom—obviously an essentially temporal phenomenon which penetrated our being like a viscous mass, threatening to reduce it to nothing.[7]

For the officers, war time was essentially a sum of discrete, sequential units out of which the scenarios for battle were constructed, while for the soldiers in the trenches it was a seemingly endless flux, a composition in time that had neither a beginning nor an end.

The sense of the direction of time also separated the officer-planners, who scheduled ambitious offensives that looked forward in space and time, from the troops in the trenches, who reported experiencing the reverse. Eric J. Leed caught the moment: "[Robert] Graves, [Charles] Carrington, and many others noted that the roaring chaos of the barrage effected a kind of hypnotic condition that shattered any rational pattern of cause and effect, allowing, even demanding, magical reversals. This state was often described in terms of a loss of coherence and the disappearance of any sense of temporal sequence."[8]

Prewar evaluation of the past emphasized its continuous and profound impact on the present. Historians, psychologists, and philosophers embraced the genetic approach for understanding human experience; and even those who, like Nietzsche, Ibsen, and Joyce, were suspicious of the enervating effect of an excessively influential past, believed that its impact was enormous, though negative. But the war ripped up the historical fabric and cut everyone off from the past suddenly and irretrievably. In his reminiscence about four weeks in the trenches, written in 1915 six months after he left the eastern front, the German musician Fritz Kreisler apologized for his inability to order events in time. A "curious indifference of the memory to the values of time and space" was, he observed, "characteristic of most people [he] met who were in the war." Leed concluded that Kreisler's experience was typical. "The invisibility of the enemy and the necessity of hiding in the earth, the layered intricacy of the defensive system, the ear-shattering roar of the barrage, and the fatigue caused by the day and night shifts, combined to shatter those stable structures that can customarily be used to sequentialize experience."[9] Blunden described with irony the weather-beaten sandbags in the trenches as "venerable" objects that "shared the past with the defenses of Troy . . . The skulls which spades disturbed about it were in a manner coeval with those of the most distant wars."[10] War creates a surrealistic sense of history that comes from confrontation with the grotesque newness of everything. In Henri Barbusse's best-selling war novel, *Under Fire* (1916), the constant shelling buried the soldiers "deep in an everlasting battlefield." They grew accustomed to the noise and only heard it when they listened—"like the ticking of the clocks at home in the days gone by—in the now almost legendary past."[11] Thomas Mann explained in the Foreword to *The Magic Mountain* that the "exaggerated pastness" of his narrative "is due to its taking place before the epoch when a certain crisis shattered its way through life and consciousness and left a deep chasm behind."[12] Proust expanded the sense of temporal distance between past and present in his novel by concluding it after the war. Then no one bothered to recall that M. Bontemps had been a Dreyfusard, Marcel explained, because "all that had been a very long time ago, a 'time' which these people affected to think longer than it was, for one of the ideas most in vogue was that the prewar days were separated from the war by something as profound, something of apparently as long a duration, as a geological period, and Brichot himself, that great nationalist, when he alluded to the Dreyfus case now talked of 'those pre-historic days.' "[13] In four years the belief in evolution, progress, and history itself was wiped out as Europeans were separated from the "pre-historic days" of the prewar years by the violence of war. In *The Great War and Modern Memory,* Paul Fussell concluded that "the image of strict division clearly dominates the Great War conception of Time Before and Time After, especially when the mind dwells on the contrast between the prewar idyll and the wartime nastiness."[14]

Minkowski's patient panicked every night from his fear of being dismembered because his sense of the flow of time from past to present was interrupted, and he was without access to a past as emotional ballast. His world therefore seemed overwhelmingly hostile. Cut off from the past by the enormous contrast between prewar civilian life and morality and the wartime violence and killing, the front-line soldier experienced panic in a world that justified the most bizarre paranoid fears. The air *was* full of deadly projectiles, skulls *did* pop out of the mud like mushrooms. Suddenly infantilized by the strangeness of everything, he was bereft of the consoling sense of continuity with what he had known and the security of expecting a recognizable world to return to in the future. The war years were excerpted from the flow of his life like a piece of calendar stuck on a Cubist collage. Only pieces of the past occasionally pierced the uniform monotony and periodic horrors of it all.

Intense and unpredictable flashes of the past did occur, as David Jones recalled: "I suppose at no time did one so much live with a consciousness of the past, the very remote, and the more immediate and trivial past, both superficially and more subtly." But as with Proust's involuntary memories, their very intensity was a function of their remoteness. Images of the former life became more fixed and idealized the more distant they were. The strange newness and overwhelming force of experience clamped the soldier in the present as if bracketed from past and future. Jones explained the title of his account of the war, *In Parenthesis,* which he published in 1937. "This writing is called 'In Parenthesis' because I have written it in a kind of space between—I don't know between quite what—but as you turn aside to do something; and because for us amateur soldiers . . . the war itself was a parenthesis—how glad we thought we were to step outside its brackets at the end of '18—and also because our curious type of existence here is altogether in parenthesis." In the beginning there was some continuity with the past—"a certain attractive amateurishness, and elbow-room for idiosyncrasy that connected one with a less exacting past." But the battle of the Somme put a stop to all that. The new recruits came in an endless stream of unfamiliar faces, and the units, decimated by losses and shuffled with new orders, kept changing their composition. Jones also recalled the dislocating effect of learning to do gas drills, to be "attuned to many newfangled technicalities," all of which required "a new and strange direction of the mind, a new sensitivity." He attempted to link the war with the past by including on the title page of each part of his book quotations from an epic poem of the sixth century that commemorated the raid of 300 Welshmen into the English kingdom of Deira. He explained that "the choice of fragments of this poem as 'texts' is not altogether without point in that it connects us with a very ancient unity and mingling of races; with the island as a corporate inheritance, with the remembrance of Rome as a European unity."[15] In fact his quotations *were* without point: the war experience had very little connection even with the years just prior to its outbreak, let alone the sixth century. The war thus contradicted the historicist thrust of the preceding century that conceived of the past as a continuous source of meaning for the present.

Several features of the war intensified the sense of the present—the disjunction from past and future, the emotionally and physically demanding nature of the ongoing experience, and the simultaneity of multiple distant events that were shared by hundreds of thousands of men in battle and witnessed by the civilian population who attempted to draw those events together into a single coherent pattern. Barbusse described the disjunction with the past. "In line from left to right fires emerge from the sky and explosions from the ground. It is a frightful curtain which divides us from the world, which divides us from the past and from the future. We stop, fixed to the ground, stupefied by the sudden host that thunders from every side."[16] Blunden recalled, "Already it seemed ages since I had last seen poor Tice, and looked at this very patch of ground with him, . . . but the gulf between this and three days before was indeed a black and lethal abyss." During a battle, "the men drowsed and yawned. Time went by, but no one felt the passage of it, for the shadow of death lay over the dial." Fixation on the present was one response to the imminence of death, intensifying the moments of peace between the bombardments. At such times, Blunden observed, one learns "that light is sweet, that a day in peace is a jewel whose radiances vary and frolic innumerably as memory turns it in her hand, infinitude of mercy. Here is this jewel; kind Nature will shield it from the corrosions of yesterday; yield yourself to this magical hour."[17] It is likely that Joyce's insistence on the supreme reality of the present in *Ulysses* was in part suggested by the overwhelming force of it during the war. He repeated variations of one obvious reference to the war like a refrain: "I hear the ruin of all space, shattered glass and toppling masonry, and time one livid final flame."[18] Time seemed to burn brightly in the war, riveting consciousness in an eternal present. In an essay on psychical phenomena in the war Hereward Carrington described a contraction of consciousness that took place as one approached the front, fixing spatially on an ever narrower visual sphere and focusing temporally in the present. When a soldier leaves the civilian world everything begins to change and contract. "Everyone he meets thinks as he does, about the same subjects, in the same way; every one is dressed alike; every one's thought runs in the same narrow groove. There is no longer the clash of opinion, the interchange of rival thoughts. Gradually, imperceptibly the images and thoughts of ordinary civilian life begin to fade; thoughts of home, wife, friends, even begin to grow dim and recede in the memory. The present, the vital present, occupies and grips the mind."[19]

One distinctive feature of the prewar sense of the present was a thickening of its temporal length beyond a "knife edge" between the past and future into an extended interval that included part of the past and future. Bergson's *durée,* James's specious present, Husserl's halos and fringes of retentions and protentions, and Gertrude Stein's continuous present all implied that intervals of present time involved streamings from the past and into the future. These concepts also implied that our life in days, weeks, and years included what came before and after and expanded its reach across time as we grew older. The war contradicted such notions of an extended present on a grand scale by isolating the present moment from the flow of time. However, the other extension of the present that we observed—a spatial extension that included a multiplicity of distant events—was dramatically embodied in the war experience.

As in a vast simultaneous drama, countless different events were pulled together under such single rubrics as "the battle of the Marne," even though no one ever experienced all of them directly at the same time. On August 23, 1914, in one episode of the initial German advance, the French were still attacking in some places and trying to throw the Germans back across the Sambre; in other places they were holding their own; and in still others retreating in broken disorder.[20] The shadows of victory and defeat fell every which way over what came to be known as a single battle, unified in historical consciousness like faceted forms of a Cubist landscape reflecting light from a multiplicity of sources. One of the most influential simultaneous poets of the prewar period, Apollinaire, applied his technique to the war itself. In a poem of 1917, *Merveille de la guerre,* he envisioned a sublime simultaneity of events emerging from the chaos.

. . .
I leave to the future the story of Guillaume Apollinaire
Who was in the war and knew how to be everywhere
In the happy cities behind the front
In the whole rest of the universe
In those who died trampling in barbed wire
In the women in the canons in the horses
From zenith to nadir to the 4 cardinal points
. . .
And without doubt it would be even more beautiful
If I could suppose that all these things everywhere
In which I am could also inhabit me
But in this sense nothing can be done about it
Because I am everywhere at this hour
. . .[21]

The sense of simultaneity experienced by diplomats during the July Crisis was magnified a thousand times during the war, as millions of soldiers were united by chain of command, electronic communication, and synchronized watches and united in spirit by the commonality of events. And their struggle was in turn witnessed by the millions at home, who learned about these multifarious events almost at the same time as they were happening, read about them in newspapers, viewed them in movie houses, and discussed them incessantly. Europe became a communications network that processed more information than ever before about more people involved in more events in widely distant places at the same time. World War I was *the* simultaneous drama of the age of simultaneity.

The sense of the future depended on rank. The officers attempted to appropriate the future actively with carefully devised battle plans. One of the most widely discussed pieces of literature in British military circles before the war was a short story, "A Sense of Proportion," about a general, apparently patterned on Moltke, who made such elaborate preparations for a battle that just before its commencement he was confident enough of the outcome to spend the time fly-casting for trout. John Keegan cited the story as illustrative of a distinctive sense of the future among military planners in the First World War. The plan for the British army's Thirteenth Corps (the same size as Wellington's Waterloo army) for the first day of the battle of the Somme was thirty-one pages. Wellington, in comparison, issued no written plans for the battle of Waterloo. Such careful planning, Keegan argued, was done in a spirit of "attempting to preordain the future; a spirit borne out by the language of the orders: 'infantry and machine-guns will be pushed forward at once . . .'; 'the siege and heavy artillery will be advanced . . .'; 'After the capture of their final objective the 30th Division will be relieved by the 9th Division . . .' "[22]

In war or peace the rich and powerful have a stronger and more active sense of the future than the poor and powerless. Great wealth is a bridge to the future—it has the power to control people and events, to support oneself and one's family in difficult times, to create trusts and inheritances that insure the well-being of future generations, to build monuments and endow institutions as stakes for immortality. The great hubris of the generals during the war, the conviction that their next offensive was bound to break the deadlock, came from generations born in a class system separating the powerful and confident from those who were traditionally obliged to carry out commands passively and obediently. The expectation of the officer class that the future was theirs to make broke down during the war along with the older rigidities of a class system that originally gave the generals their blind and persistent faith in their future. In *Tender Is the Night,* F. Scott Fitzgerald sketched the extent of that breakdown.

> This Western-front business couldn't be done again, not for a long time. The young men think they could do it again but they couldn't. They could fight the first Marne again but not this. This took religion and years of plenty and tremendous sureties and the exact relation that existed between the classes. The Russians and Italians weren't any good on this front. You had to have a whole-souled sentimental equipment going back further than you could remember. You had to remember Christmas, and postcards of the Crown Prince and his fiancée, and little cafés in Valence and beer gardens in Unter den Linden and weddings at the Mairie, and going to the Derby, and your grandfather's whiskers. . . . This was a love battle—there was a century of middle-class love spent here.[23]

After his initial confidence was shattered, the front-line soldier experienced the immediate future in the passive mode, waiting for the next shell burst and the next scream. Barbusse wrote: "In a state of war, one is always waiting. We have become waiting-machines. For the moment it is the food we are waiting for. Then it will be the post. But each in its turn. When we have done with dinner we will think about the letters. After that, we shall set ourselves to wait for something else."[24] In all war, as Barbusse implied, there is waiting, but the sense of passive waiting was particularly strong in this one, which became a protracted defensive struggle. Leed concluded that "trench warfare, perhaps more than any kind of war before or after, eroded officially sponsored conceptions of the soldierly self as an agent of aggression."[25] Long-range artillery, machine guns, trenches, barbed wire, and gas immobilized men for long periods of time in cramped quarters under circumstances of great stress. The normal response to mortal danger is active aggression, but the front-line soldiers were forced to be passive. Their humiliating circumstances produced a kind of "defensive personality" that became a distinctive characteristic of war neurosis. One psychiatrist observed that nervous tension was especially destructive among men who had to remain inactive while being shelled.[26] In a study of war neurosis, W. H. R. Rivers found that of all the soldiers, pilots had the fewest mental breakdowns, a fact which he attributed to their active sense of control over their fate. Their medical record contrasted sharply with that of observers in the balloon service, passive targets who hung suspended from long ropes over the front, whose mental breakdowns exceeded the number of physical wounds. Leed also concluded that the high incidence of sexual impotence among soldiers was another manifestation of the generally passive attitude they were forced to adopt in a war so dominated by machines and chemicals and so uniquely characterized by waiting. When everything went underground, the men had to resign themselves to their fate and internalize normal outward manifestations of aggression as well as libido.[27]

While the immediate future was dominated by a sense of passive waiting, the distant future, a time after the war, came to seem ever more remote. Already in 1916 Barbusse anticipated the moral and experiential gap that was opening up between the present and future. "The future, the future!" one of his characters exclaimed. "The work of the future will be to wipe out the present, to wipe it out more than we can imagine, to wipe it out like something abominable and shameful."[28] Blunden remembered that "one of the first ideas that established themselves in my inquiring mind was the prevailing sense of the endlessness of the war. No one here appeared to conceive any end to it." One chilling portent of that endless horror stood out in his memory: "as for the future, one of the first hints that came home to me was implied in a machine-gun emplacement stubbornly built in brick and cement, as one might build a house."[29] There was truth in the irony, since machine-gun pits and concrete bunkers remained intact throughout the war and long after as monuments to it. The anticipated remoteness from the future was confirmed years later in published reminiscences, reporting that the time between the end of the war and the resumption of a "normal" peacetime life was often very long and some veterans failed to reintegrate. More than twenty years after the armistice, Henri Massis recalled: "Yes, the war was the home of our youth. We were born of the war, the war was immediately upon us, and in truth we have never done anything else."[30] In the 1920s some American veterans migrated back and forth between Europe and America, trying to forge a new sense of connection and restore the continuity of their life in time. The restlessness that Hemingway lived and recounted in *A Moveable Feast* was the search of the "lost generation" for *temps perdu* and for reintegration in the flow of time.[31]

The war shut off direct access to the immediate future and opened an abyss between the present and the distant future for everyone, even the Futurists. At first they welcomed the coming war as the fulfillment of their aspirations to destroy the past and realize a new and endlessly innovative mode of existence. In March of 1914 Marinetti wrote that he first perceived the new beauty of "geometrical and mechanical splendor" on the deck of a dreadnaught. "The ship's speed, its trajectories of fire from the height of the quarterdeck in the cool ventilation of warlike probabilities, the strange vitality of orders sent down from the admiral had suddenly become autonomous, human no longer, in the whims, impatiences, and illnesses of steel and copper. All of this radiated geometric and mechanical splendor. I listened to the lyric initiative of electricity flowing through the sheaths of the quadruple turret guns, descending through sheathed pipes to the magazine, drawing the turret guns out to their breeches, out to their final flights."[32] He sustained this wild enthusiasm into the first months of war in 1915, when he published a collection of pro-war Futurist manifestos and added some tracts urging Italian intervention. In *War, The World's Only Hygiene* (written between 1911 and 1915) Marinetti rhapsodized about his hopes for the fulfillment of the Futurists' goals through the war. But his shrill voice was eventually drowned out by the din of battle, the exultation of the "hygiene" of war was tempered by the filth of the trenches, and soon the group was thinned out and scattered. Boccioni and Sant'Elia were killed, Marinetti and Russolo were seriously wounded, and others were lost in the chaos. Ironically the war cut them off from the future even more than most—it destroyed their movement by realizing their ideals, fulfilling their demand for an ephemeral art and the repudiation of all orthodoxy.

World War I was the apotheosis of the prewar sense of speed. The acceleration of the pace of events during the July Crisis and the imperative of speed during mobilizations continued as armies deployed on the field. On August 6, 1914, 11,000 trains began the transport of 3,120,000 German troops across the Rhine. One reason for the failure of the Schlieffen Plan was its initial success, as the troops overran their own timetable and supplies failed to keep pace.[33] And although the French initially suffered some rapid defeats, they also fielded an army with unprecedented speed. For the first engagements in August about 2,000,000 Frenchmen were deployed in 4,278 trains, and only nineteen ran late.[34] Although once in the trenches the men frequently perceived the war as a protracted and monotonous struggle, the actual fighting was far faster than anything in history, revolutionized by the widespread use of magazine loading rifles, rapid fire artillery guns, and—that symbol of speedy killing—machine guns, which caused about 80 percent of all casualties. On July 1, the first day of the battle of the Somme, the British took about 60,000 casualties, of whom 21,000 were killed. John Keegan has speculated that most of them were killed in the first hour of the attack, perhaps even in the first minutes.[35] The war set some gruesome speed records.

∞

Stein's Cubist metaphor is most appropriate for characterizing the spatiality of the war as a function of prewar conceptions of space and its modes of form, distance, and direction. Her observation that the war was "not a composition in which there was one man surrounded by a lot of other men" suggests the prewar philosophy of perspectivism. There was a front in Turkey and the Mediterranean, an Eastern and a Western front, naval encounters in the English Channel and the Atlantic, bombings and dogfights in the skies. The Western front snaked from Switzerland to the Channel, and along it vast armies squared off, now penetrating enemy lines, now being penetrated, as in a Cubist landscape. Schlieffen's own anticipation of the role of the commander was a partially accurate prediction of what happened and an uncanny prevision of Stein's image:

> The modern commander-in-chief is no Napoleon who stands with his brilliant suite on a hill. Even with the best binoculars he would be unlikely to see much, and his white horse would be an easy target for innumerable batteries. The commander is farther to the rear in a house with roomy offices, where telegraph and wireless, telephone and signalling instruments are at hand, while a fleet of automobiles and motorcycles, ready for the longest trips, wait for orders. Here, in a comfortable chair before a large table, the modern Alexander overlooks the whole battlefield on a map. From here he telephones inspiring words, and here he receives the reports from army corps commanders and from balloons and dirigibles which observe the enemy's movements and detect his positions.[36]

Schlieffen hoped that the multiplicity of different events, observed from numerous different perspectives, could be relayed to the commanding officer electronically and that even while battles were in progress he could send back effective orders to his soldiers. In reality the situation proved to be quite different. Commanders received spotty reports often long after the situation had substantially changed, creating massive confusion and making effective responses impossible. But Schlieffen accurately foresaw that in the new composition of war there would be disjointed encounters that could not be viewed from a single spot.

The centripetal force of multiple fronts made it impossible for one man to process everything. Gustave Le Bon, who in 1895 had emphasized the potency of the single leader in crowd behavior, saw in the Great War "a complete disappearance of the older type of battle which, from the days of Hannibal and Caesar to those of Napoleon, was the personal work of a General."[37] In the current war, battlefields stretched for hundreds of miles, men could scarcely be distinguished from the ground, artillery and trenches were hidden from view, and the general was usually located far from the battle zone. Schlieffen, Le Bon, and Stein concurred that comprehension of modern war required a multiplicity of perspectives.

Stein's comment that in the composition of the war "one corner was as important as any other" may have referred to the lack of a single focus among all the battles and events that were taking place at the same time. For the individual soldier, by contrast, there was a vital demarcation between safety and danger, and the corners where shells fell were certainly more important than others.[38] But from another perspective, in this war, as in a Cubist painting, all spaces were of equal value as even empty space took on special importance as a constituent element—no-man's-land was its positive negative space.

The several thousand yards between enemy trenches, narrowing at points to fifty or a hundred yards, became ever more important as the fighting made them ever more desolate. No-man's-land became a synonym for the void—a place where no man ought to be—pitted with shell holes, stinking from decaying bodies, puddled with mud and gas, a poisonous wasteland, a lifeless and threatening expanse of nothingness, and yet a space that acquired extraordinary value, reckoned by the dead piled up fighting for it. It was alternately a place of maddening noise and unnerving quiet. Telephone lines ran to the front and stopped: once there a soldier could be suddenly lost in silence. In *The Void of War,* Reginald Farrer described the "crowded emptiness" that he saw in the Somme battlefields, where every trace of habitation intensified his sense of a "huge haunted solitude." "Perhaps I ought not exactly to call it empty," he wrote, "it is more 'full of emptiness' that I mean: an emptiness that is not really empty at all. There is something personal about it throughout and I come to think of this piece of country, not as a scene, but as a person." Farrer justified the irony of his title. "The void of war is a splendid name! It means these places completely battered and smashed and apparently quite lifeless and abandoned: yet really with military life buzzing in them indefatigably."[39] For the soldier under fire, no-man's-land was chopped up like an irregular checkerboard of safety and danger, and dislocations of a centimeter could determine life or death. But from afar one corner was as important as another. Houses, trees, roads were blown away, creating a uniform expanse of wreckage. It was the great empty space that corrupted men and armies as Conrad's "heart of darkness" had corrupted Kurtz. It was a horror, but it was also what the fighting seemed to be all about. On it class, rank, and nation were leveled; World War I assaulted far more of the hierarchical structures of privilege than its participants had ever expected.

∞

No-man's-land was also a frontier, a constantly reforming line between two sides of a world at once highly polarized and tightly knit by battle. From a reading of hundreds of war memoirs, Leed reported that "astonishing numbers of those who wrote about their experience of war designate No Man's Land as their most lasting and disturbing image. This was a term that captured the essence of an experience of having been sent beyond the outer boundaries of social life, placed between the known and the unknown, the familiar and the uncanny."[40] The psychological fragmentation experienced in no-man's-land during the war was but one of a series of shattered forms—national boundaries, political systems, social classes, family life, sexual relations, privacy, moral imperatives, religious convictions, human sensibilities. The form of the fighting itself was unlike that of any previous wars, as trenches replaced fixed fortifications, battles expanded into the third dimension, and camouflaged machines blended into the surrounding countryside.

Gertrude Stein recorded another memorable incident that suggests once again the suitability of her Cubist metaphor. "I very well remember at the beginning of the war being with Picasso on the Boulevard Raspail when the first camouflaged truck passed. It was at night, we had heard of camouflage but we had not yet seen it and Picasso amazed looked at it and then cried out, yes it is we who made it, that is cubism."[41] The armies of the nineteenth century were outfitted in bright colors as a display of the pomp of the governments that fielded them. The striking blues and reds and whites were supposed to highlight the wealth, polish, and discipline of the advancing army and intimidate the enemy. But with the extension of the range of accuracy from the one or two hundred yards of a musket to the two thousand yards of the breech-loading rifle and the increased fire power of sweeping lines of bullets from a machine gun blast, bright, colorful uniforms and tight, neat formations were suicidal. The British switched to khaki after the Boer War, and by the outbreak of World War I the Germans had changed from Prussian blue to field gray. But in 1914 the French soldier still wore the red kepi and pantaloons of the Second Empire. General Messimy tried to change the uniform, but the army at first refused to dress its soldiers in some drab, earthy color. "Eliminate the red trousers?" one-time War Minister M. Etienne exploded, "Never! *Le pantalon rouge c'est la France.*"[42] After the August and September massacres, the proud and hierarchically minded officer corps that at first protested against making French soldiers blend into the surroundings, now eagerly looked for some way to make them invisible. Guirand de Scévola, working as a telephonist for an artillery unit at Pont-à-Mousson, thought of concealing an artillery gun under a net splashed with earth colors. Shortly after the battle of the Marne, Guirand interested Marshall Joffre and President Poincaré in his idea, and the first *section de camouflage* was created in the French Army to begin systematically developing techniques for concealing soldiers and military equipment. The red kepi and pantaloons were also abandoned for new uniforms of horizon blue.

It is not known whether Picasso knew of Guirand de Scévola, but Guirand did know of the work of Picasso. He was quoted as saying: "In order to totally deform objects, I employed the means Cubists used to represent them—later this permitted me, without giving reasons, to hire in my [camouflage] section some painters, who, because of their very special vision, had an aptitude for denaturing any kind of form whatsoever."[43] By the end of the war camouflage sections employed three thousand *camoufleurs* (including such prominent artists as Forain and Segonzac) to dissimulate the big guns and other conspicuous objects. Their insignia was a chameleon.

If Picasso's observation was not made with an understanding of the direct influence that Cubism had had on the invention and development of camouflage, it was at least a recognition that there might be some indirect link between the two phenomena that occurred at almost the same time in history and had such a similar cultural function. Side by side they implied that the traditional ways are not necessarily the best ways of ordering objects in pictorial space or men and guns on a battlefield or, with a bit of interpretive stretching, classes in society. The abandonment of the old military uniform, so intimately associated with aristocratic society, was a repudiation of the convention of deference to rank in the army and in the civilian world. Henceforth troops and artillery guns, like pictorial objects, would be given prominence only if the situation required, not because of outmoded conventions. Cubism and camouflage leveled the older hierarchies in order to re-hierarchize the world in ways that suited the real exigencies of the current situation.

The idea of camouflage spread rapidly. England had camouflage factories by the summer of 1917, sometimes assisted by the French. The French painter André Mare kept notebooks with drawings of camouflage patterns and sent them to assist the English. In 1917 the British Naval Commander Norman Wilkinson devised a *trompe-l'oeil* camouflage technique of painting the sides of ships with geometric patterns in sharply contrasting colors, making it difficult to determine their size and direction of travel when seen through a periscope. German camouflage (*Tarnung*) was in use by early 1916 and the German expressionist artist Franz Marc was employed painting some of the camouflage nets and canvases that were used to cover German guns for the battle of Verdun.[44] In America camouflage was used mostly on ships and was directly inspired by recent studies of natural protective coloring among animals.[45] Gertrude Stein commented on the distinctive national characteristics of the camouflage patterns but concluded that despite national differences they collectively showed the "inevitability" of the "whole theory of art"—by which she meant Cubism.[46]

While camouflage broke up the conventional visual borders between object and background, the tremendous fire power unleashed during the great battles broke up the terrain as no previous war had done. Fortresses were demolished, houses blown away, craters opened, hillocks flattened, rivers dammed and diverted in their course, roads transformed into bogs. Barbed wire stitched the fabric of the land. In the cities bodies were pounded under the shattered glass and toppling masonry. In the countryside pieces of men stuck in trees, and pieces of trees stuck into men, as in a nightmarish Cubist landscape. Barbusse has left a vivid picture of the war terrain.

> The trees bestrew the ground or have disappeared, torn away, their stumps mangled. The banks of the road are overturned and overgrown by shell-fire. All the way along . . . are trenches twenty times blown in and re-hollowed . . . The more we go forward, the more is everything turned terribly inside out . . . We walk on a surface of shell fragments, and the foot trips on them at every step. We go among them as if they were snares, and stumble in the medley of broken weapons or bits of kitchen utensils, of water-bottles, fire-buckets . . .

Further on the soldiers encounter a village that has disappeared.

> Here, within the framework of slaughtered trees that surrounds us as a spectral background in the fog, there is no longer any shape. There is not even an end of wall, fence, or porch that remains standing; and it amazes one to discover that there are paving-stones under the tangle of beams, stones, and scrap-iron. This—here—was a street . . . The bombardment has so changed

> the face of things that it has diverted the course of the mill-stream, which now runs haphazard and forms a pond.[47]

Trenches zig-zagged through the land, outlining the ever-changing facets of the front and linking areas of military operation behind.[48] Nearest the enemy was the front line, from which shallow "saps" ran into no-man's-land and led to the most forward observation posts, grenade-throwing posts, and machine-gun emplacements. About two hundred yards behind that was the support line in which the men spent most of the time living in dugouts cut into the dirt walls and in deeper chambers tunneled down as far as thirty feet, where they were safe from all shelling. Several hundred yards behind that was the reserve line, and communication trenches ran perpendicular to, and connected, all three. The frequent traverses every few yards created kinks in the lines that made it impossible for an enemy soldier, who might get into a trench during an attack, from firing down a long stretch of it. From an aerial perspective the trench system gave the terrain a Cubist-like composition of irregular geometricized forms.

Another breakdown occurred in the old formal concept of the line of battle. In past wars one measure of victory was the extent to which one side held the battle line. It was as precise a frontier for fighting as class lines were for ordering the social hierarchy of aristocratic Europe. There was a place for every soldier behind the line, and in the well-ordered army every soldier knew his place and remained in it. But the modern front line eventually collapsed under the tremendous pressure of shelling and massed offensives. The determination of tradition-minded generals to hold the line was responsible for a good deal of the slaughter during the first two years, but by 1916 they began to use a more flexible strategy of "defense in depth." Armies would defend the front line, but if the enemy reached the trenches, the defenders would immediately retreat to a support trench and regroup for a counterattack. And instead of the entire army surging in massed units under a single command, defense in depth also broke up the authority of "one man in the center" and depended more on a dispersal of authority and initiative.

There is a striking analogy between this structural change in the art of war—the defense in depth—and the shift in painting from single vanishing point perspective to the multiple perspectives of Cubism. Leed's description of the new strategy suggests some other parallels with Cubism. "Defense in depth," he wrote, "meant the fragmentation of coherence, the shattering of any clear, geometrical structure, the dissolution of the company into small, independent squads and pockets of defenders."[49] Yet there was reluctance to give up the old structure. As late as 1918 the German soldier Ernst Jünger pleaded: "We must come to an irrevocable break with the idea of the line from which, for historical and disciplinary reasons, we were never truly able to detach ourselves during the entire war . . . The correct picture is that of a net, into which the enemy may be capable of penetrating here and there only to be immediately crushed to earth from all sides by fiery meshes."[50] The idea of the line lost its inviolability as a frontier separating two distinct realms in war and in painting. The two arts took on a new composition that incorporated the ambiguities and irregular contours of reality. The Cubists had sought a new unification of the aesthetic value of the entire picture surface; the war drew together disparate elements of class, rank, profession, and nation, leveling traditional hierarchical distinctions. Uniform crosses threw geometric shadows across the mass graves—a final commemoration of the social leveling of the war.

In one respect there was deep international disunity. Passports were reinstituted, borders closed, armies squared off, and national consciousness was polarized by propaganda. But the nations of Europe were never more intimately and totally involved with each other than during the Great *European* War. Ironically there was even more international travel—not with permission, but travel nevertheless. More Germans went to France and Russia in August and September of 1914 than ever before in such a short period of time, and there was communication of all sorts between the men of the invading armies and their opponents, civilian and military. The consciousness of Europe was locked together like its armies, sometimes even with unambiguous positive feelings as in the fraternizing of enemy troops across no-man's-land during Christmas of 1914. Mud, fatigue, pain, and death wiped out the features that distinguished soldiers along national lines, and among the civilians in the separate countries there was an intense, climactic sense of unity, especially in the first weeks of August 1914, such as Zweig described in Austria:

> As never before, thousands and hundreds of thousands felt what they should have felt in peace time, that they belonged together. A city of two million, a country of nearly fifty million, in that hour felt that they were participating in world history, in a moment which would never recur, and that each one was called

> upon to cast his infinitesimal self into the glowing mass, there to be purified of all selfishness. All differences of class, rank, and language were flooded over at that moment by the rushing feeling of fraternity. . . . Each individual experienced an exaltation of his ego, he was no longer the isolated person of former times, he had been incorporated into the mass, he was part of the people, and his person, his hitherto unnoticed person, had been given meaning.[51]

As the war dragged on it released other powerful and dislocating forces that broke up the old dividers and forged new unities. Masses of civilians in Serbia and Belgium were directly affected by the fighting as their cities were bombarded by artillery, and the distinction between front and home disappeared. Later in the war Germans bombed English cities and aimed at civilian targets. Their intention was to weaken the will of the enemy, but the scattered destruction had the opposite effect of breaking down differences and misunderstandings that separated front-line soldier from civilian and of unifying the English in their resolve to pursue the war on the Continent. In the trenches distinctions of class, rank, and profession became as muddied as the uniforms. Barbusse recalled: "Our callings? A little of all—in the lump. In those departed days when we had a social status, before we came to immure our destiny in the molehills . . . what were we? Linked by a fate from which there is no escape, swept willy-nilly by the vast adventure into one rank, we have no choice but to go as the weeks and months go—alike. The terrible narrowness of the common life binds us close, adapts us, merges us one in the other. It is a sort of fatal contagion."[52] Charles de Gaulle saw the war as the culmination of a "collective spirit" that had been building for generations as a consequence of universal suffrage, compulsory education, industrialization, city life, the press, mass political parties, trade unions, and sport. "The mass movements and mechanization to which men and women were subjected by modern life had preconditioned them for mass mobilization and for the brutal, sudden shocks which characterized the war of peoples."[53]

The high level of unification required to mobilize mass armies and fight a war on many fronts was made possible by the new technology. Napoleon marched across Europe at the head of his Grand Army, and at one time his empire stretched from Lisbon to Moscow, but the Napoleonic wars did not have the scope of the Great European War that came to be known as the First *World* War. That distinction came from the enormous number of men involved and the vast distances over which operations were conducted. In an effort to break the deadlock in the west, military leaders opened fronts in distant places, as when the British attacked Gallipoli through the Dardanelles in 1915. Commenting on such an extension of the scope of the war, the combatant and military historian, Captain B. H. Liddell Hart, wrote that "modern developments had so changed conceptions of distance and powers of mobility, that a blow in some other theater of war would correspond to the historic attack on an enemy's strategic flank."[54] The contraction of distances that was effected before the war by means of telephones, the wireless, automobiles, and airplanes (in addition to the telegraph and railroad) transformed fighting during the war and the experience of distance in it. New weapons made it possible to increase the space between enemy soldiers, while electronic communication could extend the distance between sectors of coordinated operations and between commanding officers and their men. In several respects World War I was conducted at unprecedented long range.

The large-bore, muzzle-loading musket of the Napoleonic era had an effective lethal range of between one and two hundred yards. The small-bore, breach-loading rifle of the First World War shot a small-calibre bullet that was lethal at up to 2,000 yards and that traveled in a low trajectory and so could kill over a greater portion of its path. Heavy artillery in the war had a maximum range of up to 9,000 yards, although its effective range was limited to about 4,000 to 5,000 yards, beyond which it was generally impossible for the spotters at advanced positions to observe the location of impact and to telephone corrections back to their batteries. The shells therefore fell and killed far beyond the vision of soldiers in the trenches and the artillery men behind the lines. During an offensive the advancing men were blinded by the moving wall of smoke and dirt tossed up by the barrage. There was an eerie anonymity about this fighting, as Carrington observed: "The artilleryman rarely sees the object of his fire; he has no personal contact with the enemy, but suddenly finds himself in a scorching fire, from a source which he cannot ascertain, from an enemy he cannot see. It is like quarreling by telegraph."[55]

At the battle of the Marne military headquarters on both sides were far from the front. On August 30 the supreme command of the German offensive was located in Luxembourg, and wireless communication with the front was frequently disrupted by interference from the Eiffel Tower in Paris, several hundred miles away. The French headquarters were at Châtillon-sur-Seine, about 120 miles from the battlefield. On the morning of September 4 French aviators determined that the German army under General von Kluck, the "last man on the right" of Schlieffen's grand sweeping plan, had curved to the southeast and offered a vulnerable flank to the French and British. Galliéni spoke with Joffre through an intermediary (Joffre did not like to use the telephone directly), and in a series of urgent conversations that day received permission to counterattack. Later Galliéni said that the battle of the Marne was won by "coups de téléphone."[56]

The distance between the fighting and the decision-making created an experiential and emotional gap between the generals and the men at the front that enabled commanders to continue to spin table-top plans for offensives and be shielded from direct contact with the disastrous consequences. In a memoir the British Field Marshal Alexander complained that he never saw any officers above brigadier commander at the front. General Haig refrained from visiting casualty-clearing stations because it made him physically ill, and Joffre never got near front-line soldiers except as they marched past in a decoration parade. After pinning a medal on one blinded soldier, Joffre protested that he did not want to see any more of the wounded because he "would no longer have the courage to attack."[57] In earlier wars generals drew courage from proximity to battle; in this war they preserved their courage by moving far away from it.

The wireless not only expanded the range of military communication but also made possible the transmission and reception of orders in many widely scattered theaters of war. One such dramatic transmission was the message sent from the British Admiralty to all ships at 5:30 P.M. on August 3, 1914, which read: "The war telegram will be issued at midnight authorising you to commence hostilities against Germany but in view of our ultimatum they may decide to open fire at any moment. You must be ready for this." At midnight the second message went out, ordering the commencement of hostilities against Germany. At 5:00 P.M. the next day German wireless stations with a radius of around 2,000 miles were able to order German ships all over the world to make for a neutral port, thus saving a great part of the German merchant marine.[58]

Sending secret information across great distances by wireless could also be dangerous. As Russian armies moved farther from their base in August 1914, they were reduced to sending uncoded wireless messages because their staffs lacked codes and cryptographers. While maneuvering near Tannenberg, the Germans began to intercept those messages, and General Ludendorff was emboldened to attack one section of the Russian army after he learned that it could be surrounded and taken without danger of interference from the rest of the Russian army in that sector. The intercepted Russian messages were reported to Ludendorff every night at 11:00 P.M., and when they were late he went to the decoding office and got them directly.[59] His spectacular victory at Tannenberg (92,000 men captured) would not have been possible without access to this unique source of information.

For the soldiers the direction of the front became invested with magical, devilish properties as the entire landscape radiated into zones of safety and danger. The British philosopher T. E. Hulme observed this from the front in 1915: "In peacetime each direction on the road is as it were indifferent, it all goes on ad infinitum. But now you know that certain roads lead, as it were, up to the abyss."[60] The horizontal bearings were not qualitatively different from those of previous wars, but the vertical ones were. The war at sea was for the first time fought on an up-down axis with submarines, and the land war pushed down into the dugouts and mines and galleries of the trenches. But the most distinctive new shift in the directional orientation of the war was upwards.

In July 1917 Lovat Fraser, a writer for the London *Times,* reflected: "If I were asked what event of the last year has been of most significance to the future of humanity, I should reply that it is not the Russian Revolution, nor even the stern intervention of the United States in a sacred cause, but the appearance of a single German aeroplane flying at high noon over London last November." On November 28, 1916, a German airplane dropped six bombs on London's West End. The chimney was blown off a bakery, a stable was wrecked, the dressing room in a music hall was demolished. A newspaper reported also that "one cobblestone was cracked in Eccleston Mews."[61] Two years into the war such damage was negligible, but the exactitude of the report reveals something else—the anxiety generated from this new kind of war against people and cities for which no defense seemed possible. Germany had started aerial bombings on August 6, 1914, when a Zeppelin dropped thirteen bombs on Liège, killing nine people. On August 29, a German Taube commenced the bombing of Paris, a practice that continued sporadically throughout the war and inflicted minor damage. German airplanes dropped a few small bombs along the British coast regularly from 1914 on, but the London raid inaugurated an intensification of the number and effectiveness of bombings.

The year 1917 marked the commencement of regular heavy bombings of cities from airplanes. Anti-aircraft guns, barrage balloons, observer posts, sound locators, searchlights, maroons (sound bombs), sirens, and two-way air-to-ground wireless equipment (developed in 1917) engaged the attention of civilian and military personnel alike. Londoners braved the danger and remained in the open, staring up at the enormous Gotha bombers even as they were dropping their bombs. One report of a raid on June 13, 1917, described the fascination as "enemy aeroplanes journeyed through the clouds like little silver birds and their passage was watched by thousands of men and women . . . It was amazing because it was so beautiful." That particular display of beauty killed 162 people, marking a new level of destruction of civilian life and property. By the end of the war the British had suffered 835 killed and 1,972 wounded from German raids. Between 1915 and 1918 French and British bombings killed 746 Germans and wounded 1,843.[62] Although the total casualties were negligible compared to the losses in the trenches (the British had 2,500 casualties a day at the front in 1917), the bombings had a psychological impact that could not be reckoned in numbers alone. This form of war blurred the distinction between soldier and civilian, front and home, safety and danger. Walls and fortresses, rivers and channels, perhaps even oceans could not stop the airplanes that created a new vulnerability among those at home who had formerly felt safe. The airplane was a great leveler.

In the description of her flight over America, Gertrude Stein reflected on the structure of reality in the twentieth century and alluded once again to an underlying connection between Cubism, the new vision of aerial flight, and the composition of World War I.

> . . . yes I saw and once more I knew that a creator is contemporary, he understands what is contemporary when the contemporaries do not yet know it, but he is contemporary and as the twentieth century is a century which sees the earth as no one has ever seen it, the earth has a splendor that it never has had, and as everything destroys itself in the twentieth century and nothing continues, so then the twentieth century has a splendor which is its own and Picasso is of this century, he has that strange quality of an earth that one has never seen and of things destroyed as they have never been destroyed.[63]

Picasso's remark that the camouflaged trucks were Cubism also implied that even before his contemporaries had understood, he had seen through the old conception of the object and space as distinct kinds of being and understood that a truly contemporary art must recompose them as two modes of the same thing. Cubism was a creative response to broad cultural pressures that bore down on all sorts of traditional forms and necessitated new compositions and new perspectives. Stein's observation embraced the inevitable destruction and the splendor of new creations that emerge from sweeping cultural and material transformations. If an artist, aviator, or anyone should actually see the earth as no one has ever seen it before, then the old world must go smash. To record that new vision she fractured traditional syntax, recomposing thoughts from new and multiple perspectives, shading each variation as if turning it in her hand to see it and show it from all sides. Constantly looking for new points of view she gasped with pleasure during her first flight over America—the gasp of an artist who has finally found a long-sought-after composition. Her vision juxtaposed the new art and the new war, and, writing in 1937 just before German bombers leveled Guernica, she suggested that, like Picasso, she had understood what is contemporary even before her contemporaries knew it. Later that year Picasso began his epic painting, which commemorated the visitation of twentieth-century aerial warfare to a small civilian town in his native Spain. Using the Cubist techniques he had pioneered thirty years before, Picasso depicted things destroyed as they had never been destroyed before and made his unforgettable condemnation of the war from the sky.

CONCLUSION

No single thesis can explain all of the technological, scientific, literary, artistic, and philosophical currents that shaped the experience of time and space from 1880 to 1918. It is possible, however, to draw some general conclusions about the more important developments and their relation to the larger historical picture.

The introduction of World Standard Time had an enormous impact on communication, industry, war, and the everyday life of the masses; but explorations of a plurality of private times were the more historically unique contributions of the period. The assault on a universal, unchanging, and irreversible public time was the metaphysical foundation of a broad cultural challenge to traditional notions about the nature of the world and man's place in it. The affirmation of private time radically interiorized the locus of experience. It eroded conventional views about the stability and objectivity of the material world and of the mind's ability to comprehend it. Man cannot know the world "as it really is," if he cannot know what time it really is. If there are as many private times as there are individuals, then every person is responsible for creating his own world from one moment to the next, and creating it alone.

Among the three modes of time, the sense of the past was not qualitatively different from older notions, although its influence on the present was given more weight. The sense of the future, largely a reconstruction of past experience projected ahead in time, also resembled older experiential modes. The sense of the present was the most distinctively new, thickened temporally with retentions and protentions of past and future and, most important, expanded spatially to create the vast, shared experience of simultaneity. The present was no longer limited to one event in one place, sandwiched tightly between past and future and limited to local surroundings. In an age of intrusive electronic communication "now" became an extended interval of time that could, indeed must, include events around the world. Telephone switchboards, telephonic broadcasts, daily newspapers, World Standard Time, and the cinema mediated simultaneity through technology. The sinking of the *Titanic* dramatized it with S.O.S. messages beamed across the entire Atlantic world. In contrast no new technology transformed the experience of the past or future so fully. In the cultural sphere no unifying concept for the new sense of the past or future could rival the coherence and the popularity of the concept of simultaneity. Psychoanalysis was perhaps the most systematic and collective exploration of the past, but it was limited to a small group of *cognoscenti.* The Futurists formalized a philosophy of the past, but only negatively with their aesthetic of *passéisme*—an outlook that was as irresponsible as their affirmation of the future was impulsive. In spite of the bombast with which they proclaimed their love of the future, the subject of their art and the true focus of their manifestos was the present. And in addition to the Futurists, a growing number of playwrights, novelists, critics, musicians, painters, and even sculptors identified their work under the general concept of simultaneity, conceived of it to be unique to their age, and acknowledged the direct causal impact of recent technology.

As an experience that had spatial as well as temporal aspects, simultaneity had an extensive impact, since it involved many people in widely separate places, linked in an instant by the new communications technology and by the sweeping ubiquity of the camera eye. The cultural effects of the new temporality however, were generally not as extensive as those of the new spatiality, because the private nature of private time limited them to the phenomenal world of the individual and precluded a public or collective restructuring of experience. In contrast the emerging modes of space had extensive social, political, and religious manifestations.

One common theme among the several changes in those modes was the leveling of traditional hierarchies. The plurality of spaces, the philosophy of perspectivism, the affirmation of positive negative space, the restructuring of forms, and the contraction of social distance assaulted a variety of hierarchical orderings. While the plurality of lived spaces envisioned by the geometers, physicists, and biologists and those created by the painters and novelists did not always aim directly at the social structure of the aristocracy, they energized a general cultural challenge to all outmoded hierarchies.

The affirmation of positive negative space rejected the conventional view of space as less important than the objects contained in it. The reconstruction of forms rejected the conventional hierarchical orderings of those objects. Samuel Smiles's homily, "A place for everything and everything in its place," was an apt formulation of the older ordering. It was comforting to know that everything had a right place, even if the rationale for that placing seemed unjust. The challenge to hereditary privileges of class that began in seventeenth-century England spread eastward but slowly. The French Revolution of 1789 renewed that challenge, but even in nineteenth-century France the mystique of aristocracy continued to dictate social groupings. In eastern Europe rule by the noble class had the force of law. Up to the beginning of World War I, the Austro-Hungarian Empire preserved intact respect for royalty and the privileges of nobility. An aristocracy of birth set itself above everyone lacking sixteen noble quarterings. Many aristocrats lived secluded in hundreds of castles throughout the empire. They monopolized the higher posts in the army and diplomatic corps, controlled the conservatiye politics of the empire, upheld the power of the Catholic Church, set fashion in dress and furniture, and dictated proper decorum for the entire European world. That overbearing hierarchical world became the target of numerous artists and intellectuals, who challenged its metaphysical foundation and its concrete social, political, and religious institutions.

Dramatic transformations in the sense of distance were created by the new transportation and communications technology. Recognition of the potential threat of that technology to the aristocratic society of the Austro-Hungarian Empire is illustrated by the Emperor's reluctance to allow the newfangled gadgetry into the royal palace. Reared under the rigid formalism of military life and the exacting requirements of one of the oldest surviving royal dynasties, convinced of his divine right to rule, hostile to the incursions of popular government, isolated socially in a circle of high nobility, and contemptuous of everyone of low birth, Francis Joseph was an embodiment of the hierarchical world of the European aristocracy. In the Hofburg in Vienna, the favorite Habsburg palace for six hundred years, he allowed no electric lights, and kerosene lamps provided illumination. He shunned the use of typewriters and automobiles and refused to install telephones. The telephone in particular was incompatible with the aristocratic principle that certain persons, by virtue of their position in society—generally propinquity to the monarch—have special importance. Telephones break down barriers of distance—horizontally across the face of the land and vertically across social strata. They make all places equidistant from the seat of power and hence of equal value. The elaborate protocol of introductions, calling cards, invitations, and appointments is obviated by their instantaneity; and the protective function of doors, waiting rooms, servants, and guards is eliminated by the piercing of their intrusive ring. Telephones penetrate and thus profane all places; hence there are none in churches. The ancient frontiers of the Austro-Hungarian Empire (an empire that abounded in both horizontal and vertical frontiers) were incompatible with the universality, the irreverence, and the pugnacity of the telephone.

While the telephone most obtrusively penetrated the sanctuaries of privilege, much of the other new technology had a similar effect, leveling traditional hierarchies and creating new social distances. Already in 1913 the cinema was tagged a "democratic art," as the camera eye penetrated everywhere and as its cheap admission prices and mixed seating arrangements brought the highbrow culture of the theater to the working classes. The bicycle was a "great leveler" that bridged social space and made travel over longer distances accessible to the middle and lower classes who could not afford a carriage or automobile. The democratizing effect of the automobile was recognized immediately, even before it was inexpensive enough for the masses. The crowding of people in cities created a tangible drama of modern democracy. As the status of the rural setting of the aristocratic world gave way to the new status of the urban setting of the bourgeoisie, the value of distance in the maintenance of social prestige was diminished. The bourgeoisie still aspired to get away from the crowd and retire to provincial estates, but the urban crowd remained, dictating values and new social forms as it flattened social hierarchies based on distance.

Modern technology also collapsed the vault of heaven. Never before the age of the wireless and airplane did the heavens seem to be so close or so accessible—a place of passage for human communication and for human bodies in man-made machines. The omnipresence and penetrating capacity of wireless waves rivaled miraculous action and reversed the direction of divine intervention. Planes invaded the kingdom of heaven, and their exhaust fumes profaned the realm of the spirit. Upwards was still the direction of growth and life, but in this period it lost much of its sacred aspect.

As Arthur Lovejoy warned, the historian ought not try to make the thought of an individual, let alone that of an age, cohere "all-of-a-piece" but must try to identify the fluctuations between opposing ideas or the "embracing of both sides of an antithesis." I have identified those opposing ideas and antitheses and reconstructed how they occurred in isolation, in debates, and in the paired clusterings of opposites that appear most clearly only in historical perspective. If, in the interest of literary unity and in defiance of Lovejoy's prudent warning, I were to suggest a drawstring for this multiplicity of developments, it would be (and here technology supplies the metaphor) the miles of telephone wire that criss-crossed the Western world. They carried signals for World Standard Time and the first public "broadcasts"; revolutionized newspaper reporting, business transactions, crime detection, farming, and courting; made it possible for callers to control the immediate future of anyone they wished and intrude upon the peace and privacy of homes; accelerated the pace of life and multiplied contact points for varieties of lived space; leveled hierarchical social structures; facilitated the expansion of suburbs and the upward thrust of skyscrapers; complicated the conduct of diplomacy; forced generals to leave their lofty promontories and retire behind the front lines to follow battles from telephone headquarters; brought the voices of millions of people across regional and national boundaries; and worked to create the vast extended present of simultaneity.

NOTES
INDEX

NOTES

Introduction

1. Roger Shattuck, *The Banquet Years: The Origins of the Avant Garde in France, 1885 to World War I* (New York, 1958; rev. 1967); Carl E. Schorske, *Fin-de-siècle Vienna: Politics and Culture* (New York, 1980); Stuart Hughes, *Consciousness and Society: The Reorientation of European Social Thought 1890–1930* (New York, 1958).

1. The Nature of Time

1. Arthur O. Lovejoy, *Essays in the History of Ideas* (1948; rpt. New York, 1970), xiv, xv.

2. Sanford Fleming, "Time-Reckoning for the Twentieth Century," in *Smithsonian Report* (1886), 345–366.

3. "General von Moltke on Time Reform," in *Documents Relating to the Fixing of a Standard of Time* (Ottawa, 1891), 4.

4. Harrison J. Cowan, *Time and Its Measurements* (Cleveland, 1958), 45, estimates that there were over three hundred local times in the United States. According to a pamphlet by Charles Ferdinand Dowd, *A System of National Time for Railroads* (1870), there were at that time eighty different time standards on the American railroads. See Derek Howse, *Greenwich Time and the Discovery of the Longitude* (New York, 1980), 121.

5. Hugh Mill, "Time Standards of Europe," *Nature* (June 23, 1892): 175.

6. John Milne, "Civil Time," *The Geographical Journal*, 13 (January–June 1899): 179.

7. L. Houllevigue, "Le Problème de l'heure," *La Revue de Paris* (July–August 1913): 871. See also M. Ferrié, "La Télégraphie sans fil et le problème de l'heure," *Revue scientifique* (1913): 70–75; and *Conférence internationale de l'heure* (Paris, 1912).

8. Houllevigue, "Problème de l'heure," 875.

9. Charles W. Super, "Time and Space," *The Popular Science Monthly* (March 1913): 283.

10. Paul Delaporte, *Le Calendrier universel* (Paris, 1913).

11. Alexander Philip, *The Reform of the Calendar* (London, 1914), 34.

12. Johannes C. Barolin, *Der Hundertstundentag* (Vienna, 1914).

13. Henry Olerich, *A Cityless and Countryless World* (1893; rpt. New York, 1971), 173.

14. "Recording Time of Employees," *Scientific American* (August 12, 1890).

15. E. P. Thompson has identified a number of sources on the link between Puritan discipline, bourgeois exactitude, and capitalism in the West from the seventeenth to the nineteenth centuries, in "Time, Work-Discipline, and Industrial Capitalism," *Past and Present* (February 1968).

16. George M. Beard, *American Nervousness* (New York, 1881), 103.

17. Marcel Proust, *Within a Budding Grove* (1918; rpt. New York, 1970), 290.

18. Kafka wrote most of *The Trial* in 1914–15, but it was published posthumously for the first time in 1925. Quoted in Theodor Ziolkowski, *Dimensions of the Modern Novel* (Princeton, 1969), 41.

19. Franz Kafka, *Tagebücher, 1910–23* (Frankfurt, 1951), 552.

20. The interpretation of the rhythm of each chapter comes from Stuart Gilbert's *James Joyce's Ulysses* (New York, 1930), 30. Joyce himself helped Gilbert to unravel the keys to *Ulysses*, including a chart of the "technic" or rhythmic structure of each chapter.

21. James Joyce, *Ulysses* (1922; rpt. New York, 1961), 668–669.

22. Ernst Mach, *The Science of Mechanics* (1883; rpt. New York, 1919), 223.

23. Hendrick Lorentz, "Michelson's Interference Experiment," in *The Principle of Relativity*, ed. A. Sommerfeld (New York, 1952).

24. Albert Einstein, *Relativity: The Special and the General Theory* (1916; rpt. New York, 1952), 26.

25. Albert Einstein and Leopold Infeld, *The Evolution of Physics* (New York, 1938), 181.

26. Franz Lukas, *Die Grundbegriffe in den Kosmogonien der alten Völker* (Leipzig, 1893), 238–263, distinguished two basic cosmological forms: those that begin with matter and those that begin with principles like time or space. In the second category time is socially relative and becomes embodied in a deity such as Chronos, who in his turn creates things and originates history. Christian Pflaum, "Prolegomena zu einer völkerpsychologischen Untersuchung des Zeitbewusstseins," *Annalen der Naturphilosophie,* I (1902), laid down principles for studying the social origins of time and comparing them with the child's sense of time. Also relevant is Gustav Bilfinger, *Untersuchungen über die Zeitrechnung der alten Germanen* (Stuttgart, 1899–1901).

27. Emile Durkheim, *The Elementary Forms of the Religious Life* (1912; rpt. New York, 1965), 22, 32.

28. Karl Jaspers, *General Psychopathology* (1913; rpt. Chicago, 1963).

29. Jean Guyau, *La Genèse de l'idée du temps* (Paris, 1890); Pierre Janet, *L'Évolution de la mémoire et de la notion du temps* (Paris, 1928), 47.

30. Charles Blondel, *La Conscience morbide; essai de psychopathologie générale* (Paris, 1914), 214–225.

31. Henry Ellis Warren invented the modern electric clock driven by alternating current in 1916. See Brooks Palmer, *The Romance of Time* (New Haven, 1954), 47.

32. J. Marey, "The History of Chronophotography," *Annual Report of the Board of Regents of the Smithsonian Institution* (1902), 317; J. Marey, *La Chronophotographie* (Paris, 1899).

33. Anton Bragaglia, "Futurist Photodynamism" (1911), in *Futurist Manifestos,* ed. Umbro Apollonio (New York, 1973), 38.

34. The original reads: "Es bleibt dabei: die Zeitfolge ist das Gebiete des Dichters, so wie der Raum das Gebiete des Malers," Gotthold Lessing, *Laokoon oder über die Grenzen der Malerei und Poesie* (1776), chap. 18.

35. Richard W. Murphy, *The World of Cézanne 1839–1906* (New York, 1968), 58. George Heard Hamilton, "Cézanne, Bergson, and the Image of Time," *College Art Journal* (Fall 1956): 2–12, contrasts the Impressionist effort to depict a separate moment in time with Cézanne's "pictorial equivalent of the Bergsonian concept of space as known only in and through time." Hamilton concludes, boldly, that Cézanne "was the first modern artist to create an image of time."

36. Edward Fry, ed., *Cubism* (London, 1966), 57, 60, 62, 66–67.

37. Ernst Te Peerdt, *Das Problem der Darstellung des Moments der Zeit in den Werken der malenden und zeichnenden Kunst* (Strassburg, 1899), 40.

38. Albert Gleizes, "Portrait of the Publisher Figuière," (first exhibited in 1913); see Daniel Robbins, *Albert Gleizes 1881–1953* (New York, 1964), 31.

39. William James, "On Some Omissions of Introspective Psychology," *Mind* (January 1884): 2, 6, 11, 16.

40. William James, *Principles of Psychology* (New York, 1890), I, 239.

41. Henri Bergson, *An Introduction to Metaphysics* (1903; rpt. New York, 1955), 23–26.

42. Henri Bergson, *Creative Evolution* (1907; rpt. New York, 1944), 335.

43. Georges Sorel, *Réflexions sur la violence* (Paris, 1908).

44. Charles Péguy, *Oeuvres en prose de Charles Péguy, 1909–1914* (Paris, 1957), 1259–1286.

45. Wyndham Lewis, *Time and Western Man* (London, 1927), 428, 120.

46. In *Stream of Consciousness in the Modern Novel* (Berkeley, 1954), Robert Humphrey distinguishes between stream of consciousness and various literary techniques to render it such as direct interior monologue.

47. Edouard Dujardin, *We'll to the Woods no More* [*Les Lauriers sont coupés*], tr. Stuart Gilbert (1888; rpt. New York, 1935), 6.

48. Joyce, *Ulysses*, 750.

49. Richard Ellmann, *Ulysses on the Liffey* (New York, 1972), 163. Rayner Banham, *The Architecture of the Well-Tempered Environment* (London, 1969), 64.

50. Ellen Glasgow, *Phases of an Inferior Planet* (New York, 1898), 4.

51. Georges Méliès, "Les Vues cinématographiques" (1907), in *Intelligence du cinématographe*, ed. Marcel L'Herbier (Paris, 1946), 186–187.

52. Edgar Monin, *Cinéma ou l'homme imaginaire* (Paris, 1958), 50–54; and Arnold Hauser, *The Social History of Art* (New York, 1958), IV, 240.

53. Hugo Münsterberg, *The Film: A Psychological Study* (1916; rpt. New York, 1970), 77.

54. Charles B. Brewer, "The Widening Field of the Motion-Picture," *Century Magazine*, 86 (1913): 75.

55. Joseph Conrad, Preface to *Nigger of the Narcissus* (London, 1897).

56. Ford Madox Ford, *Joseph Conrad: A Personal Reminiscence* (Boston, 1924), 192–195.

57. Virginia Woolf, *A Writer's Diary* (New York, 1954), 136, 93. Malcolm Bradbury and James McFarlane concluded that one major concern of the modern novel was to free narrative art from "the determination of an onerous plot" and to question "linear narrative, logical and progressive order." See their *Modernism 1890–1930* (New York, 1976), 393.

58. Sigmund Freud, Letter to Wilhelm Fliess, May 25, 1897, in *The Standard Edition of the Complete Psychological Works of Sigmund Freud* (London, 1953) I, 252.

59. Freud, *Standard Edition*, XVIII, 28. Marie Bonaparte, "Time and the Unconscious," *The International Journal of Psycho-Analysis*, 21 (1940): 427–468, gives a full elaboration of the psychoanalytic approach to the "timelessness" of unconscious processes observed in children, dreams, fantasies, love, drug-induced states, mystic ecstasy, psychoses, and fairy tales.

60. Henri Hubert and Marcel Mauss, "Étude sommaire de la représentation du temps dans la religion et la magie," in their *Mélanges d'histoire des religions* (Paris, 1909), 201, 207, 209, 211–212.

61. Paul Langevin speculated how the theory might create a paradoxical situation from the dilation of time. If time slowed as a result of motion, then

so would biological processes. If a man could be sent into space at a speed sufficiently close to that of light and return after he had aged two years, he would find that two hundred years had passed on earth. See his "L'Evolution de l'espace et du temps," *Revue de métaphysique et de morale,* 19 (1911): 466.

62. Samuel Alexander, *Space, Time and Deity* (London, 1890); Camile Vettard, "Proust et le temps," *Les Cahiers Marcel Proust,* I (1927): 194; Lewis, *Time and Western Man.*

2. The Past

1. Eugène Minkowski, *Lived Time: Phenomenological and Psychopathological Studies* (Evanston, 1970), 148–168.

2. William Thompson, "On the Age of the Sun's Heat," *Macmillan's Magazine* (March 1862).

3. Archibald Geikie, "Twenty-five Years of Geological Progress in Britain," *Nature,* 51 (1895): 369.

4. Cited in Joe D. Burchfield, *Lord Kelvin and the Age of the Earth* (New York, 1975), 11. See also Francis C. Haber, "The Darwinian Revolution in the Concept of Time," in *The Study of Time,* ed. J. T. Fraser (New York, 1972).

5. J. Joly, "Radium and the Geological Age of the Earth," *Nature* (October 1, 1903): 526.

6. L. Azoulay, "L'Ère nouvelle des sons et des bruits—musées et archives phonographiques," *Revue scientifique,* 13 (1900): 712–715.

7. G. S. Lee, *The Voice of the Machines* (New York, 1906).

8. James Joyce, *Ulysses* (1922; rpt. New York, 1961), 114.

9. Hugo Münsterberg, *The Film: A Psychological Study* (New York, 1970), 77.

10. Nicolas Pevsner, *Pioneers of Modern Design* (London, 1964), 33.

11. Marcel Proust, *Pleasures and Days* (1913; rpt. New York, 1957), 286.

12. Marcel Proust, *Swann's Way* (1914; rpt. New York, 1928), 85.

13. Georg Simmel, "Die Ruine" (1911), in Kurt Wolff, *Georg Simmel: 1858–1918* (Columbus, Ohio, 1959), 265–266.

14. Henry Maudsley, *Physiology and Pathology of the Mind* (London, 1867), 182.

15. Ewald Hering, *Ueber das Gedächtnis als eine allgemeine Funktion der organisierten Materie* (Vienna, 1870).

16. Sigmund Freud and Joseph Breuer, "Studies in Hysteria," in *Standard Edition* (1895; rpt. London, 1953), II, 21–47.

17. Théodule Ribot, *Les Maladies de la mémoire* (Paris, 1895), 164.

18. On the childhood origin of these mental phenomena see Bernard Perez, *L'Enfant de trois à sept ans* (Paris, 1886), 275; Friedrich Scholz, *Die Charakterfehler des Kindes* (Leipzig, 1891), 101–102; and Friedrich Scholz, *Schlaf und Traum* (Leipzig, 1887), 34–36. On the childhood origins of adult

sexual pathology see Jean-Martin Charcot and Valentin Magnan, "Inversion du sens génital," *Archives de Neurologie*, III–IV (1882): 315; Jules Dallemagne, *Dégénérés et déséquilibrés* (Brussels, 1894), 525–527; Anton von Schrenck-Notzing, *The Use of Hypnosis in Psychopathia Sexualis* (1896; rpt. New York, 1956), 1, 154–155; Havelock Ellis, *Sexual Inversion* (Philadelphia, 1908), 156.

19. Freud, *Standard Edition*, XX, 33.

20. Ernst Kris, ed., *The Origins of Psycho-Analysis: Letters to Wilhelm Fliess, Drafts and Notes: 1887–1902* (New York, 1954), 246.

21. Freud, *Standard Edition*, VI, 274.

22. For a discussion of this theory of memory and perception and of Kant's philosophy of time generally see Charles M. Sherover, *The Human Experience of Time: The Development of its Philosophic Meaning* (New York, 1975), 112.

23. Henri Bergson, *Matter and Memory* (1896; rpt. New York, 1959), 52, 142.

24. Henri Bergson, *Time and Free Will* (1889; rpt. New York, 1960), 101.

25. Bergson, *Matter and Memory*, 52–53, 143.

26. Henri Bergson, *Creative Evolution* (1907; rpt. New York, 1944), 7, 52.

27. Edmund Husserl, *The Phenomenology of Internal Time-Consciousness* (Bloomington, Indiana, 1964), 71.

28. Wilhelm Dilthey, *Pattern and Meaning in History* (New York, 1961), 85, 97, 67.

29. Henri Bergson, *Essai sur les données immédiates de la conscience* (Paris, 1961), 73, 74.

30. Bergson, *Time and Free Will*, 11, 12.

31. Bergson, *Creative Evolution*, 219.

32. Bergson, *Time and Free Will*, 133–134; syntax altered slightly.

33. Marcel Proust, *By Way of Sainte-Beuve* (1954; rpt. London, 1958), 17, 18.

34. Proust, *Swann's Way*, 535, 529.

35. Marcel Proust, *Within a Budding Grove* (1918; rpt. New York, 1970), 72.

36. Ibid., 132.

37. Ibid., 133. See also Roger Shattuck, *Proust's Binoculars: A Study of Memory, Time and Recognition in À la recherche du temps perdu* (New York, 1963), on the function of metaphor and the perspective of time regained.

38. Marcel Proust, *The Letters of Marcel Proust* (New York, 1966).

39. Relevant here is Martin Buber's observation that "The Jew of antiquity was more an audient than a visual being and felt more in terms of time than of space." In Martin Buber, ed., *Jüdische Künstler* (Berlin, 1903), 7.

40. Translated as *The Use and Abuse of History* (New York, 1957), 3–49.

41. Friedrich Nietzsche, "Thus Spoke Zarathustra," in *The Portable Nietzsche*, ed. and tr. Walter Kaufmann (New York, 1954), 250–253.

42. My discussion of the function of the past in Ibsen and of the sense of the past generally in this period owes much to a number of discussions with Rudolph Binion, who shared ideas and sources from a work of his about the

power of the past on the present in the nineteenth and twentieth centuries (to be published under the title *The Present Past*).

43. The translation of the title as "Ghosts" from the Norwegian, *Gengangere,* is a bad one. Ibsen himself remarked to Magnus Hirschfeld that the German translation, *Gespenster,* was not right, and they both agreed that the French translation, *Les Revenants,* gave a better sense of the original which meant "those who return"; Magnus Hirschfeld, "Literarische Selbsterkenntnis. Zu meinem 60. Geburtstag," in *Literarische Welt,* IV, no. 21–22 (May 25, 1928).

44. Henrik Ibsen, *Ghosts and Three Other Plays* (New York, 1962), 163.

45. The dramatic representation of the destructive action of a past remembered or suddenly recalled in Ibsen's plays was the subject of a book by a German critic, who interpreted Ibsen's "vampire-like sucking memory" (*vampirartig saugende Erinnerung*) as part of a general movement in European thought beginning around 1880 when a number of intellectuals including Strindberg, Zola, Sudermann, and Hardy abandoned their optimism about transcending the past and settled into a fatalistic pessimism based on biological, historical, and psychological theories to the effect that we are fated to inherit our parents' diseases and repeat their vices. Kurt K. T. Wais, *Henrik Ibsen und das Problem des Vergangenen im Zusammenhang der gleichzeitigen Geistesgeschichte* (Stuttgart, 1931), 246 *ff.*

46. James Joyce, "The Dead," *Dubliners* (1916; rpt. New York, 1961).

47. Joyce, *Ulysses,* 421, 189, 38, 45, 24, 583, 186.

48. I am indebted to Carl E. Schorske's *Fin-de-siècle Vienna* for my discussion of Wagner and the Ringstrasse.

49. R. W. Flint, ed., *Marinetti: Selected Writings* (New York, 1971), 42, 46, 55–56, 60–64, 66–67, 97.

50. Proust, *Swann's Way,* 61.

51. Proust, *Letters,* 226.

52. Marcel Proust, *The Past Recaptured* (1927; rpt. New York, 1971), 139.

53. There is a passive component in psychoanalysis, the method of free association, which allows the therapist to follow capricious leaps of consciousness. Freud believed that this technique would enable the therapist to break down the defenses built into our use of logic and grammar and allow repressed material from the past to surface to consciousness and effectuate therapeutic relief. But despite this one element of chance, the basic structure of psychoanalytic inquiry is methodical, persistent, and incessant. In psychoanalysis the royal road to the unconscious is open and heavily trafficked the year around.

54. Henry James, *The Sense of the Past* (New York, 1917), 66.

55. Stephen Toulmin and June Goodfield, *The Discovery of Time* (New York, 1965), 232. In *Political Philosophy and Time* (Middletown, Connecticut, 1968), 252, John G. Gunnell concluded that the nineteenth century accomplished "the complete historicization of existence." Donald M. Lowe's *History of Bourgeois Perception* (Chicago, 1982), 40, identifies a unique bourgeois

sense of the past, which became fully developed in the nineteenth century when new disciplines such as anthropology, archaeology, and mythology came into being and when there was a special interest in identifying art styles and motifs with new terms such as "Neoclassicism," "Romanticism," "Medievalism," and "Primitivism."

56. Hayden V. White, "The Burden of History," *History and Theory,* 5 (1966): 119.

57. My dissertation, "Freud and the Emergence of Child Psychology: 1880–1910" (Columbia University, 1970), surveys many works that anticipated Freud by reconstructing the personal past and establishing its effect on adult behavior.

58. In *Studies in Human Time* (Baltimore, 1956), 35, Georges Poulet evaluated Bergson's most important contribution to the thought of the twentieth century to be his affirmation of freedom in the face of determinism and historicism: "Not in his conception of memory, nor in his philosophy of the continuous, but in his affirmation that duration is something other than history or a system of laws; that it is a free creation."

3. The Present

1. Walter Lord, *A Night to Remember* (New York, 1955); Richard O'Connor, *Down to Eternity* (New York, 1956); Peter Padfield, *The Titanic and the Californian* (London, 1965); Geoffrey Marcus, *The Maiden Voyage* (New York, 1969).

2. Lawrence Beesley, *The Loss of the SS Titanic* (New York, 1912), 101.

3. U. N. Bethell, *The Transmission of Intelligence by Electricity* (New York, 1912), 6; Smith quote cited by Wyn Craig Wade, *The Titanic: End of a Dream* (New York, 1979), 399–400.

4. Lord Salisbury's speech was printed in *The Electrician,* November 8, 1889, and cited by Asa Briggs, "The Pleasure Telephone: A Chapter in the Prehistory of the Media," in *The Social Impact of the Telephone,* ed. Ithiel Pool (Cambridge, 1977), 41.

5. G. E. C. Wedlake, *SOS: The Story of Radio-Communication* (London, 1973), 18–74.

6. Sylvester Baxter, "The Telephone Girl," *The Outlook* (May 26, 1906): 235.

7. Julien Brault, *Histoire du téléphone* (Paris, 1888), 90–95.

8. Jules Verne, "In the Year 2889," *The Forum,* 6 (1888): 664.

9. "The Telephone Newspaper," *Scientific American* (October 26, 1896); Arthur Mee, "The Pleasure Telephone," *The Strand Magazine,* 16 (1898): 34; and Asa Briggs, "The Pleasure Telephone," 41.

10. "The Telephone and Election Returns," *Electrical Review* (December 16, 1896): 298.

11. Max Nordau, *Degeneration* (1892; rpt. New York, 1968), 39.

12. Paul Claudel, "Connaissance du temps," in *Fou-Tcheou* (1904), quoted in Pär Bergman, *"Modernolatria" et "Simultaneità": Recherches sùr deux tendances dans l'avant-garde littéraire en Italie et en France à la veille de la première guerre mondiale* (Uppsala, Sweden, 1962), 23.

13. Lewis Jacobs, *The Rise of the American Film* (1939; rev. ed. New York, 1967).

14. Hugo Münsterberg, *The Film: A Psychological Study* (1916; rpt. New York, 1970), 14.

15. Filippo Marinetti, Bruno Carrà, Emilio Settimelli, Arnaldo Ginna, Giacomo Balla, "The Futurist Cinema," (manifesto of September 1916), in *Marinetti: Selected Writings,* ed. R. W. Flint (New York, 1971), 207.

16. It is ironic that these poets, who insisted that in the modern world things are experienced all at the same time, that simultaneity was a distinctive feature of their age and a compelling subject of modern art, should fight so bitterly over priority. The fireworks began when an anonymous article in *Paris-Journal* of October 24, 1913, contended that "the first simultaneous book" was a "mongrel plagiarism" modeled on Barzun's poetry. Cendrars denied that he owed anything to Barzun, and for several weeks a debate raged between Barzun and Cendrars. Barzun's poetry was published in parallel columns to be read sequentially. The critics charged that if the poems were read that way, they would create nothing but cacophony. Also there were precedents in the poetry of Jules Romains, who, in an article of 1907, had deplored the fact that a single voice may recite a poem expressing contrary passions. In 1908 Romains staged a reading of a poem, "L'Église," with four voices responding to each other and occasionally mixing in authentic simultaneity. The debate erupted again in June 1914, when Cendrars published the "Open Letter to Barzun," claiming that his own *Prose du Transsibérien* was indeed the first simultaneous book. In October 1913, after Barzun had accused Apollinaire of imitating Marinetti, Apollinaire entered the fray with an article that compared Barzun's poetry with "Frère Jacques." To discredit Barzun's claims to originality Apollinaire surveyed a variety of earlier simultaneous productions in poetry, drama, and art, and argued that the concept was, as Delaunay had said, a *terme de métier* for all the modern arts. See Guillaume Apollinaire, "Simultanisme-Librettisme," *Les Soirées de Paris,* June 15, 1914, 322–325. For the details of this dispute see Bergman, *"Modernolatria,"* 291–323, 362–369; Michel Decaudin, *La Crise des valeurs symbolistes: 20 ans de poésie française 1895–1914* (Toulouse, 1960), 477–483; Volker Neumann, *Die Zeit bei Guillaume Apollinaire* (Munich, 1972), 122–140.

17. Henri-Martin Barzun's poem in his journal *Poème et Drame,* 3 (March 1913): 54.

18. Henri-Martin Barzun, *L'Ère du drame: essai de synthèse poétique moderne* (Paris, 1912), 15–35.

19. Henri-Martin Barzun, *Voix, rythmes et chants simultanés* (Paris, 1913), 25–46. These ideas were summarized in Ernst Florian-Parmentier, *La Littérature et l'époque* (Paris, 1914), 291–303.

20. "Le Futurisme," *Revue synthétique illustrée* (January 11, 1924).

21. Walter Albert, ed., *Selected Writings of Blaise Cendrars* (New York, 1962), 67–99.

22. Cited by Bergman, *"Modernolatria,"* 8.

23. Ibid., 392.

24. H. H. Stuckenschmidt, *Twentieth-Century Music* (New York, 1969), 72, is the major source for my discussion of simultaneity in music.

25. Filippo Marinetti, "Geometrical and Mechanical Splendor and the Numerical Sensibility," in Flint, *Marinetti,* 97.

26. Richard Ellmann, *James Joyce* (New York, 1965), 310–313; Craig Wallace Barrow, *Montage in James Joyce's Ulysses* (Madrid, 1980).

27. A number of scholars have noted Joyce's "spatial form." Harry Levin observed that Joyce's mind is not temporal, but spatial: "His characters move in space, but they do not develop in time"; see his *James Joyce: A Critical Introduction* (New York, 1960), 134. Joyce's friend and chronicler, Frank Budgen, observed that the writer's view of life "is that of a painter surveying a still scene rather than that of a musician following a development through time"; *James Joyce and the Making of Ulysses* (Bloomington, Indiana, 1973), 153. Joseph Frank, "Spatial Form in the Modern Novel," in *Critiques and Essays on Modern Fiction 1920–1951,* ed. John W. Aldridge (New York, 1952), 46, concluded that Joyce can only be grasped upon rereading and that he "proceeded on the assumption that a unified spatial apprehension of his work would ultimately be possible." Edmund Wilson conceived of *Ulysses* as "something solid like a city which actually existed in space and which could be entered from any direction—as Joyce is said, in composing his books, to work on different parts simultaneously." See his *Axel's Castle* (New York, 1950), 210.

28. There is a mountain of critical literature. Most helpful was Stuart Gilbert, *James Joyce's Ulysses* (New York, 1955) and A. Walton Litz, *The Art of James Joyce* (London, 1961), 62–74.

29. Joyce, *Ulysses,* 359–360, 361, 363.

30. In 1913 Delaunay wrote that his generation had been inspired by "the poetry of the [Eiffel] Tower which communicates mysteriously with the whole world [and] the factories, the bridges, iron constructions, dirigibles, the countless movements of airplanes, the windows simultaneously seen by crowds." See Pierre Francastel, ed., *Du Cubisme à l'art abstrait* (Paris, 1957), 111–112.

31. Albert Einstein, "On the Electrodynamics of Moving Bodies" (1905), in *The Principle of Relativity,* ed. H. A. Lorentz (New York, 1952), 42–43.

32. Pär Bergman, *"Modernolatria,"* x.

33. Albert Heim, "Notizen über den Tod durch Absturz," *Jahrbuch des schweizerischen Alpenclubs,* 27 (Bern, 1964): 168. Victor Egger, "La Durée apparente des rêves," *Revue philosophique* (July 1895): 41–59; Pierre Janet, *Les Obsessions et la psychasthénie* (Paris, 1903), I, 481.

34. William James, *Principles of Psychology* (1890; rpt. New York, 1950), II, 613–614.

35. David Hume, *Treatise on Human Nature,* 1730–1740, part 2, sec. 1.

36. The work by E. R. Clay is cited as "The Alternative." James gives no more bibliographical information and I have not been able to locate it. See William James, *Principles,* II, 608–609.

37. Josiah Royce, *The World and the Individual* (New York, 1901), 111–149. See also Milič Čapek, "Time and Eternity in Royce and Bergson," *Revue internationale de philosophie,* 70–80 (1967): 23–45.

38. Edmund Husserl, *The Phenomenology of Internal Time-Consciousness* (Bloomington, Indiana, 1964), 41, 43, 52, 76, 149.

39. Anton Bragaglia, "Futurist Photodynamism" (1911) and Gino Severini, "The Plastic Analogies of Dynamism" (1913), in *Futurist Manifestos,* ed. Umbro Apollonio (New York, 1973), 47, 121.

40. Guillaume Apollinaire, *The Cubist Painters* (1913; rpt. New York, 1949), 10.

41. James Joyce, *Stephen Hero* (New York, 1944), 211. Theodore Ziolkowski has argued that these distinctive moments in modern literature create a "timeless suspension" in defiance of death. He interprets Rilke's *The Notebooks of Malte Laurids Brigge* as an affirmation of private time in which past and future become reworked by the artist into a timeless, eternal present. See his *Dimensions of the Modern Novel: German Texts and European Contexts* (Princeton, 1969), 3–35, 212.

42. F. W. Dupee, ed., *Selected Writings of Gertrude Stein* (New York, 1972), 516, 342–343. In 1908 Hugo von Hofmannsthal argued that the task of the modern poet was to achieve both a spatially and a temporally expanded present. He must respond to the myriad separate events of the modern age and he must also create a thickened present. "The concept of time, of the past and future, will be transformed into a single present." See his "Der Dichter und diese Zeit," in *Erzählung und Aufsätze* (Frankfurt, 1957), 455–456, 464.

43. "William Blake" in *The Critical Writings of James Joyce,* ed. Ellsworth Mason and Richard Ellmann (New York, 1970), 222.

44. The English Vorticists also affirmed the present in their journal, *Blast,* the first issue of which appeared on the eve of war, June 20, 1914. The editor, Wyndham Lewis, announced that the Vorticists "stand for the Reality of the Present—not for the Sentimental Future, or the sacrosanct Past." The central image of their art—the vortex—was a meeting point of the real forces of the universe in the here and now. They viewed everything absent—past or future—as a negation of life. Although they distinguished their emphatic repudiation of the past from the "sentimental" philosophy of the Futurists, in the long view the two groups fall together. See Wyndham Lewis, "Long Live the Vortex," *Blast,* 1 (June 20, 1914). No page numbers.

45. Friedrich Nietzsche, "Die fröhliche Wissenschaft," in *Werke in drei Bänden* (Munich, 1966), II, 202–203.

4. The Future

1. Eugène Minkowski, *Lived Time: Phenomenological and Psychopathological Studies* (Evanston, 1970), 6, 87–88. Walter Lippmann saw America in 1914 as having two possible modes of moving into the future: it could continue the current "drift" or struggle to achieve a new, active mode of "mastery." These contrasting modes were prominent in his title, *Drift and Mastery: An Attempt to Diagnose the Current Unrest* (New York, 1914), and were elaborated throughout the text.

2. Herbert N. Casson, *The History of the Telephone* (Chicago, 1910), 231.

3. See next chapter for a discussion of Taylorism.

4. Cited in William L. Langer, *The Diplomacy of Imperialism* (New York, 1935), I, 78.

5. Henry Adams, *The Education of Henry Adams* (1907; rpt. New York, 1931), 382.

6. Published in *Nature*, 65 (February 6, 1902): 326–331.

7. E. F. Bleiler, ed., *Three Prophetic Novels of H. G. Wells* (New York, 1960), 142, 41.

8. H. G. Wells, *Anticipations of the Reaction of Mechanical and Scientific Progress Upon Human Life and Thought* (1901; rpt. London, 1914), 2, 46, 59, 32, 184. For another unusually accurate prediction of the future of warfare see I. S. Bloch, *The Future of War in Its Technical Economic and Political Relations* (1897; rpt. New York, 1899).

9. Kenneth M. Roemer, *The Obsolete Necessity: American Utopian Writings 1888–1900* (Kent, Ohio, 1976), 4. He surveyed 160 such works that appeared between 1888 and 1900 and concluded that during this period the utopian novel was one of the most widely read types of literature in America.

10. Mark R. Hillegas, *The Future as Nightmare: H. G. Wells and the Anti-Utopians* (Carbondale, Illinois, 1967).

11. F. T. Marinetti, "The Founding Manifesto of Futurism" (1909), in *Futurist Manifestos*, ed. Umbro Apollonio (New York, 1973), 19–24.

12. Umberto Boccioni, Carlo Carrà, Luigi Russolo, Giacomo Balla, Gino Severini, "Manifesto of the Futurist Painters" (1910) in ibid., 24–25.

13. Antonio Sant'Elia, "Manifesto of Futurist Architecture" (1914), in ibid., 160–172.

14. Marianne W. Martin, *Futurist Art and Theory 1909–1915* (Oxford, 1968), 190.

15. Pierre Laplace, quoted in Milič Čapek, *Bergson and Modern Physics* (New York, 1971), 122.

16. Emile Meyerson, *Identity and Reality* (1908; rpt. New York, 1962), 215–231.

17. Jean Guyau, *La Genèse de l'idée de temps* (Paris, 1890), 44.

18. Henri Bergson, *Creative Evolution* (1907; rpt. New York, 1944), 220.

19. Georges Sorel, *Reflections on Violence* (1906; rpt. New York, 1961), 124–125.

20. William Thomson, "On a Universal Tendency in Nature to the Dis-

sipation of Mechanical Energy," *Philosophical Magazine,* 4 (1852): 304, cited by Stephen G. Brush, "Science and Culture in the Nineteenth Century: Thermodynamics and History," *The Graduate Journal* (Spring 1967): 494. See also Jerome Buckley, "The Idea of Decadence" in *Triumph of Time* (Cambridge, 1966), 67. Thomson repeated this formulation forty years later in "On the Dissipation of Energy," *Fortnightly Review* (1892): 313-321.

21. Oswald Spengler, *The Decline of the West* (1918; rpt. New York, 1929), I, 129, 134, 137, 423-424.

22. Thomas Mann, *The Magic Mountain* (1924; rpt. New York, 1966), 219.

23. Ibid., 356-357.

5. Speed

1. Robert Ensor, *England 1890–1914* (Oxford, 1936), 278–279, 505.

2. Geoffrey Marcus, *The Maiden Voyage* (New York, 1969), 289–291.

3. Lawrence Beesley, *The Loss of the SS. Titanic* (New York, 1912), 237.

4. Critics cited by Richard O'Connor, *Down to Eternity* (New York, 1956), 186–190.

5. Morgan Robertson, *Futility* (New York, 1898), 1, 3, 4, 29.

6. Karl Lamprecht, *Deutsche Geschichte der jüngsten Vergangenheit und Gegenwart* (Berlin, 1912), I, 171.

7. Georg Simmel, "The Metropolis and Mental Life" (1900), in *The Sociology of Georg Simmel,* ed. and tr. Kurt H. Wolff (New York, 1950), 409–424.

8. Joseph B. Bishop, "Social and Economic Influence of the Bicycle," *The Forum* (August 1896): 689.

9. Sylvester Baxter, "Economic and Social Influences of the Bicycle," *The Arena* (October 1892): 583.

10. Ch. Du Pasquier, "Le Plaisir d'aller à bicyclette," *Revue scientifique,* ser. 4, vol. 6 (Paris, 1896): 145.

11. Paul Adam, *La Morale des sports* (Paris, 1907), 449–450.

12. Maurice Leblanc, *Voici des ailes!* (Paris, 1898), 19, 65, 77, 108, 145, 147.

13. Octave Mirbeau, *"La 628-E8"* (Paris, 1908), 6–7, cited in Pär Bergman, *"Modernolatria" et "Simultaneità"* (Uppsala, Sweden, 1962), 17.

14. Cited in William Plowden, *The Motor Car and Politics 1896–1970* (London, 1971), 47.

15. *The Times,* London, April 11, 1911, 6.

16. Edward W. Byrn, *The Progress of Invention in the Nineteenth Century* (New York, 1900), 56.

17. Harold I. Sharlin, "Electrical Generation and Transmission," in *Technology in Western Civilization,* ed. Melvin Kranzburg and Carroll W. Pursell Jr. (New York, 1967), I, 583.

18. John Brooks, *Telephone: The First Hundred Years* (New York, 1975), 115. In a short story, "In the Year 2889," Jules Verne envisioned an electronic computer, a "Piano Electro-Reckoner," *The Forum,* 6 (1888): 676. Georg Simmel, in "Die Bedeutung des Geldes für das Tempo des Lebens," *Neue*

Deutsche Rundschau, 8 (1897): 111–122, discusses how the introduction of paper money accelerated the pace of business transactions and the tempo of life.

19. William Crookes, "Some Possibilities of Electricity," *The Fortnightly Review,* 5 (1892): 179.

20. Ernest Solvay, "Rôle de l'électricité dans les phénomènes de la vie," *Revue scientifique,* 52 (1893): 769–778.

21. John B. Huber, "Arrhenius and His Electrified Children," *Scientific American* (April 13, 1912): 334.

22. *New York Times,* August 7, 1890, 1–2.

23. Robert Lincoln O'Brien, "Machinery and English Style," *Atlantic Monthly* (October 1904): 464–472.

24. Samuel Haber, *Efficiency and Uplift: Scientific Management in the Progressive Era 1890–1920* (Chicago, 1964).

25. Frederick W. Taylor, *The Principles of Scientific Management* (New York, 1911), 94.

26. Frederick W. Taylor, "A Piece-Rate System, Being a Step Toward a Partial Solution of the Labor Problem," read at American Society of Mechanical Engineers in 1895. An early study of the pace of competitive activity was made by Norman Triplett, "The Dynamogenic Factors in Pacemaking and Competition," *The American Journal of Psychology* (July 1898). He concluded that the presence of another contestant in a bicycle race increased the pace on an average of 5.15 seconds per mile.

27. Frederick W. Taylor, "Shop Management," reprinted in his *Scientific Management* (New York, 1947), 150–154.

28. Frank B. Gilbreth and Lillian M. Gilbreth, *Fatigue Study* (New York, 1916), 121.

29. Frank B. Gilbreth, "Motion Study in the Household," *Scientific American* (April 13, 1912): 328.

30. Frank B. Gilbreth, Jr., and Ernestine Gilbreth Carey, *Cheaper by the Dozen* (New York, 1948), 3.

31. F. Gilbreth and L. Gilbreth, *Fatigue Study,* 159.

32. Standish D. Lawder documents the influence of the cinema on the Cubists and their subsequent work with it in *The Cubist Cinema* (New York, 1975), 21–25.

33. Cited by Katherine Kuh, *Break-Up* (New York, 1966), 48.

34. A wildly speculative article entitled "Does Everything Go by Jerks?" suggested that all processes in the universe might occur by means of a series of infinitesimally small jerks rather than continuously. "There are 'atoms' of energy as well as of matter, and possibly also 'atoms' of time, causing all duration to be jerky instead of smooth, as it appears to be." And nature might therefore be "one vast cinematograph." See *The Literary Digest* (April 13, 1912).

35. Rudolf Arnheim, *Film as Art* (New York, 1933), 165–166.

36. E. A. Baughan, "The Art of Moving Pictures," *The Fortnightly Review,* 12 (1919): 450–454.

37. Horace M. Kallen, "The Dramatic Picture Versus the Pictorial Drama: A Study of the Influences of the Cinematograph on the Stage," *The Harvard Monthly* (March 1910): 28.

38. Jules Guiart, "La Vie révélée par le cinématographe," *Revue scientifique* (1914): 749. See also Charles B. Brewer, "The Widening Field of the Moving Picture," *The Century Magazine,* 86 (1913): 72.

39. Hugo Münsterberg, *The Film: A Psychological Study* (New York, 1970), 10.

40. Erwin Panofsky, "Style and Medium in the Motion Pictures," *Critique* (January-February 1947); reprinted in *Film: An Anthology,* ed. Daniel Talbot (Berkeley, 1969), 16.

41. Frank Norris, *McTeague* (1899; rpt. New York, 1964), 85.

42. Fernand Léger, "The Origins of Painting and Its Representational Value" (1913), in *Cubism,* ed. Edward F. Fry (New York, 1966), 121.

43. Fernand Léger, "Contemporary Achievements in Painting," in *Functions of Painting,* ed. Edward F. Fry (New York, 1973), 11.

44. Luigi Pirandello, *Shoot: The Notebooks of Serafino Gubbio, Cinematograph Operator* (1916; rpt. New York, 1926), 4, 10, 86.

45. Filippo Marinetti, "The Founding Manifesto of Futurism," *Le Figaro,* February 20, 1909, in *Marinetti: Selected Writings,* ed. R. W. Flint (New York, 1971), 41.

46. Filippo Marinetti, "The New Religion-Morality of Speed," *L'Italia Futurista,* May 11, 1916; in Flint, *Marinetti,* 94–95.

47. Umberto Boccioni, Carlo Carrà, Luigi Russolo, Giacomo Balla, Gino Severini, "Futurist Painting: Technical Manifesto" (1910), in *Futurist Manifestos,* ed. Umbro Apollonio (New York, 1973), 27–30.

48. Umberto Boccioni, "Absolute Motion + Relative Motion = Dynamism," in ibid., 150–154.

49. Marianne W. Martin cited this passage from Marinetti's 1915 interventionist tract, *War, the World's Only Hygiene* and made the connection with Boccioni, *Futurist Art Theory 1909–1915* (Oxford, 1968), 172.

50. In a general study of technology and culture, Werner Sombart linked the two-step with the rhythm of a machine; the "nervous and hurried" pace of music with urban life; and the city's "hard, cold, loveless" quality with the "rush and racket" of his age; see "Technik und Kultur," *Archiv für Sozialwissenschaft und Sozialpolitik,* 23 (1911): 342–347.

51. There were also studies of the origin of our sense of rhythm and its application to work. A pioneer article of 1894 identified several possible sources: the cosmic rhythms of the earth's rotation and orbiting; the living rhythms of gestation, menstruation, pulse, breathing, and sleeping; and the rhythm of walking or a horse's hoofbeat; see Thaddeus Bolton, "Rhythm," *The American Journal of Psychology,* 6 (1894): 145–238. See also Margaret Kiever Smith, "Rhythmus und Arbeit," *Philosophische Studien,* 16 (1900): 71–133.

52. William J. Schafer and Johannes Riedl, *The Art of Ragtime* (Baton

Rouge, 1973), 9, 10, 58–59; Rudi Blesh and Harriet Janes, *They All Played Ragtime* (London, 1958), 3–23.

53. Hiram Kelly Moderwell, "Ragtime," *New Republic* (October 16, 1915): 286, cited by Edward A. Berlin, *Ragtime: A Musical and Cultural History* (Berkeley, 1980), 51; Walter Lippmann, *Drift and Mastery: An Attempt to Diagnose the Current Unrest* (New York, 1914), 211.

54. William Morrison Patterson, *The Rhythm of Prose* (New York, 1916), 50–51; R. W. S. Mendl, *The Appeal of Jazz* (London, 1927), 46; William W. Austin, *Music in the 20th Century* (New York, 1966).

55. Igor Stravinsky, *An Autobiography* (New York, 1936), 47.

56. George M. Beard, *American Nervousness: Its Causes and Consequences* (New York, 1881), 116.

57. Sir James Crichton-Browne, "La Vieillesse," *Revue scientifique*, 49, (1892): 168–178.

58. Max Nordau, *Degeneration* (New York, 1968), 37–42.

59. John H. Girdner, *Newyorkitis* (New York, 1901), 119.

60. Gabriel Hanotaux, *L'Energie française* (Paris, 1902), 355.

61. Willy Hellpach, *Nervosität und Kultur* (Berlin, 1902), 12.

62. For a survey of the *psychische Spannung* of the age in medical and imaginative literature see Andreas Steiner, *Das nervöse Zeitalter: der Begriff der Nervosität bei Laien und Ärzten in Deutschland und Österreich um 1900* (Zurich, 1964).

63. Henry Adams, *The Education*, 499.

64. William Dean Howells, *Through the Eye of the Needle* (New York, 1907), 10–11.

65. In 1913 there were 4,200 total traffic deaths in the United States. United States Bureau of the Census, *Historical Statistics of the United States Colonial Times to 1970* (Washington, D.C., 1975), 720.

66. Robert Musil, *The Man Without Qualities* (1930; rpt. New York, 1966), 6, 7, 30.

67. Stefan Zweig, *The World of Yesterday* (Lincoln, Nebraska, 1964), 25–26.

68. Charles Féré, "Civilisation et névropathie," *Revue philosophique*, 41 (1896): 400–413.

69. Octave Uzanne, *La Locomotive à travers le temps, les moeurs et l'espace* (Paris, 1912), vi–vii, 244–247, 304. Émile Magne surveys a number of sources which, after Zola's *La Bête humaine* in 1885, affirmed the new aesthetic of the machine: "Le Machinisme dans la littérature contemporaine," *Mercure de France*, LXXXIII (January 16, 1910): 202–217.

6 The Nature of Space

1. Albert Einstein, "Autobiographical Notes," in *Albert Einstein: Philosopher-Scientist*, ed. Paul Arthur Schilpp (Evanston, 1949), 9–11.

2. Max Jammer, *Concepts of Space: The History of Theories of Space in Physics*

(Cambridge, Massachusetts, 1969), 144–146; A. d'Abro, *The Evolution of Scientific Thought from Newton to Einstein* (New York, 1927), 35–48.

3. Lawrence Beesley, *The Loss of the SS. Titanic* (New York, 1912), 105.

4. Henri Poincaré, *Science and Hypothesis* (1901; rpt. New York, 1952), 50–58. See also his article, "On the Foundations of Geometry," *The Monist*, 9 (1898): 42.

5. Ernst Mach, *Space and Geometry in Light of Physiological, Psychological, and Physical Inquiry* (1901; rpt. Chicago, 1906), 9, 94.

6. V. I. Lenin, *Materialism and Empirio-Criticism: Critical Comments on a Reactionary Philosophy* (1908; rpt. New York, 1927), 176–189.

7. On the political context of this issue see "Lenin and the Partyness of Philosophy," in David Joravsky, *Soviet Marxism and Natural Science 1917–1932* (London, 1961), 24–44.

8. Cited by Lenin, *Materialism*, 189.

9. Albert Einstein, *Relativity* (New York, 1961), 9.

10. Ibid., 139.

11. E. de Cyon, "Les Bases naturelles de la géométrie d'Euclide," *Revue philosophique*, 52 (July-December 1901): 1–30.

12. Louis Couturat, "Sur les bases naturelles de la géométrie d'Euclide," in ibid., 540–542.

13. E. von Cyon, *Das Ohrlabyrinth als Organ der mathematischen Sinne für Raum und Zeit* (Berlin, 1908), chap. 7.

14. Jacob von Uexküll, *Umwelt und Innenwelt der Tiere* (Berlin, 1909), 195. He extended these findings in *Bausteine zu einer biologischen Weltanschauung* (Munich, 1913).

15. The other categories discussed included cause, class, substance, number, and force. There is an analysis of these arguments in Steven Lukes, *Émile Durkheim: His Life and Work* (New York, 1973), 436–445.

16. Émile Durkheim and Marcel Mauss, *Primitive Classification* (New York, 1970), 43–44, 82, 86.

17. Émile Durkheim, *The Elementary Forms of the Religious Life* (New York, 1965), 22, 32, 489–492.

18. In *Consciousness and Society: The Reconstruction of European Social Thought 1890–1930* (New York, 1958), H. Stuart Hughes interpreted this generation as having discovered "the subjective character of social thought," the necessary mediation of consciousness in the study of man and society. According to Hughes, Durkheim was one of many who "found themselves inserting between the external data and the final intellectual product an intermediate stage of reflection on their own awareness of these data"; see pp. 16, 17. For a discussion of the polarization of space between the right and left hand in the religious practices of different societies see Robert Hertz, "La Prééminence de la main droite: étude sur la polarité religieuse," *Revue philosophique* (December 1909): 553–580. The Marxist historian Henri Lefebvre has analyzed the social "production of space" (especially unique capitalistic and socialistic forms), and he has identified a breakdown of the older uniform

space around 1910, when "Euclidean and perspectival space disappeared as referents along with the other common places (the city, history, paternity, the tonal system in music, the moral tradition, etc.)." See *La Production de l'espace* (Paris, 1974), 34 *ff.*

19. Oswald Spengler, *The Decline of the West* (New York, 1926), I, 174–178; 188–190; H. Stuart Hughes, *Oswald Spengler* (New York, 1952), 78–79.

20. Spengler, *Decline,* 337.

21. A pioneer study of the cultural impact of the introduction of perspective is Erwin Panofsky's "Die Perspektive als 'symbolische Form'," *Vorträge der Bibliothek Warburg* (1924–25).

22. Samuel Y. Edgerton Jr., *The Renaissance Rediscovery of Linear Perspective* (New York, 1975), 30–40.

23. L. Keith Cohen, "The Novel and the Movies: Dynamics of Artistic Exchange in the Early Twentieth Century" (Ph.D diss., Princeton University, 1974), 49–50.

24. For a general discussion of Cézanne's role in the breakdown of scientific perspective see Fritz Novotny, *Cézanne und das Ende der wissenschaftlichen Perspektive* (1938; rpt. Vienna, 1970), 184 and *passim.*

25. John Rewald, ed., *Paul Cézanne Letters* (Oxford, 1946), 262.

26. Maurice Merleau-Ponty, *Sense and Non-Sense* (1948; rpt. Evanston, 1964), 14.

27. Georges Matoré, *L'Espace humain: l'expression de l'espace dans la vie, la pensée et l'art contemporains* (Paris, 1962), 236–242; Standish Lawder, *The Cubist Cinema* (New York, 1975), 12.

28. Charles B. Brewer, "The Widening Field of the Moving-Picture," *The Century Magazine,* 86 (1913): 73–74.

29. Roger Allard, "At the Paris Salon d'Automne" (1910), in *Cubism,* ed. Edward Fry (New York, 1966), 62.

30. Jean Metzinger, "Cubism and Tradition" (1911), in ibid., 66.

31. Guillaume Apollinaire, *Cubist Painters* (1913; rpt. New York, 1944), 13.

32. E. Jouffret, *Traité élémentaire de géometrie à quatre dimensions* (Paris, 1903), 153. This connection was made by Linda Dalrymple Henderson in an article on possible influences of physics and geometry on the Cubists: "A New Facet of Cubism: 'The Fourth Dimension' and 'Non-Euclidean Geometry' Reinterpreted," *The Art Quarterly* (Winter 1971): 411–433. She refutes the facile connections between Cubism and science made by Paul M. Laporte and others and shows that although the Cubists could not have known about Einstein's relativity or Minkowski's space-time theory, they might have learned about the fourth dimension and non-Euclidean geometry from their friend, the insurance actuary Maurice Princet. Although Picasso denied having discussed the fourth dimension with Princet, she speculates that it is possible that he or the other Cubists picked up a suggestion of it from him indirectly.

33. See Judith Wechsler, ed., *Cézanne in Perspective* (Englewood Cliffs, New Jersey, 1975), 7.

34. Albert Gleizes and Jean Metzinger, "Cubism," in *Modern Artists on Art,* ed. Robert L. Herbert (New York, 1964), 7–8.

35. Pablo Picasso, "Statements to Marius de Zayas" (1923), in Fry, *Cubism,* 168.

36. Novotony, *Cézanne,* 141–143, 188.

37. Siegfried Giedion, *Space, Time, and Architecture* 5th ed. (1941; rpt. Cambridge, Mass., 1967), 435.

38. Pierre Francastel, *Peinture et société: naissance et destruction d'un espace plastique de la Renaissance au cubisme* (Lyon, 1951), 247.

39. Max Kozloff, *Cubism/Futurism* (New York, 1973), 70.

40. Wylie Sypher, *Rococo to Cubism in Art and Literature* (New York, 1960), 263–277.

41. Marcel Proust, *Swann's Way* (1914; rpt. New York, 1928), 258–261. On the Cubist nature of the steeples of Martinville see Matoré, *L'Espace humain,* 206. On multiple perspective in Proust and Einstein see Camille Vettard, "Proust et Einstein," *La Nouvelle revue française* (August 1922): 246–252.

42. Cited in Georges Poulet, *Studies in Human Time* (Baltimore, 1956), 319.

43. Proust, *Swann's Way,* 611.

44. Joyce, *Ulysses,* 698–699, 736.

45. Edmund Wilson, *Axel's Castle* (New York, 1931), 221. R. M. Kain wrote of *Ulysses:* "In its cubistic arrangement of contrasting planes and perspectives it is a perfect art form for the modern era"; see *Fabulous Voyager* (New York, 1959), 240. For another aspect of multiple viewpoint in Joyce and its possible connection with cinema see Paul Deane, "Motion Picture Technique in James Joyce's 'The Dead,'" *James Joyce Quarterly* (Spring 1969): 231–236.

46. Friedrich Nietzsche, *Thus Spoke Zarathustra* (New York, 1954), 237.

47. Friedrich Nietzsche, *On the Genealogy of Morals* (New York, 1967), 119.

48. José Ortega y Gasset, "Adám en el Paraíso," in his *Obras Completas* (1910; rpt. Madrid, 1946), I, 471. See Julian Marías, *José Ortega y Gasset,* n.p. (1970), 325–378.

49. José Ortega y Gasset, *Meditations on Quixote* (1914; rpt. New York, 1963), 44.

50. José Ortega y Gasset, "Verdad y perspectiva," *El Espectador,* 1 (1916), 10 *ff.*

51. José Ortega y Gasset, *The Modern Theme* (New York, 1961), 143.

52. Gasset, "Verdad y perspectiva," 116.

53. José Ortega y Gasset, "Doctrine of Point of View," *The Modern Theme,* 94.

54. J. J. Thomson, "Cathode Rays" (1897), in *The World of the Atom,* ed. Henry A. Boorse and Lloyd Motz (New York, 1966), 426.

55. Jean Perrin, *Les Atomes* (Paris, 1914), 226.

56. William Clifford, "On the Space-Theory of Matter" (1876), cited by Max Jammer, *Concepts of Space: The History of Theories of Space in Physics* (Cambridge, Mass., 1954), 161.

57. Hiram M. Stanley, "Space and Science," *The Philosophical Review* (November 1898): 616–617.

58. Richard Herr, *Wireless Telegraphy Popularly Explained* (London, 1898), 17.

59. Harriet Prescott Spofford, "The Ray of Displacement," *The Metropolitan Magazine* (October 1903).

60. Albert Einstein, *Relativity*, 150; "The Problem of Space, Ether, and the Field in Physics" in *Ideas and Opinions* (1934; rpt. New York, 1976), 274.

61. Bruno Zevi, *Architecture as Space* (New York, 1957).

62. Study cited by Reyner Banham, *The Architecture of the Well-Tempered Environment* (Chicago, 1969), 55 and *passim*, on the history of lighting and ventilation.

63. Peter Collins, *Concrete: The Vision of a New Architecture* (London, 1959); Giedion, *Space, Time, and Architecture*, 326–330.

64. Banham, *Well-Tempered Environment*, 81–2, 171–78.

65. Siegfried Giedion, *Mechanization Takes Command* (1948; rpt. New York, 1969), 301, 390.

66. Charles Voysey, "The Aims and Conditions of the Modern Decorator" (1895), cited by David Gebhard, *Charles F. A. Voysey Architect* (Los Angeles, 1975), 52.

67. Friedrich Naumann, "Die Kunst im Zeitalter der Maschine" (1904), cited by Nicolas Pevsner, *Pioneers of Modern Design* (London, 1964), 35.

68. Adolf Loos, "Ornament und Verbrechen" (1908), cited by Reyner Banham, *Theory and Design in the First Machine Age* (New York, 1960), 93–94. Loos's friend Karl Kraus fought to simplify language, politics, and society along the same lines. In the first issue of his journal *Die Fackel* (April 1899) Kraus announced that his program will be a "drainage system for the vast marches of phraseology." In Musil's *The Man Without Qualities* Ulrich finds himself suffocating under the elaborate conventions and traditions of Viennese society. He systematically withdraws from human relationships and tries to divest himself of all the "qualities" that that ornate society had imposed upon him.

69. Hendrick P. Berlage, *Gedanken über Stil in der Baukunst* (Leipzig, 1905); *Grundlagen und Entwicklung der Architektur* (Berlin, 1908), 46, 68, 115.

70. Frank Lloyd Wright, "A Testament" and "An Autobiography" in Edgar Kaufmann and Ben Raeburn, *Frank Lloyd Wright: Writings and Buildings* (New York, 1960), 314, 76, 313.

71. Theodor Lipps, *Grundlegung der Aesthetik* (Hamburg, 1903), chap. 3, "Raumaesthetik."

72. Geoffrey Scott, *The Architecture of Humanism* (1914, rpt. London, 1924), 226–228.

73. For a discussion of their works see George R. Collins and Christiane Crasemann Collins, *Camillo Sitte and the Birth of Modern City Planning* (New York, 1965), 17–18.

74. Camillo Sitte, *Der Städtebau nach seinen künstlerischen Grundsätzen* (1889; rpt. Vienna, 1909), chap. 2.

75. Adolphe Appia, *Die Musik und die Inscenierung* (n.p., 1899); Lee Simonson, "The Ideas of Adolphe Appia," in Eric Bentley, *The Theory of the Modern Stage* (New York, 1968).

76. Gordon Craig, *The Art of the Theatre* (1911); Joachim Hintze, *Das Raumproblem im modernen deutschen Drama und Theater* (Marburg, 1969), 94 *ff.*

77. George L. Mosse, *The Nationalization of the Masses: Political Symbolism and Mass Movements in Germany from the Napoleonic Wars Through the Third Reich* (New York, 1975), 62–66.

78. In a textbook of sociology published in 1908, Georg Simmel identified the specific social functions of empty space (*leerer Raum*). It can function as a neutral zone between rivals or as a meeting place or common ground. It can be a place for trading or a place where individuals or groups at war meet under peaceful conditions to negotiate a truce. Although Simmel's examples were drawn from the historical record, his elaboration of a theory of the social function of empty space was unique to this period. See his *Soziologie* (Leipzig, 1908), 703–708.

79. Boccioni, "Technical Manifesto of Futurist Sculpture" (1912), in *Futurist Manifestos,* ed. Apollonio, 61–65.

80. Alexander Archipenko, *Archipenko: Fifty Creative Years 1908–1958* (New York, 1960), 51–56.

81. I am indebted to Sean Shesgreen for this example.

82. Fernand Léger observed, "From the day the impressionists liberated painting, the modern picture set out at once to structure itself on contrasts; instead of submitting to a subject, the painter makes an insertion and uses a subject in the service of purely plastic means." "Contemporary Achievements in Painting" in *Functions of Painting,* ed. Edward Fry (New York, 1973), 14.

83. Pierre Francastel has assessed this contribution: "Cézanne introduced the notion of the reality of the 'motif,' that is to say the positive character of voids constituted by the intervals between objects. With him appeared the unity of every part of the figurative image." See his *Art et technique* (Paris, 1956), [illegible]

84. Carl E. Schorske, *Fin-de-siècle Vienna* (New York, 1980), 289.

85. Dora Vallier, "Braque, la peinture et nous: Propos de l'artiste recueillis," *Cahiers d'art,* 29 (October 1954): 15–16.

86. John Golding concluded, "For the first time in the history of art, space had been represented as being as real and as tangible, one might say as 'pictorial,' as the objects which it surrounded (here one must distinguish between an Impressionist depiction of atmosphere and the painting of empty, clear space, such as it is found before Cubism only in the work of

Cézanne)," in his *Cubism: A History and an Analysis, 1907–1914* (London, 1959), 185.

87. For a discussion of this poem and the rest of Williams's early poetry in the context of developments in art and photography see Bram Dijkstra, *Cubism, Stieglitz, and the Early Poetry of William Carlos Williams: The Hieroglyphics of a New Speech* (Princeton, 1969), 66 *ff.*

88. Meyer Shapiro, *Cézanne* (New York, n.d.), 125.

89. Boccioni et al., "Futurist Painting," in *Futurist Manifestos,* ed. Apollonio, 27.

90. Preface to the Bernheim-Jeune catalog of the first Futurist exhibition in Paris in 1912, cited by Marianne W. Martin, *Futurist Art Theory 1909–1915* (Oxford, 1968), 111.

91. I will discuss the development of geopolitics in Chapter 8.

92. Frederick Jackson Turner, "The Significance of the Frontier in American History," paper read at a meeting of the American Historical Association in Chicago, July 12, 1893; published in Frederick Jackson Turner, *The Frontier in American History* (New York, 1920), 30.

93. Frederick Jackson Turner, "Contributions of the West to American Democracy," in ibid., 259.

94. Roderick Nash, "The American Invention of National Parks," *American Quarterly* (Fall 1970): 726–735.

95. Susanne Howe, *Novels of Empire* (New York, 1949), 85.

96. Leonidas Andreiyeff, *Silence* (Philadelphia, 1910), 16, 29, 32.

97. Maurice Maeterlinck, "Silence," in his *The Inner Beauty* (London, 1911), 36, 47.

98. Marcel Proust, *The Guermantes Way* (1920; rpt. New York, 1970), 85.

99. Suzanne Bérnard, "Le 'Coup de dés' de Mallarmé replacé dans la perspective historique," *Revue d'histoire littéraire de la France* (April-June 1951): 183.

100. Aimé Patri, "Mallarmé et la musique du silence," *La Revue musicale* (January 1952): 101–111; for a historical interpretation of his contribution to composing with blanks and the connection with silences see Camille Mauclair, *L'Art en silence* (Paris, 1901) and Jean Voellmy, *Aspects du silence dans la poésie moderne* (Zurich, 1952), 24–32 and *passim.*

101. Stéphane Mallarmé, "La Musique et les lettres," *Oeuvres complètes* (Paris, 1945), 635–657.

102. Stéphane Mallarmé, "Sur Poe," *Oeuvres complètes* (Paris, 1954), 872.

103. Stéphane Mallarmé, "Mystery in Literature" (1895), in *Mallarmé: Selected Prose Poems, Essays, and Letters,* ed. Bradford Cook (Baltimore, 1956), 33.

104. Stéphane Mallarmé, "Un Coup de dés," *Cosmopolis* (May 1897): 419–427.

105. Original proofs of the projected Lahure edition of "Un Coup de dés," with marginalia by Mallarmé, in the Houghton Library Collection of Harvard University.

106. Paul Valéry, "Le Coup de dés," *Variété II* (Paris, 1929), 173.

107. Archipenko, *Fifty Creative Years,* 58.

108. Roger Shattuck, "Making Time: A Study of Stravinsky, Proust, and Sartre," *The Kenyon Review* (Spring 1963): 258.

109. Otto Deri, *Exploring Twentieth-Century Music* (New York, 1968), 358–359. In an obituary, "Le Silence de Anton von Webern," René Liebowitz wrote that "if a silence surrounds Webern's music, it also forms a part of it"; *Labyrinthe* (November 1945): 14.

110. Edgar Rubin, *Synsoplevede Figur* (Copenhagen, 1915). The first German translation appeared under the title *Visuell wahrgenommene Figur* (Berlin, 1920).

111. William James, *The Principles of Psychology* (New York, 1950), I, 240.

112. William James, *The Letters of William James* (Boston, 1926), II, 277–278.

113. Horace Meyer Kallen, *William James and Henri Bergson: A Study in Contrasting Theories of Life* (Chicago, 1914), 11, 30, 105.

114. J. Hillis Miller, *The Disappearance of God: Five Nineteenth-Century Writers* (Cambridge, Mass., 1963); " 'Thou art indeed just, Lord'," in W. H. Gardner, ed., *Poems and Prose of Gerard Manley Hopkins* (New York, 1979), 67.

115. Friedrich Nietzsche, *The Gay Science* (1882; rpt. New York, 1974), 181.

116. Nietzsche, *Genealogy of Morals,* 163.

7. Form

1. Walter E. Houghton, *The Victorian Frame of Mind 1830–1870* (New Haven, 1957), 10–15, 162.

2. Samuel Smiles, *Thrift* (London, 1876), 70.

3. Hugo von Hofmannsthal, "Der Dichter und diese Zeit" in his *Erzählungen und Aufsätze* (1905; rpt. Frankfurt, 1957), 445. Carl E. Schorske interpreted this passage in "Politics and Psyche in *fin de siècle* Vienna: Schnitzler and Hofmannsthal," *The American Historical Review* (July 1961): 930–946.

4. Mabel Dodge, *Camera Work* (June 1913): 7. Cited in Bram Dijkstra, *Cubism, Stieglitz, and the Early Poetry of William Carlos Williams* (Princeton, 1969), 25.

5. Robert Musil, *The Man Without Qualities* (New York, 1965), I, 62.

6. Georg Simmel, "The Conflict in Modern Culture" (written in 1914, first published in 1918), in *Georg Simmel: The Conflict in Modern Culture and Other Essays,* ed. and tr. K. Peter Etzkorn (New York, 1968), 11–25. Walter Lippmann, *Drift and Mastery* (New York, 1914), xvii.

7. José Ortega y Gasset, "Signs of the Times" in *The Modern Theme* (1921–22; rpt. New York, 1961), 79. Ortega wrote it after the war but documented it with examples from the prewar period. W. B. Yeats, "Introduction," *The Oxford Book of Modern Verse* (New York, 1936), xxviii, cited by Edward Engelberg, "Space, Time, and History: Towards the Discrimination of

Modernisms," *Modernist Studies: Literature + Culture, 1920–1940,* I, 1 (1975): 21.

8. Virginia Woolf, "Mr. Bennett and Mrs. Brown" (1924), in *The Captain's Bed and Other Essays* (New York, 1956), 96, 115–117.

9. Henri Bergson, *Matter and Memory* (1896; rpt. New York, 1959), 192, 196–197.

10. "The Disappearing Line Between Matter and Electricity," *Current Literature,* 41 (1906): 98–99. S. L. Bigelow, "Are the Elements Transmutable, the Atoms Divisible, and Forms of Matter but Modes of Motion?" *The Popular Science Monthly* (July 1906): 38–51.

11. H. G. Wells, *Tono-Bungay* (1909; rpt. New York, 1961), 299.

12. Albert Einstein, "On the Electrodynamics of Moving Bodies" in *The Principle of Relativity,* ed. H. A. Lorentz (New York, 1952), 48.

13. Albert Einstein, *Relativity* (New York, 1961), 99. Gerald Holton concluded that the "hierarchically ordered, harmoniously arranged cosmos, rendered in sharply delineated lines" that reigned until the mid-nineteenth century, gave way to an unbounded "restless" universe in which "the clear lines of the earlier mandala [were] replaced by undelineated, fuzzy smears." See his *Thematic Origins of Scientific Thought: Kepler to Einstein* (Cambridge, Mass., 1973), 35–36.

14. Edward Byrn, *The Progress of Invention in the Nineteenth Century* (New York, 1900), 319.

15. Umberto Boccioni, "Futurist Painting: Technical Manifesto," in *Futurist Manifestos,* ed. Umbro Apollonio (New York, 1973), 28.

16. Thomas Mann, *The Magic Mountain* (1924; rpt. New York, 1966), 218–219.

17. For an imaginative semiotic reading of the Eiffel Tower as "une sorte de degré zéro du monument" see Roland Barthes, *La Tour Eiffel* (Paris, 1964).

18. Paul Scheerbart, *Glasarchitektur* (1914; rpt. New York, 1972). See Rosemarie Haag Bletter, "Paul Scheerbart's Architectural Fantasies," *Journal of the Society of Architectural History* (May 1975): 83–97.

19. Morgan Brooks, "The Relation of Lighting to Architectural Interiors," *Scientific American Supplement,* 83 (June 2, 1917): 367.

20. Reyner Banham, *Architecture of the Well-Tempered Environment* (London, 1969), 70.

21. Frank Lloyd Wright, "An Autobiography," in *Frank Lloyd Wright: Writings and Buildings,* ed. Edgar Kaufmann and Ben Raeburn (New York, 1960), 82.

22. Julien Brault, *Histoire du téléphone et exploitation des téléphones en France et à l'étranger* (Paris, 1888), 86–87.

23. *New York Times,* August 23, 1902. For a discussion see Alan F. Westin, *Privacy and American Law* (New York, 1970), 338.

24. G. S. Lee, *The Voice of the Machines* (New York, 1906), 53, 56.

25. Arnold Bennett, "Your United States," *Harper's Monthly Magazine* (July 1912): 191.

26. Harriet Prescott Spofford, *The Elder's People* (Boston, 1920), 57–76.

27. "The Telephone Newspaper," *Scientific American* (October 26, 1895): 267; Robert Morton, "Curbing the Wireless Meddler," *Scientific American* (March 23, 1912): 226.

28. Charles Mulford Robinson, *The Improvement of Towns and Cities,* 4th rev. ed. (1901; rpt. New York, 1913), 63–79. See also Lawrence Baron, "Noise and Degeneration: Theodor Lessing's Crusade for Quiet," *Journal of Contemporary History* (January 1982): 165–187, for an account of Lessing's articles on noise pollution, published in 1901 and 1902; the founding of The Society for the Suppression of Unnecessary Noise by Mrs. Julia Barnett-Rice in New York in 1906; and the German counterpart, founded by Lessing in 1908—the Deutscher Lärmschutzverband.

29. E. L. Godkin, "The Rights of the Citizen—to His Own Reputation," *Scribner's Magazine* (July 1890): 58–67. On the history of this law see Westin, *Privacy and American Law,* 338–349, and P. Allan Dionisopoulos and Craig R. Ducat, *The Right to Privacy* (St. Paul, Minnesota, 1976), 20–25.

30. Samuel D. Warren and Louis B. Brandeis, "The Right to Privacy," *Harvard Law Review* (December 15, 1890): 195–196 and *passim.*

31. "Union Pacific Ry. Co. v. Botsford," *Supreme Court Register,* 11, 1891 (141 U.S. 250).

32. For an account of *Roberson v. Folding Box* and the dissenting opinion see Morris L. Ernst and Alan Schwartz, *Privacy: The Right to Be Left Alone* (New York, 1962): 108–127.

33. *Pavesich v. New England Life Insurance Co.;* see Dionisopolous and Ducat, *Right to Privacy,* 25.

34. See Gerald N. Izenberg, *The Existentialist Critique of Freud: The Crisis of Autonomy* (Princeton, 1976): 329.

35. David Daiches, *The Novel and the Modern World* (Chicago, 1960), 25, 36 *ff.* Leon Edel offered a similar interpretation in *The Modern Psychological Novel* (New York, 1964) especially in chap. 2, where he argued that the modern psychological novel pioneered by James, Proust, Joyce, and Richardson turned fiction away from external to internal reality and reflected "the deeper and more searching inwardness of our century." Ian Watt regarded Bloom as the "supreme culmination" of a formal trend, begun in the eighteenth century, that involved the exploration of inner consciousness. See his *The Rise of the Novel* (Berkeley, 1957), 206–207.

36. The theory of David Riesman, who observed a general shift from the inner-oriented to the outer-directed individual in the modern age, and Richard Sennett, who traced "the fall of public man" in the same period, offer opposing and one-sided interpretations of the history of privacy. The two phenomena occur simultaneously and must be understood as a historical and dialectical interaction. Richard Sennett contrasts his theory with Riesman's in *The Fall of Public Man* (New York, 1975), 5.

37. See Sam B. Warner, Jr., *Streetcar Suburbs: The Process of Growth in Bos-*

ton, 1870–1900 (New York, 1973) for a model study of this general transformation.

38. Henry Olerich, *A Cityless and Countryless World: An Outline of Practical Co-Operative Individualism* (1893; rpt. New York, 1971), 4.

39. Frank T. Carlton, "Urban and Rural Life," *The Popular Science Monthly* (March 1906): 260.

40. Ebenezer Howard, *To-Morrow: A Peaceful Path to Real Reform* (London, 1898). In 1902 and thereafter it was published under the title, *Garden Cities of To-morrow.*

41. Theodor Fritsch, *Die Stadt der Zukunft* (1896; rpt. Leipzig, 1912), 14.

42. For a discussion of Taut see Wolfgang Pehnt, *Expressionist Architecture* (New York, 1973), 78–82.

43. Arthur Meyer, *Forty Years of Parisian Society* (London, 1912), 111.

44. Edward Alsworth Ross, *Changing America* (New York, 1912), 8.

45. Philip Gibbs, *The New Man* (London, 1913), 144–148.

46. Marcel Proust, *The Past Recaptured* (1927; rpt. New York, 1970). For the social milieu of the novel see Seth L. Wolnitz, *The Proustian Community* (New York, 1971), 81 and *passim.*

47. Carl E. Schorske, *Fin-de-siècle Vienna* (New York, 1980), 296, 285.

48. Egidio Reale, *Le Régime des passeports et la société des nations* (Paris, 1930), 25–29; Jean Heimweh, *Le Régime des passeports en Alsace-Lorraine* (Paris, 1890), 12; and, for a history of passports from the German point of view, Werner Bertelsmann, *Das Passwesen: eine völkerrechtliche Studie* (Strassburg, 1914).

49. Stefan Zweig, *The World of Yesterday* (Lincoln, Nebraska, 1964), 410.

50. Heinrich Wöfflin, *Principles of Art History* (1915; rpt. New York, 1950), especially chap. 3. Two more recent studies have interpreted the whole of modern art as essentially a "break-up" or "disintegration" of traditional forms. See Katherine Kuh, *Break-Up: The Core of Modern Art* (London, 1965) and Erich Kahler, *The Disintegration of Form in the Arts* (New York, 1968). Both studies are a good source of examples but neglect the reconstruction of new forms that occurred along with the disintegration of forms.

51. Gertrude Stein, *Picasso* (1938; rpt. New York, 1959), 12.

52. Georg Simmel, "Der Bildrahmen," first published in *Der Tag* (1902); reprinted in *Zur Philosophie der Kunst* (Potsdam, 1922), 46–54.

53. Paul Souriau, *La Beauté rationnelle* (Paris, 1904), 352–373. See also Georg Simmel, "Brücke und Tür," *Der Tag* (September 15, 1909); and his "Soziologie des Raumes," *Jahrbuch für Gesetzgebung, Verwaltung und Volkswirtschaft,* 27 (1903) on the function of borders in social life and the significance of objects that bridge different spatial realms. While he found little to indicate that borders were changing in his own time, these studies represent an original scrutiny of their functions.

54. Boccioni, "Futurist Manifesto," in *Futurist Manifestos,* ed. Apollonio, 28.

55. Anton Bragaglia, "Futurist Photodynamism," in ibid., 50.

56. Umberto Boccioni, "Technical Manifesto of Futurist Sculpture" (1912), in ibid., 63; on "Sculpture's Vanishing Base" see Jack Burnham, *Be-*

yond Modern Sculpture: The Effects of Science and Technology on the Sculpture of this Century (New York, 1967), 19–28.

57. Review from the *London Standard* quoted in Gay Morris, "La Loie," *Dance Magazine* (August 1977): 39.

58. For a discussion of this and of the Harz mountain theater see George L. Mosse, *Nationalization of the Masses,* (New York, 1975), 111–112.

59. Walter R. Volbach, *Adolphe Appia: Prophet of the Modern Theater, A Profile* (Middletown, Connecticut, 1968), 47, 59.

60. Jocza Savits, *Von der Absicht des Dramas* (Munich, 1908), 40.

61. Joachim Hintze, *Das Raumproblem im modernen deutschen Drama und Theater* (Marburg, 1969), 97; see also Georg Fuchs, *Die Revolution des Theaters* (n.p., 1908), 109 *ff;* and Sybil Rosenfeld, *A Short History of Scene Design in Great Britain* (London, 1973), 166–167, for a drawing of the staging of *The Miracle* at the Olympia Theater in London.

62. Colonel William D'Alton Mann, *Town Topics* (January 18, 1912): 1, cited in Lewis A. Erenberg, *Steppin' Out: New York Nightlife and the Transformation of American Culture, 1890–1930* (Westport, Connecticut, 1981), 131. In a chapter entitled "Breaking the Bonds," Erenberg shows how new luxury hotels such as the Waldorf-Astoria (opened in 1897) enabled the elite of old wealth to abandon its former private life for a more public one that absorbed the rapid influx of the newly wealthy from the Midwest in the 1880s and 1890s. A discussion of the pre-World War I dance craze explains how the controlled, regular, and patterned movement of nineteenth-century dances was displaced by the sexually suggestive, irregular, and individualistic dances of the 1910s that included the foxtrot, bunny hug, Texas tommy, and tango.

63. Filippo Marinetti, "The Variety Theater" (1913), and Marinetti, Emilio Settimelli, Bruno Corra, "The Futurist Synthetic Theater," in *Marinetti: Selected Writings,* ed. R. W. Flint (New York, 1971), 118, 121, 128.

64. See Michael Kirby, *Futurist Performance* (New York, 1971), 48–49, for a discussion of these productions.

65. August Strindberg, "A Dream Play," in *Six Plays of Strindberg* (New York, 1955), 193. For a discussion of "the subjectivization of space" in Hauptmann, Ibsen, and Strindberg see Hintze, *Das Raumproblem,* 68–73.

66. The first reference in the *O. E. D.* was to *The Century Dictionary,* 1889–1891.

67. Werner Haftmann, *Painting in the Twentieth Century* (New York, 1965), I, 136–137.

68. Bruno Corra, "Abstract Cinema—Chromatic Music" (1912), in *Futurist Manifestos,* ed. Apollonio, 66–70.

69. Carlo Carrà, "The Painting of Sounds, Noises and Smells," in ibid., 111–115.

70. Haftmann, *Painting in the Twentieth Century,* 135.

71. Wassily Kandinsky, *Concerning the Spiritual in Art* (1911; rpt. New York, 1977), 2, 38, 47.

72. Friedrich Nietzsche, *Beyond Good and Evil* (New York, 1966), 18.

73. William James, *Essays in Radical Empiricism and A Pluralistic Universe*, ed. Ralph Barton Perry (New York, 1971), 243, 248. In *Our Knowledge of the External World* (London, 1914), 9–11, Bertrand Russell described Bradley's monism as the "straight-waistcoated benevolent institution which idealism palms off as the totality of being." The stiff sartorial image was apt. The form and content of Bradley's philosophy suggest the constriction of a corset, the tight tailoring of a nineteenth-century gentleman, who was as certain of the moral categories of good and bad as of the suitability of black tie for the opera, confident that each of the "apparent" diverse experiences of life had its proper place in the totality of the Absolute as a watch had a proper pocket in a properly made suit. John Higham, "The Reorientation of American Culture in the 1890's," in *The Origin of Modern Consciousness*, ed. John Weiss (Detroit, 1965), 42–47, identified a deep affinity of William James's pluralism with Frank Lloyd Wright's open architectural forms and Frederick Jackson Turner's frontier hypothesis. They all showed "a common opposition to all closed and static patterns of order" and affirmed a distinctively open and fluid consciousness.

74. For an account of this notion in Husserl see Quentin Lauer, *Phenomenology: Its Genesis and Prospect* (New York, 1958), 17.

75. Sigmund Freud, *Standard Edition* (London, 1953), XIV, 73–82.

76. Freud, *Standard Edition*, XV, 127.

77. The "memoir" in which he introduced this theory appeared in 1907. I have relied on Banesh Hoffmann, *Albert Einstein: Creator and Rebel* (New York, 1972) and Milič Čapek, "The Fusion of Space with Time and Its Misinterpretation," chap. 11 in his *Bergson and Modern Physics* (New York, 1971).

78. Hermann Minkowski, "Space and Time," in Lorentz, *Principle of Relativity*, 75–76.

79. Einstein, *Relativity*, 150.

80. Wyndham Lewis, *Time and Western Man* (London, 1927), 434.

81. In 1896 in New York City phone service cost $20 a month, while the average income of a worker was $38.50 a month. See Ithiel de Sola Pool, "Retrospective Technology Assessment of the Telephone" (A Report to the National Science Foundation) (1977), I, 252.

82. I am indebted to Larry May for these sources on the democratic meaning of the cinema in his *Screening Out the Past: The Birth of Mass Culture and the Motion Picture Industry* (New York, 1980): "A Democratic Art," *The Nation* (August 28, 1913): 193; D. W. Griffith, "Radio Speech," Griffith File, Museum of Modern Art Film Library, New York; Herbert Francis Sherwood, "Democracy and the Movies," *Bookman* (March 1918): 238. See also George Walsh, "Moving Picture Drama for the Multitude," *The Independent* (February 6, 1908): 306–310, who notes the democratic function of the cinema, which, along with the phonograph, "brings grand opera in a way down to the level of the poorest."

83. Louis H. Sullivan, *Kindergarten Chats and Other Writings* (New York, 1979), 105, 163, 39, 73.

8. Distance

1. Peter Costello, *Jules Verne: Inventor of Science Fiction* (London, 1978), 118–121.

2. "Le Tour du monde en quarante jours," *Revue scientifique,* 18 (1902): 602.

3. See Leo Marx, *The Machine in the Garden: Technology and the Pastoral Ideal in America* (New York, 1964); Walter Ledig, *Über den Einfluss der Eisenbahnen auf Kultur und Volkswirtschaft* (Leipzig, 1896); and Wolfgang Schivelbusch, "Railroad Space and Railroad Time," *New German Critique,* 14 (Spring 1978): 31–40.

4. Emile Zola, *The Human Beast* (New York, 1948), 51.

5. Frank Norris, *The Octopus* (New York, 1964), 42, 205, 458.

6. John Brooks, *Telephone: The First Hundred Years* (New York, 1975), 93; Friedrich Ludwig Vocke, *Die Entwickelung des Nachrichtenschnellverkehrs und das Strassenwesen* (Heidelberg, 1917), 92.

7. Herbert N. Casson, *The History of the Telephone* (Chicago, 1910), 256; Julien Brault, *Histoire du téléphone* (Paris, 1888), 154, 227–229; J. H. Robertson, *The Story of the Telephone* (London, 1947), 88; U. N. Bethell, *The Transmission of Intelligence by Electricity* (New York, 1912), 3.

8. H. R. Mosnart, "The Telephone's New Uses in Farm Life," *The World's Work* (April 1905): 6104.

9. Gerald Stanley Lee, *Crowds: A Moving-Picture of Democracy* (New York, 1913), 65.

10. Marcel Proust, *Letters of Marcel Proust,* ed. and tr. Mina Curtiss (New York, 1966), 73.

11. Marcel Proust, *The Guermantes Way* (1920; rpt. New York, 1970), 93–94.

12. Sylvester Baxter, "The Telephone Girl," *The Outlook* (May 26, 1906): 232.

13. "The Telephone: A Domestic Tragedy," *Temple Bar,* 107 (1896): 106–110.

14. Peter Cowan, *The Office* (New York, 1969), 29–30.

15. "Action at a Distance," *Scientific American,* 77 (1914): 39.

16. Gary Allan Tobin, "The Bicycle Boom of the 1890's: The Development of Private Transportation and the Birth of the Modern Tourist," *Journal of Popular Culture* (Spring 1974), 838–849.

17. Maurice Leblanc, *Voici des ailes!* (Paris, 1898), 147.

18. R. J. Mecredy, "Cycling," *Fortnightly Review,* 56 (1891): 76.

19. Articles cited by Robert A. Smith, *A Social History of the Bicycle* (New York, 1972), 112.

20. Joseph B. Bishop, "Social and Economic Influence of the Bicycle," *The Forum* (August 1896): 683–689.

21. Alfred C. Harmsworth, *Motors and Motor-Driving* (London, 1902), 28.

22. Siegfried Sassoon, *Memoirs of a Fox-Hunting Man* (London, 1928), 14.

23. Stefan Zweig, *The World of Yesterday* (Lincoln, Nebraska, 1964), 193–194.

24. Paul Adam, *La Morale des sports* (Paris, 1898), 115–124.

25. Henry Adams, *The Education of Henry Adams* (1907; rpt. New York, 1931), 469–470.

26. Marcel Proust, *The Past Recaptured* (1927; rpt. New York, 1970), 142.

27. Marcel Proust, *Within a Budding Grove* (1918; rpt. New York, 1970), 161.

28. Proust, *Past Recaptured*, 147.

29. José Ortega y Gasset, "Le Temps, la distance et la forme chez Proust," *Les Cahiers Marcel Proust*, 1 (1927): 288.

30. Lewis Jacobs recorded that the episode "brought a torrent of criticism" and that "everybody in the Biograph studio was shocked." The close-up was too close, too intimate, and violated the going sense of personal privacy. See his *Rise of the American Film* (New York, 1967), 42, 294–296.

31. Hugo Münsterberg, *The Film: A Psychological Study* (New York, 1970), 16, 38, 91. On early microcinematography in France see Standish Lawder, *The Cubist Cinema* (New York, 1975) and Jules Guiart, "La Vie révélée par le cinématographe," *Revue scientifique* (1914): 745–748.

32. Rémy de Gourmont, "Cinématographe," *Mercure de France* (September 1, 1907): 124–127; René Doumic, "L'Age du cinéma," *Revue des deux mondes* (1913): 919–922.

33. Münsterberg, *The Film*, 46.

34. Filippo Marinetti, "Technical Manifesto of Futurist Literature" (1912), in *Marinetti: Selected Writings*, ed. R. W. Flint (New York, 1971), 84–85.

35. Marinetti, "Destruction of Syntax—Imagination Without Strings—Words-in-Freedom" (1913), in *Futurist Manifestos*, ed. Umbro Apollonio (New York, 1973), 98.

36. In "The Plastic Analogies of Dynamism" (1913) Severini proclaimed that Marinetti was the first person to use analogies systematically in accord with the dynamics of the new technology, and he concluded that by using analogies "we can penetrate the most expressive part of reality and render simultaneously the subject and the will at their most intensive and expansive"; see ibid., 122.

37. Alexander Mercereau, "Introduction to the Catalogue of the Forty-fifth Exhibition of the Mánes Society, Prague, February-March, 1914, in *Cubism*, ed. Edward Fry (New York, 1966), 133.

38. Edward R. Tannenbaum, *1900: The Generation Before the Great War* (New York, 1976), 14, 20, 24; the quotation is from Allan Janik and Stephen Toulmin, *Wittgenstein's Vienna* (New York, 1973), 51.

39. Gabriel Tarde, *The Laws of Imitation* (1890; rpt. New York, 1903), 370, 17.

40. Émile Durkheim, *The Division of Labor in Society* (1893; rpt. New York, 1933), 157, 257, 262–266.

41. Scipio Sighele, *Psychologie des Auflaufs und der Massenverbrechen* (Dresden, 1897).

42. Gustav Le Bon, *The Crowd* (1895; rpt. New York, 1960), 14, 15.

43. The concepts "social contagion" and "social epidemics" began to appear in sociological literature in the 1870s, following some dramatic discoveries about infectious disease, especially in the work of Pasteur, who traced the life cycles and methods of transmission of various pathogenic microorganisms. As people came into greater proximity, the likelihood of contagious transmission increased, and by the 1890s the problem of crowding intensified fears of this means of communicating social as well as physical pathology. Prosper Despine, *De la contagion morale* (Paris, 1870); Paul Moreau de Tours, *De la contagion du suicide à propos de l'epidémie actuelle* (Paris, 1875); Jean Rambosson, *Phénomènes nerveux, intellectuels et moraux, leur transmission par contagion* (Paris, 1883); Paul Aubry, *La Contagion du meurtre: étude d'anthropologie criminelle* (Paris, 1887).

44. Robert K. Merton's introduction to Le Bon, *The Crowd*, xvii–xviii.

45. One of the most frantic critics of the crowd, Edward Ross, was particularly attentive to the role of the new technology in creating the "mob mind." In former times, he argued, a shock might agitate people within a hundred miles in the course of a day, but by the time it passed beyond them they would have cooled down somewhat. "Now, however, our space-annihilating devices, by transmitting a shock without loss of time, make it all but simultaneous. A vast public shares the same rage, alarm, enthusiasm, or horror." It swamps the individual and creates dehumanized mobs capable of lynchings. See "The Mob Mind," *Appleton's Popular Science Monthly* (July 1897): 394.

46. G. S. Lee, *Crowds: A Moving Picture of Democracy* (Garden City, New York, 1913), 4, 19, 274–278.

47. Stefan Zweig, *Émile Verhaeren* (1910; rpt. London, 1914), 5–6, 95.

48. Jules Romains, *La Vie unanime* (1908; rpt. Paris, 1913), 26, 47–48, 58; on the cultural context of his philosophy see P. J. Morrish, *Drama of the Group: A Study of the Unanimism in the Plays of Jules Romains* (Cambridge, England, 1958).

49. Wilhelm Götz, *Die Verkehrswege im Dienste des Welthandels: Eine historisch-geographische Untersuchung samt einer Einleitung für eine 'Wissenschaft von den geographischen Entfernungen'* (Stuttgart, 1888). For the same focus by an eminent British geographer see Sir George S. Robertson, "The Science of Distances," *Appleton's Popular Science Monthly* (March 1901): 526–539.

50. Friedrich Ratzel, *Anthropogeographie* (1882; rpt. Stuttgart, 1899, 2, 3, 9, 236–238, 253.

51. Friedrich Ratzel, "Die Gesetze des räumlichen Wachstums der Staaten," in his *Petermans Mitteilungen* (1896).

52. Friedrich Ratzel, *Politische Geographie,* first published in 1897. The sec-

ond edition was revised under the title *Politische Geographie oder die Geographie der Staaten, des Verkehrs und Krieges* (Munich, 1903), 1, 32, 227, 363, 371–372, 381–389, 398–400.

53. Friedrich Ratzel, *Das Meer als Quelle der Völkergrösse* (Munich, 1900), 1, 5.

54. Ellen Churchill Semple, *Influences of Geographic Environment on the Basis of Ratzel's System of Anthropo-Geography* (New York, 1911), 1–2, 175.

55. Camille Vallaux, *Géographie sociale: le sol et l'état* (Paris, 1911), 154–163.

56. H. J. Mackinder, "The Geographical Pivot of History," *The Geographical Journal,* 23 (April 1904): 422, 436.

57. Marshall McLuhan, *Understanding Media: The Extensions of Man* (New York, 1966), 23, 24, 19.

58. Charles Richet, "Dans cent ans," *Revue scientifique,* 48 (1891): 780. For the political impact of such changes see Charles A. Fisher, "The Changing Dimensions of Europe," *Journal of Contemporary History* (July 1966): 3–20.

59. H. G. Wells, *Anticipations* (1901; rpt. London, 1914), 216–217, 267, and the chapter on "The Larger Synthesis."

60. J. Novicow, *La Fédération de l'Europe* (Paris, 1901), 152–159, 502.

61. George S. Morison, *The New Epoch as Developed by the Manufacture of Power* (1903; rpt. New York, 1972), 6

62. On the Hague Conferences see F. S. L. Lyons, *Internationalism in Europe: 1815–1914* (Leyden, 1963), 342–358.

63. G. S. Lee, *The Voice of the Machines* (Northampton, Mass., 1906), 166–167.

64. The French critic Jean Cassou discussed the displacement of the "voyage" by "tourism" in the late nineteenth century, when improvements in transportation and the growth of travel agencies made it possible for the masses to travel. The classic voyage emerged in the eighteenth century, as the rich and the adventurous went off to explore historical monuments and take note of the sensibilities and customs of different nations. The golden age of the voyage was the age of Romanticism. In France travel accounts by Chateaubriand, Hugo, Mérimée, Stendhal, and Gobineau provided texts for experiences shared by others who traveled on their own to seek love and adventure in exotic, uncharted places. But the tourist agency brought the voyage to the masses as a denatured experience, stripped of the possibility of discovery. The modern tourist was locked into a collective consciousness of the group and saw everything passively, according to a preconceived itinerary that excluded the possibility of error and adventure. He experienced only what the tour guide had already experienced—looked when and where he was told, felt prescribed feelings, ordered recommended foods. See "Du voyage au tourisme," *Communications* 10 (1967): 25–34.

65. K. Mühl, *Weltreise* (Leipzig, 1907), i: "Zu den 'Globetrottern,' früher meist nur ein Typ der reisemutigen Engländer, gehört jetzt auch der deutsche Tourist, und schon seit Jahren unternimmt das Reisebureau der Hamburg-Amerika-Linie Gesellschaftsreisen um die Erde, auf denen die Teil-

nehmer zu den Hauptschaustücken Asiens und Nordamerikas geführt werden."

66. "Poèmes par un riche amateur," *Valéry Larbaud oeuvres* (Paris, 1957), 1192–1193.

67. Karl Lamprecht, *Deutsche Geschichte der jüngsten Vergangenheit und Gegenwart* (Berlin, 1912), I, 172.

68. Oron James Hale, *Publicity and Diplomacy: With Special Reference to England and Germany* (New York, 1940), 40.

69. Herbert Feis, *Europe the World's Banker 1870–1914* (New York, 1965), 5.

70. V. I. Lenin, "Imperialism, the Highest Stage of Capitalism" (written in 1916, first published in 1917), in *Selected Works* (Moscow, 1977), I, 730.

71. Silvanus P. Thompson, "Le But et l'oeuvre de la Commission Electrotechnique Internationale," *La Vie internationale* (1914): V, 10–13.

72. Lyons, *Internationalism*, 14, 229–233.

73. Hendrik Christian Anderson, *Creation of a World Centre of Communication* (Paris, 1913), x.

74. "La Deuxième session du Congrès Mondial," *La Vie internationale* (1913): iii, 524.

75. In *The Great Illusion* (1910) Angell attempted to refute the validity of that assumption, but the book proved to be, ironically, a classic example of the great delusion that war had become impossible because it was so contrary to the commercial interest of all nations. He maintained this pacifist hope even though there was a "universal assumption that a nation, in order to find outlets for expanding population and increasing industry . . . is necessarily pushed to territorial expansion and the exercise of political force against others." 4th ed. rev. (New York, 1913), ix.

76. J. R. Seeley, *The Expansion of England* (1883; rpt. Chicago, 1971), 234.

77. This was the title of a famous early imperialist tract: Charles Dilke, *Greater Britain*, 2 vols. (London, 1866–67).

78. James Froude, *Oceania or England and Her Colonies* (London, 1885), 387, 389, 392.

79. Charles W. Dilke, *Problems of Greater Britain* (London, 1890), II, 506–507, 582.

80. Ronald Robinson, John Gallagher, Alice Denny, *Africa and the Victorians: The Climax of Imperialism* (New York, 1961), 3.

81. Paul Leroy-Beaulieu, *De la colonisation chez les peuples modernes* (1874; rpt. Paris, 1886), 748–749.

82. Louis Vignon, *L'Expansion de la France* (Paris, 1891), 354.

83. Eugène Poiré, *L'Emigration française aux colonies* (Paris, 1897), 331–332; Jules Harmand, *Domination et colonisation* (Paris, 1910), chap. 2, "L'Expansion naturelle," 28–52.

84. Cited by Imanuel Geiss, ed., *July 1914: The Outbreak of the First World War, Selected Documents* (New York, 1975), 29–31.

85. Fritz Fischer, *Griff nach der Weltmacht*, tr. as *Germany's Aims in the First*

World War (New York, 1967), 7–49; *War of Illusions: German Policies from 1911 to 1914* (New York, 1975).

86. Kurt Riezler, *Die Erforderlichkeit des Unmöglichen* (Munich, 1913), 229 ff.

87. J. J. Ruedorffer (pseud. for Kurt Riezler), *Grundzüge der Weltpolitik in der Gegenwart* (Stuttgart, 1914), 4, 10, 27, 216, 226.

88. John L. O'Sullivan, "The True Title," *New York Morning News*, December 27, 1845.

89. Charles A. Conant, "The Economic Basis of 'Imperialism,'" *North American Review* (September 1898): 326–327.

90. Cited in William Appleman Williams, *The Contours of American History* (New York, 1966), 368.

91. William Appleman Williams, *The Tragedy of American Diplomacy* (New York, 1962), 24.

92. Brooks Adams, *America's Economic Supremacy* (New York, 1900), 29.

93. Erich Marcks, "Die imperialistische Idee in der Gegenwart" (1903), in his *Männer und Zeiten: Aufsätze und Reden zur neuen Geschichte* (1911), II, 271.

9. Direction

1. H. G. Wells, *Anticipations* (London, 1914), 35n.

2. H. G. Wells, *The War in the Air* (London, 1908), 96, 98, 99.

3. Marginal note on a report from the German Ambassador Tschirschky at Vienna to the German Chancellor Bethmann Hollweg on June 30, 1914. Imanuel Geiss, ed., *July 1914, The Outbreak of the First World War: Selected Documents* (New York, 1974), 64.

4. Wells, *War in the Air*, 105, 243–244.

5. C. F. G. Masterman, *The Condition of England* (London, 1909), 239, 182.

6. Paul Scheerbart, *Die Entwicklung des Luftmilitarismus und die Auflösung der europäischen Land-Heere, Festungen und Seeflotten* (Berlin, 1909).

7. Stefan Zweig, *The World of Yesterday* (Lincoln, Nebraska, 1964), 196.

8. Serrano Villard, *Contact! The Story of the Early Birds, Man's First Decade of Flight from Kitty Hawk to World War I* (New York, 1968), 73–84.

9. Victor Laugheed, *Vehicles of the Air* (Chicago, 1909), 36, 40–41.

10. G. S. Lee, *Crowds: A Moving Picture of Democracy* (Garden City, New York, 1913), 88, 60.

11. Marcel Proust, *The Past Recaptured* (New York, 1970), 80.

12. Filippo Marinetti, "Technical Manifesto of Futurist Literature," in *Marinetti: Selected Writings*, ed. R. W. Flint (New York, 1971), 88.

13. Gertrude Stein, *Picasso* (1938; rpt. New York, 1959), 49–50.

14. Henry Harrison Suplee, "The Peace Makers," *Cassier's* (September 1913): 97, 102.

15. H. Brougham Leech, "The Jurisprudence of the Air," *The Fortnightly Review* 98 (1912): 234–251.

16. William A. Robson, *Aircraft in War and Peace* (London, 1916), 166–167, 175.

17. See Rudyard Kipling, *The Ballad of East and West,* 1889.

18. Ernest E. Williams, *"Made in Germany"* (London, 1896), 10–13.

19. The German historian Ernst Robert Curtius analyzed that strong Parisian core in *The Civilization of France* (1930; rpt. New York, 1932), 51–64. Jacques Attalie and Yves Stourdze, "The Birth of the Telephone and Economic Crisis: The Slow Death of Monologue in French Society," in *The Social Impact of the Telephone,* ed. Ithiel de Sola Pool (Cambridge, Mass., 1977), 97–111, traced how the introduction of the telephone in France in the 1880s challenged the traditional direction of communication in the centralized state that flowed primarily from the capital to outlying areas in a one-way monologue. The telephone created reciprocity, equality, easy access, and two-way dialogue.

20. Marc Ferro, *The Great War 1914–1918* (London, 1973), 28.

21. Georg Wegener, *Die geographischen Ursachen des Weltkrieges* (Berlin, 1920), 113–126.

22. Friedrich von Bernhardi, *Germany and the Next War* (1912; rpt. New York, 1914), 76; Alfred Tirpitz, *My Memoirs* (New York, 1919), I, 77.

23. Hajo Holborn, "Moltke and Schlieffen," in his *Germany and Europe: Historical Essays* (New York, 1970), 74 *ff.*

24. Von Bernhardi, *Germany and the Next War,* 76.

25. J. J. Ruedorffer (pseud. for Kurt Riezler), *Grundzüge der Weltpolitik in der Gegenwart* (Stuttgart, 1914), 103.

26. Rudolf Kjellén, *Die Grossmächte der Gegenwart* (Berlin, 1914), 59.

27. Imanuel Geiss, ed., *July 1914, The Outbreak of the First World War* (New York, 1974), 294–295.

28. Georg Wegener's conclusion was typical of the geopolitical determinism that fueled *Einkreisung:* "Geographical considerations lead us almost irrefutably to the conviction that the prevailing naturally determined tensions could *only* have been resolved by war." See *Die geographischen Ursachen,* 129. In 1929 German legal historian Hermann Kantorowicz argued that *Einkreisung* was a "national myth . . . invented by Holstein and Bülow, and adapted for the mentality of children." See his *The Spirit of British Policy and the Myth of the Encirclement of Germany* (London, 1931), 365. Since national consciousness is shaped around myths and symbols, the abundance of popular versions he presents confirms how widely held was this view of the German position.

29. Colmar von der Goltz, *Die deutsche Bagdadbahn* (Vienna, 1900), cited by Henry Cord Meyer, *Mitteleuropa in German Thought and Action 1815–1945* (The Hague, 1955), 96.

30. Paul Rohrbach, *Die Bagdadbahn* (Berlin, 1902).

31. Charles Sarolea, *The Bagdad Railway and German Expansion as a Factor in European Politics* (Edinburgh, 1907), 3; Morris Jastrow, *The War and the Bagdad*

Railway: The Story of Asia Minor and its Relation to the Present Conflict (Philadelphia, 1917), 114–115.

32. Cited by Fritz Fischer, *Germany's Aims in the First World War* (New York, 1967), 9.

33. On Julius Wolf and Friedrich Naumann see H. C. Meyer, *Mitteleuropa*, 64–5, 88–95; Friedrich Naumann, *Patria: Bücher für Kultur und Freiheit* (Berlin, 1910), 6–7.

34. Oscar Jászi, *The Dissolution of the Habsburg Monarchy* (1929; rpt. Chicago, 1964), 34.

35. Wegener, *Geographischen Ursachen*, 96; Jászi, *Dissolution*, 4, 12.

36. Wegener, *Geographischen Ursachen*, 98. For an analysis at that time of the connection between domestic and foreign policy see Rudolf Goldscheid, *Das Verhältnis der äussern Politik zur innern* (Vienna, 1914).

37. J. R. Seeley, *The Expansion of England* (1883; rpt. Chicago, 1971), 237.

38. Albert J. Beveridge, *The Russian Advance* (New York, 1903), 1, 70.

39. Henry Adams, *The Education of Henry Adams* (1907; rpt. New York, 1931), 439–440.

40. Wolf von Schierbrand, *Russia: Her Strength and Her Weakness* (New York, 1904), iv., 28.

41. Friedrich Nietzsche, *Beyond Good and Evil* (New York, 1966), 188.

42. Thomas Mann, *The Magic Mountain* (New York, 1966), 243.

43. Leon Trotsky, *The Russian Revolution* (New York, 1959), 16.

44. Ladis K. D. Kristof, "The Russian Image of Russia: an Applied Study in Geopolitical Methodology," in *Essays in Political Geography*, ed. Charles A. Fisher (London, 1968), 356–364, 369–373.

45. Edward Said, *Orientalism* (New York, 1978), 1.

46. Halford J. Mackinder, "The Geographical Pivot of History," *The Geographical Journal*, 23 (April 1904): 423, 432–437.

47. Ladis K. D. Kristof, "Mackinder's Concept of Heartland and the Russians," paper delivered at the XXIII International Geographical Congress (Leningrad, July 22–26, 1976).

10. Temporality of the July Crisis

1. Oswald Spengler, *Der Untergang des Abendlandes: Umrisse einer Morphologie der Weltgeschichte* (1918; rpt. Munich, 1923), I, 176.

2. Sidney B. Fay listed "an intensely bitter press campaign of vilification between Austria and Serbia" during the three weeks after the Archduke's murder as one of the major causes of the breakdown of peace; *The Origins of the World War* 2nd ed., rev., 1930 (1928; rpt. New York, 1966), II, 332, 558.

3. In January of 1900 one popular magazine commented that "The American Biograph is taking a prominent part in the two wars now occupying the center of the world's stage [Boer and Spanish American] . . . We

are promised some vivid, soul-stirring pictures of actual gruesome war." *Leslie's Weekley* (January 6, 1900).

4. René Doumic, "L'Age du cinéma," *Revue des deux mondes* (1913): 923.

5. Stefan Zweig, *The World of Yesterday* (Lincoln, Nebraska, 1964) 210.

6. Conrad von Hötzendorf, *Aus meiner Dienstzeit 1906–1918* (Vienna, 1921–1925), IV, 30–31, cited by Luigi Albertini, *The Origins of the War of 1914* (Oxford, 1965), II, 123.

7. Imanuel Geiss, ed., *July 1914, The Outbreak of the First World War* (New York, 1974), 64–65.

8. Ibid., 63, 77, 79.

9. Herbert Butterfield, "Sir Edward Grey in July 1914," in *Historical Studies,* ed. J. L. McCracken (London, 1965), V, 11. "The original mood—the original expectation of a sudden, angry reaction on the part of Austria-Hungary—explains why Grey could be so unprotesting when he learned on 6 July that Austria might even attack Serbia and that Germany might not be prepared to hold her back." In opposition to others such as Fritz Fischer, who have attacked the Kaiser for his bellicosity and recklessness, Butterfield thinks that the Kaiser's reaction in early July reflected his accurate sense that European powers would see speedy retribution by Austria as a moment of justifiable anger that would not draw all of Europe into war.

10. Geiss, *July 1914,* 81, 106–107, 108.

11. Karl Dietrich Erdmann, *Kurt Riezler: Tagebücher, Aufsätze, Dokumente* (Göttingen, 1972), 185.

12. Albertini, *The Origins,* II, 173.

13. See Geiss, *July 1914,* 134.

14. Albertini, *The Origins,* II, 285.

15. Geiss, *July 1914,* 175, 184, 188, 195.

16. Fay, *Origins of World War,* II, 287.

17. Geiss, *July 1914,* 201.

18. Tisza to Francis Joseph, July 25, in Albertini, *The Origins,* II, 386.

19. Ibid., 373–374. Fay observed that Giesl's departure "certainly established the speed record for the rupture of diplomatic relations." *Origins of World War,* II, 349.

20. Albertini, *The Origins,* II, 375.

21. Conrad, cited in ibid., 131

22. Ibid., 456.

23. Geiss, *July 1914,* 222–223, 256.

24. Ibid., 261, 260, 287, 290, 291, 304, 323, 344, 347.

25. Lt.-Col. Philip Neame, *German Strategy in the Great War* (London, 1923), 2.

26. Victor Derrécagaix, *Modern Warfare* (1885; rpt. Washington, 1888–1890), I, 212.

27. Major-General Sir Edward Spears, cited by A. J. P. Taylor, *War by Time-Table: How the First World War Began* (New York, 1969), 16, 26.

28. "Schlieffen's Memorandum of December 1905," in Gerhard Ritter, *The Schlieffen Plan: Critique of a Myth* (London, 1958), 139.

29. In an interpretation of the outbreak of the war that focused on the impact of military timetables, A. J. P. Taylor concluded: "When cut down to essentials, the sole cause for the outbreak of war in 1914 was the Schlieffen plan—product of the belief in speed and the offensive . . . No one had time for a deliberate aim or time to think. All were trapped by the ingenuity of their military preparations, the Germans most of all." *War by Time-Table,* 121.

30. Albertini, *The Origins,* II, 480.

31. Fay, *Origins of World War,* II, 470–472.

32. Sergi Dobrorolski, *Die Mobilmachung der russischen Armee, 1914* (Berlin, 1921), cited in Fay, ibid., 473.

33. Norman Stone, *The Eastern Front 1914–1917* (London, 1975), 41, 48.

34. Dobrorolski, cited by Fay, *Origins of World War,* II, 481.

35. Ibid., 531.

36. Ibid., 535–546.

37. Harold Nicolson, *The Evolution of Diplomatic Method* (Oxford, 1953), 76.

38. Report from the Select Committee on the Diplomatic Service, 1861, cited in D. P. Heatley, *Diplomacy and the Study of International Relations* (Oxford, 1919), 252.

39. P. Pradier-Fodéré, *Cours de droit diplomatique* (Paris, 1899), I, 214.

40. Sir Horace Rumbold, *Recollections of a Diplomatist* (London, 1902), II, 111–112.

41. Charles Mazade, *La Guerre en France 1870–1871* (Paris, 1875), I, 37.

42. Pierre Granet, *L'Evolution des méthodes diplomatiques* (Paris, 1939), 85, 88–90.

43. Sir Ernest Satow, *A Guide to Diplomatic Practice* (London, 1917), I, 157.

44. J. H. Robertson, *The Story of the Telephone* (London, 1947), 116.

45. Marshall McLuhan, *Understanding Media* (New York, 1964), 101.

46. A classic example of dogged obtuseness is a passage from a British Cavalry Training Manual of 1907: "It must be accepted as a principle that the rifle, effective as it is, cannot replace the effect produced by the speed of the horse, the magnetism, and the terror of cold steel." Cited in John Ellis, *The Social History of the Machine Gun* (New York, 1975), 55.

47. A phrase of Jules Cambon, cited by Nicolson, *The Evolution,* 82.

48. Albertini, *The Origins,* II, 461.

49. Erdmann, *Kurt Riezler,* 190–191.

50. Edmund Blunden, *Undertones of War* (1928; rpt. New York, 1965), 146.

51. The diplomats of 1914 might well have felt a particularly strong sense of continuity with the age of Louis XIV, when his ministers Colbert and Richelieu centralized political authority under monarchical rule. The influential diplomat and historian Gabriel Hanotaux, although not directly involved in the negotiations of the July Crisis, exercised a strong influence on

the general conduct of diplomacy in the early twentieth century. His writings included *Instructions des ambassadeurs de France à Rome, depuis les traités de Westphalie* (1888) and a two-volume *Histoire du Cardinal de Richelieu* (1893–1903).

52. Taylor, *War by Time-Table,* 7.

53. The new Field Regulations of 1913, which were the basis for training the French Army, declared: "The French Army, returning to its tradition, henceforth admits no law but the offensive." Cited by Barbara Tuchman, *The Guns of August* (New York, 1971), 51.

54. Poincaré wanted to position French troops at the frontiers but ordered them to hold back a short distance from the frontier to avoid any accidental encounters with the enemy that might be construed by England (which had not yet declared its intention to enter the war) as a provocative act. In fact some troops approached within 4 or 5 kilometers of the frontier, but it became known as the "10 kilometer withdrawal." Fay, *Origins of World War,* II, 489–92.

55. Letter to Ellen Key, April 3, 1906, in *Letters of Rainer Maria Rilke 1892–1910* (New York, 1969), 101.

56. Report from Szögyény to Berchtold, in Geiss, *July 1914,* 110.

57. Private letter from Jagow to Lichnowsky, ibid., 122.

58. The source is Scheerbart, *Die Entwicklung.*

59. Fay, *Origins of World War,* II, 187.

60. Geiss, *July 1914,* 128.

61. Eugène Minkowski, *Lived Time: Phenomenological and Psychopathological Studies* (Evanston, 1970), 180–193.

62. A. J. P. Taylor, *The Struggle for Mastery in Europe 1848–1918* (Oxford, 1963), 521.

63. Cited by Fritz Fischer, *Germany's Aims in the First World War* (New York, 1967), 50.

64. Cited in V. R. Berghahn, *Germany and the Approach of War in 1914* (New York, 1973), 169.

65. Fischer, *Germany's Aims,* 50.

66. Geiss, *July 1914,* 65, 68.

67. Erdmann, *Kurt Riezler,* 183, 187.

11. The Cubist War

1. Gertrude Stein, *Picasso* (1938; rpt. New York, 1959), 11.

2. Edmund Blunden, *Undertones of War* (1928; rpt. New York, 1965), 171.

3. John Keegan, *The Face of Battle: A Study of Agincourt, Waterloo & the Somme* (New York, 1976), 241.

4. Cecil Lewis, *Sagittarius Rising* (New York, 1936), 86–87, cited in Paul Fussell, *The Great War and Modern Memory* (New York, 1977), 81.

5. David Jones, *In Parenthesis* (1937; rpt. London, 1955), 202n.

6. Philander Johnson, "Each Man's Army," selected by Admiral Samuel MacGowan; see *Everybody's Magazine* (May 1920): 36.

7. Eugène Minkowski, *Lived Time* (Evanston, 1970), 14.

8. Eric J. Leed, *No Man's Land: Combat and Identity in World War I* (Cambridge, England, 1979), 129. Robert Graves, *Goodbye to All That* (London, 1929); C. E. Carrington, *A Subaltern's War* (London, 1929).

9. Leed, *No Man's Land*, 124.

10. Blunden, *Undertones*, 30.

11. Henri Barbusse, *Under Fire* (1916; rpt. London, 1965), 6.

12. Thomas Mann, *The Magic Mountain* (New York, 1966), v.

13. Marcel Proust, *The Past Recaptured* (New York, 1970), 25–26.

14. Fussell, *The Great War*, 80.

15. Jones, *In Parenthesis*, xi, xv, xiv, 191–192n.

16. Barbusse, *Under Fire*, 244.

17. Blunden, *Undertones*, 186, 124.

18. James Joyce, *Ulysses* (1922; rpt. New York, 1961), 24, 583.

19. Hereward Carrington, *Psychical Phenomena and the War* (New York, 1918), 41.

20. Barbara Tuchman, *The Guns of August* (New York, 1971), 289.

21. Guillaume Apollinaire, *Oeuvres poétiques* (Paris, 1956), 272.

22. General Sir Edward Swinton, "A Sense of Proportion," in Keegan, *Face of Battle*, 261.

23. F. Scott Fitzgerald, *Tender is the Night* (1933; rpt. New York, 1962), 57.

24. Barbusse, *Under Fire*, 17–18.

25. Leed, *No Man's Land*, 105–114.

26. W. M. Maxwell, *A Psychological Retrospect of the Great War* (London, 1923), 66, cited in Leed, ibid., 181.

27. Ibid., 182–184.

28. Barbusse, *Under Fire*, 257.

29. Blunden, *Undertones*, 30.

30. Henri Massis, "The War We Fought," in *Promise of Greatness: The War of 1914–1918*, ed. George A. Panichas (New York, 1968), 204.

31. Ernest Hemingway, *A Moveable Feast* (New York, 1964); Malcolm Cowley, *Exile's Return* (New York, 1934).

32. Filippo Marinetti, "Geometrical and Mechanical Splendor and the Numerical Sensibility" (1914), in *Marinetti: Selected Writings*, ed. R. W. Flint (New York, 1971), 97–98.

33. Captain B. H. Liddell Hart, *The Real War 1914–1918* (1930; rpt. New York, 1964), 54.

34. Alistair Horne, *The Price of Glory: Verdun, 1916* (New York, 1967), 16.

35. Keegan, *Face of Battle*, 255.

36. Alfred von Schlieffen, "Der Krieg in der Gegenwart," in *Cannae* (Berlin, 1925), 278, cited by Hajo Holborn, "Moltke and Schlieffen: The Prussian-German School," in *Makers of Modern Strategy*, ed. Edward Mead Earle (Princeton, 1969), 194.

37. Gustave Le Bon, *The Psychology of the Great War* (London, 1916), 284–285.

38. In 1917 Kurt Lewin, a founder of sociological "field theory," wrote a "phenomenology" of war landscape, surveying the qualities of different parts of the entire war zone, which included the unique value attached to certain areas or directions as a function of their proximity to danger. See "Kriegslandschaft,"*Zeitschrift für angewandte Psychologie,* 12 (1917): 440–447. Paul Fussell identified the "gross dichotomy" that came to dominate war experiences: "The sharp dividing of landscape into known and unknown, safe and unsafe, is a habit no one who has fought ever entirely loses . . . One of the legacies of the war is just this habit of simple distinction, simplification, and opposition." See *The Great War,* 79.

39. Reginald Farrer, *The Void of War* (Boston, 1918), 62, 148.

40. Leed, *No Man's Land,* 15.

41. Stein, *Picasso,* 11.

42. Cited in Tuchman, *Guns of August,* 55.

43. André Ducasse, Jacques Meyer, Gabriel Perreux, *Vie et mort des Français, 1914–1918* (Paris, 1962), 510–511. I am indebted to Theda Shapiro, *Painters and Politics: The European Avant-Garde and Society, 1900–1925* (New York, 1976), 139–140, for this reference and for alerting me to the direct connection between Cubism and camouflage.

44. A number of Mare's most important drawings appeared in *Desseins faits aux armées par André Mare,* ed. J.-A. Gonon (Paris, n.d.). The source for Marc is Horne, *Price of Glory,* 13. See also Carl Nordenfalk, "Camouflage und Kubismus," *Kunstgeschichtliche Gesellschaft zu Berlin,* Sitzungsberichte, N.F. Heft 27 (1981): 9–11.

45. For a discussion of this see Roy A. Behrens, "Camouflage, Art and Gestalt," *The North American Review* (December 1980): 9–18, and also his most important source: Robert F. Sumrall, "Ship Camouflage (WWI): Deceptive Art," *U.S. Naval Institute Proceedings* (July 1971): 55–77. At the time the major source on natural protective coloring was the work of Abbott H. Thayer, especially his full-length book, *Concealing Coloration in the Animal Kingdom* (1909). Behrens also links Cubism and camouflage with developments in Gestalt psychology at that time as possible "proof of a *Zeitgeist.*"

46. Gertrude Stein, "The Autobiography of Alice B. Toklas," in *Selected Writings of Gertrude Stein* (New York, 1972), 177.

47. Barbusse, *Under Fire,* 148–150.

48. For a description of trenches see Keegan, *Face of Battle,* 209; Fussell, *The Great War,* 41; Theodore Ropp, *War in the Modern World* (New York, 1967), 247.

49. Leed, *No Man's Land,* 103.

50. Ernst Jünger, *Das Wäldchen 125. Eine Chronik aus dem Grabenkampf 1918* (Berlin, 1925), 21, cited in Leed, ibid.

51. Stefan Zweig, *The World of Yesterday* (Lincoln, Nebraska, 1964), 223.

52. Barbusse, *Under Fire,* 15, 17.

53. Charles de Gaulle, *France and Her Army* (London, 1941), 90. See also Alfred Wolff Oberlehrer, "Über Einheit und Fortschritt des Menschengeschlechts im Weltkrieg 1914/16," *Archiv für Philosophie,* 22 (1916): 104 *ff.*

54. Basil Henry Liddell Hart, *A History of the World War 1914–1918* (London, 1930), 183.

55. Carrington, *Psychical Phenomena,* 59.

56. Marius-Ary Leblond, *Galliéni parle* (Paris, 1920), 53.

57. Horne, *Price of Glory,* 22.

58. G. E. C. Wedlake, *SOS: The Story of Radio-Communication* (London, 1973), 106; Charles Bright, *Telegraphy, Aeronautics, and War* (London, 1918), 32.

59. Charles Dupont, *Le Haut commandement allemand en 1914* (Paris, 1922), 8.

60. Cited by Fussell, *The Great War,* 76.

61. Cited in Raymond H. Fredette, *The Sky on Fire: The First Battle of Britain 1917–1918* (New York, 1966), 7, 4.

62. Ibid., 220.

63. Stein, *Picasso,* 50.

INDEX